Contemporary crystallography

McGraw-Hill Series in Materials Science and Engineering

AVITZUR *Metal Forming: Processes and Analysis*
AZÁROFF *Introduction to Solids*
BARRETT AND MASSALSKI *Structure of Metals*
BLATT *Physics of Electronic Conduction in Solids*
BRICK, GORDON, AND PHILLIPS *Structure and Properties of Alloys*
BUERGER *Contemporary Crystallography*
BUERGER *Introduction to Crystal Geometry*
DeHOFF AND RHINES *Quantitative Microscopy*
ELLIOT *Constitution of Binary Alloys, First Supplement*
GILMAN *Micromechanics of Flow in Solids*
GORDON *Principles of Phase Diagrams in Materials Systems*
HIRTH AND LOTHE *Theory of Dislocations*
MURR *Electron Optical Applications in Materials Science*
PAUL AND WARSCHAUER *Solids under Pressure*
ROSENFIELD, HAHN, BEMENT, AND JAFFEE *Dislocation Dynamics*
RUDMAN, STRINGER, AND JAFFEE *Phase Stability in Metals and Alloys*
SHEWMON *Diffusion in Solids*
SHEWMON *Transformations in Metals*
WERT AND THOMSON *Physics of Solids*

Contemporary crystallography

Martin J. Buerger

Institute Professor
Massachusetts Institute of Technology

McGraw-Hill Book Company

New York St. Louis San Francisco Düsseldorf London
Mexico Panama Sydney Toronto

Contemporary crystallography

Library of Congress Catalog Card Number 74-98049

08840

1234567890 HDMM 79876543210

This book was set in Modern by The Maple Press Company, printed on permanent paper by Halliday Lithograph Corporation, and bound by The Maple Press Company. The original drawings were done by Richard M. Beger and Felix Trojer and rendered in their final form by Joseph Buchner. The editors were B. J. Clark and Antonia Stires. Stuart Levine supervised the production.

To
David Chan
and
Anne Chan

Preface

Information about the detailed nature of crystalline matter has been available for about half a century. But it is a curious fact that the many interested in the solid state make use of only a little of this fund of knowledge. At least part of the reason for this is that the scientist who furthers this knowledge—the crystallographer—has been busy with his own rapidly expanding research and the interesting sidelines uncovered by it. This preoccupation has been so intense that he has not wanted to spare the time to present his subject in perspective to others. I have realized for some time that I myself have been guilty of this in that what I have written has been directed toward the professional crystallographer and the student who wishes to become one. Accordingly, I take this opportunity to discuss some of the features peculiar to crystals in some greater perspective and for a more general audience.

This book, then, is written for the student who wants to know something of the characteristic features of crystals. It specifically outlines the theory and results concerned with the basic geometrical features of crystals, and the modern methods of experimentally determining this geometry. It is evident that in presenting the many desirable topics in a small volume one cannot discuss them in any detail, nor can the results be derived for all cases of interest—indeed, to treat the subject in that way would be to lose perspective. An attempt has been made, therefore, to display the spirit of the subject by sketching the highlights along the route, always keeping in mind the route itself.

Students in several fields are advised to take seriously the study of crystals. This admonition is especially important for those interested in solid-state physics and "materials science," but it also applies to mineralogists, chemists, metallurgists, ceramists, and biologists; for these fields deal either chiefly or in part with crystals. To scientists in these fields there is available not only the theoretical background developed in the previous century but also the results piled up in the last half century's most fertile experimental work. Those who remain unacquainted with this wealth of knowledge are less likely to make contributions to their own fields where crystals are involved.

In the writing of any book, the question arises as to the level at which the discussion should be presented. In writing this book I have taken the following view: There is little formal instruction in crystallography in this country, yet the subject matter of crystallography is fundamental to a proper understanding of anything concerned with crystalline material which occurs in chemistry, physics, metallurgy, mineralogy, and even, nowadays, biology. The study of crystals, then, ought to be started as early as possible, either in advance of or concurrent with the beginning courses in these more popular fields, and might well begin in the freshman year of college. I have, therefore, presented the discussion so that it can be understood by the student with no mathematics beyond that commonly offered to science majors in secondary schools. But I have taken the opportunity of introducing just a little vector algebra and the algebra of the complex plane, with a view to encouraging the student to look into them more thoroughly.

Crystallography is an extensive subject, and this book presents merely an elementary introduction to it. Many topics which properly belong in the range of subject matter discussed here have been considered but briefly, or have been omitted entirely for one reason or another (chiefly to avoid writing an encyclopedic work). Among the topics reluctantly omitted was the use of anomalous dispersion in symmetry determination and in the direct determination of crystal structures. The traditional stereographic and gnomonic projections were not included because they are not crystallography, but rather conventions for conveying crystallographic ideas; once popular, they are no longer required and their use is waning. Also omitted are detailed discussions of the Laue and powder methods of x-ray diffraction. These relatively weak methods for obtaining information about crystals are now largely superseded in scientific work by more powerful ones, although they continue to be used in many routine industrial applications of crystallography when some aspect of the problem (usually the character of the sample to be examined) makes more powerful methods difficult or impossible to use.

The x-ray photographs in this book have been reproduced as nearly as possible to the scale of the original films. This has been done chiefly so that the student can, by making simple measurements with a millimeter scale, obtain data for computing the unit cells.

The crystallography of an earlier era was crystal geometry. Since our present-day crystallography is a marvelous experimental realization of the theory of that subject, some discussion of it is a necessary preliminary for appreciating contemporary crystallography. Accordingly, crystal geometry is treated in outline in Chapter 2. Some readers will want to read further in this subject, yet not as extensively as those who plan to become professional crystallographers. For such readers the author has provided a more extended treatment in a companion volume entitled *Introduction to crystal geometry.*

Finally, the study of crystals is much broader than crystallography, as that subject has been limited in this book. The background established here, however, leads directly to such topics as the bonding of atoms in crystals, the relative stabilities of alternative arrangements of atoms (polymorphism), how crystals grow, imperfections in their patterns due to growth (lineage structure and dislocations), intergrowths of two kinds of crystals (topotaxy and epitaxy), intergrowths of two or more crystals of the same kind (twinning), disordered crystals, and the obvious mechanical properties of crystals. It was inappropriate to load this small book with these additional topics; instead they are being discussed in another book, now in preparation.

I am indebted to many for help in preparing this book. The manuscript was used as a text by undergraduates, who, along with graduate students and former students, called my attention to errors and obscure passages. Starting with my crude sketches, the line drawings were carefully redrawn (most of them again and again, as improvements appeared desirable) by two graduate students, Richard M. Beger and Felix Trojer. The x-ray photographs are the careful handiwork of another graduate student, Martha Redden, who did them just for the satisfaction of examining their various symmetries. I am grateful that the skills and crystallographic backgrounds of these helpers were available to me. The manuscript was reworked several times by Joyce Everitt and Joan Ellison.

Crystallography has such a fascination for a few students that, once introduced to it, they make it their scientific way of life. While this book was not intended especially for them, I will be happy if it opens that door to some.

Martin J. Buerger

Contents

Contemporary crystallography

1

Introduction

The geometrical nature of crystals
Crystals in perspective
The anisotropism of crystals

Today's physics and chemistry of solids bear little relation to those of half a century ago. Today it is known what solids are like structurally, while half a century ago little thought was given to their structures except by a small group of theoreticians, and what views there were on the subject at that time were unsupported by relevant experiments. For the most part, ignorance of the general nature of the solid state among physicists and chemists not only constituted a void in their general knowledge but also led to naïve errors, some of which persist even today.

A true, but slow, breakthrough in understanding solids began in 1912 when the physicist Max Laue, then a privatdozent at the University of Munich, conceived the experiment of attempting to diffract x-rays by the three-dimensional gratings which the theoretical crystallographers had supposed ought to characterize the structures of crystals. The experiments were carried out by the physics assistants W. Friedrich and P. Knipping, who, after an initial failure, achieved a brilliant success which marked one of the great turning points in science. The experiments proved unmistakably that crystals behave like three-dimensional gratings and that x-rays behave like waves. That crystals diffracted these waves showed that their grating periods were of the same order of magnitude as the wavelengths of x-rays. By use of chemical information (the composition of the crystal) and physical information (the density of the crystal) both the grating periods and the wavelengths were quickly established.

This experiment placed in the hands of those interested in crystals a new tool, which was quickly refined to the point where it provided data from which could be derived quite detailed information about the arrangements of atoms in crystals. In the score of years that intervened between the two world wars, the arrangements of atoms in thousands of crystals were established; these results furnished information of tremendous value to physicists, chemists, and others interested in the solid state.

The geometrical nature of crystals

Sometimes the gist of a subject can be put into a nutshell, so to speak, by the right word or phrase. For crystals the word is "ordered," because crystals, compared with other states of matter, are characterized by being geometrically ordered. If comparatively minor deviations from order, due chiefly to thermal motion of the atoms, are ignored, crystals constitute a state of matter in which the atoms are arranged in regular patterns. The most important distinguishing characteristic of a crystal is that it is based on some repeating pattern of atoms. The most fundamental data about a crystal are concerned with a description of the nature of this pattern: its symmetry and the way the pattern is repeated periodically in space.

A preliminary feeling for the pattern nature of crystals can be acquired by studying briefly some simpler patterns, specifically, patterns limited to two dimensions. It turns out that there exists only a strictly limited number of pattern types, specifically 17, in two dimensions. The geometrical unit which is repeated in a pattern is called a *motif*. The 17 pattern types are illustrated in Fig. 1, using an arbitrary scalene triangle as a motif. Given any motif, there are only 17 general ways that it can be repeated to form an ordered pattern. The same pattern type can be used to repeat a different motif, as shown in Fig. 2 (page 6).

Any specific pattern is characterized by three things:

(*a*) its symmetry—a qualitative feature describing the pattern type,

(*b*) two vectors periodically used as translating operations to repeat the contents of a properly chosen area—a quantitative feature, and

(*c*) the details of the motif—another quantitative feature.

Each of these three features, though noted here for two-dimensional patterns, has its counterpart for crystals. Crystals differ from these two-dimensional patterns in two chief ways. In the first place, the patterns in crystals exist in three dimensions, and the extra dimension not only renders the geometry of the patterns more complex but also increases the number of possible pattern types (from 17 to 230). Secondly, the motifs in crystals are not arbitrary geometrical markings, but consist of the atoms constituting the chemical composition of the crystal (not necessarily in the form of molecules).

Some examples of the patterns of atoms actually found in crystals are shown in Figs. 3 to 5 (pages 7 to 9). These are seen to differ according to characteristics (*a*) to (*c*) itemized above. The symmetries, that is, the general ways the atomic motifs are repeated, are different; the areas occupied by the motifs are different; and the numbers and kinds of atoms in the motifs are different.

The symmetry of a pattern is a most interesting feature. Without defining, at this point, what is meant by symmetry, or describing its many varieties, it may be said that the symmetry of a pattern can be designated by a symbol; those for two dimensions are given as labels in Fig. 1. The nature of symmetry and the meaning of these symbols will be treated in the following chapter.

With this brief introduction to patterns, the reader is advised to watch for patterns in his daily surroundings. These are to be found especially in wallpaper, tiling, textile prints, textile weaves, and brick walls. Having had two-dimensional patterns called to his attention, the student will see them in many places. It will prove valuable to try to identify each pattern, wherever found and whatever the motif, with one of the 17 types shown in Fig. 1.

Crystals in perspective

To appreciate why crystals constitute an ordered form of matter it is helpful to consider the nature of the several recognized states of matter. For present purposes matter may be regarded as existing in three stable states: gaseous, liquid, and crystalline. In the gaseous and liquid states there exist single atoms or small clusters of atoms which do not bond together because the average thermal energy is sufficient to disrupt any larger clusters that tend to form. If the temperature of the gas or liquid is lowered, the thermal energy is reduced, and a point is eventually reached when the bond energy is greater than the average thermal energy. Under these conditions additional bonds between single atoms or small clusters can remain intact, so that a more highly bonded state of matter develops; liquid matter normally transforms to crystalline matter, and gaseous matter may condense directly to crystalline matter. In other words, a highly bonded, and therefore condensed, state forms.

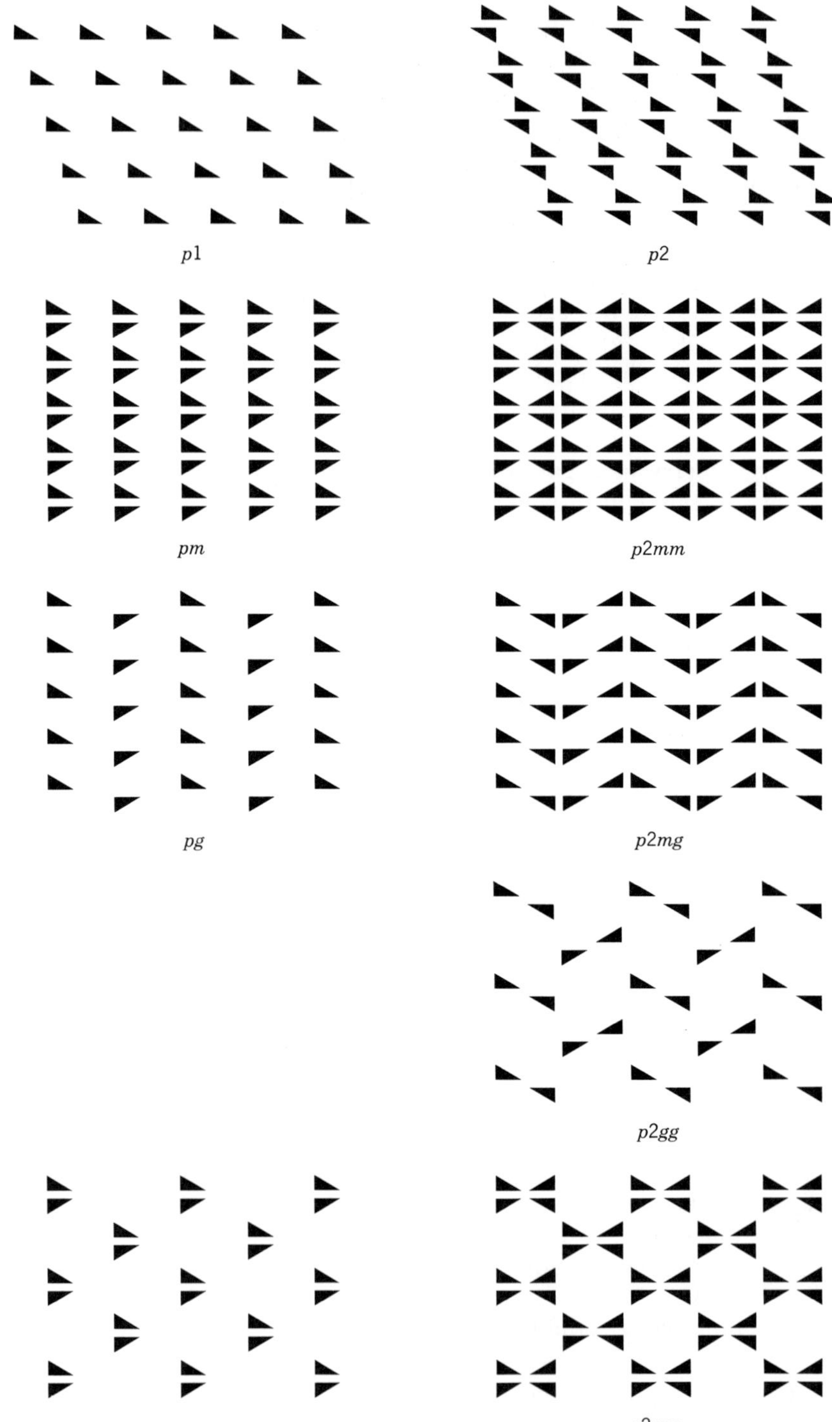

*p*1 *p*2

pm *p*2*mm*

pg *p*2*mg*

*p*2*gg*

cm *c*2*mm*

Fig. 1. The 17 types of patterns in a plane. [*From Martin J. Buerger: Elementary crystallography.* (*Wiley, New York,* 1956 *and* 1963) *end paper.*]

Fig. 2 (page 6). One of the pattern types of Fig. 1, but with a different motif than in Fig. 1.

Figs. 3 to 5 (pages 7, 8, and 9): Examples of patterns of atoms in crystal structures.

Fig. 3. Cyanuric triazide, C_3N_{12}.

Fig. 4. 1, 2, 4, 5-tetramethylbenzene, $C_6(CH_3)_4$.

Fig. 5. Marcasite, FeS_2.

The general nature of this condensation can be appreciated by considering a simplified model. In particular,

(I) consider the two-dimensional analog of a monatomic vapor;
(II) let each atom have the possibility of 4 bonds, which will be called latent bonds, at the relative azimuths of 0, 90, 180, and 270°.

At low pressures the vapor consists of independent atoms, with average kinetic energy which is proportional to the absolute temperature, that is, to kT, flying about at distances from one another which are large compared with their effective diameters. When the temperature T is reduced so that kT is just less than the bond energy, the atoms can bond together. This process does not stop with the bonding of 2 atoms to form a pair. A third atom can add itself to a pair to form a short chain, which can be either straight, as in Fig. 6*A*, or kinked, as in Fig. 6*B*. If a fourth atom condenses on one of these chain types, the energy of the resulting set of four is less if the new atom adds itself to the kinked chain to form a square array, as shown in Fig. 6*D*, than if it adds itself elsewhere on the kinked chain or anywhere on the straight chain, as shown in Fig. 6*C*. This is because in the position noted in Fig. 6*D* the atom

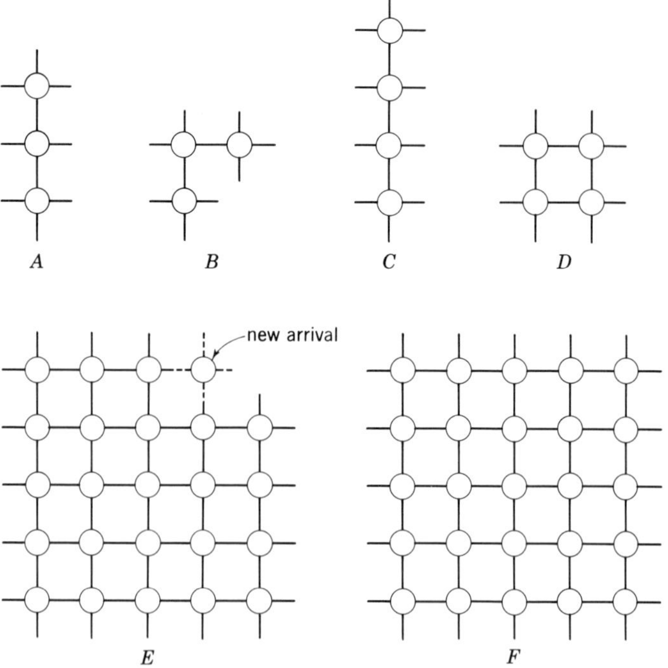

Fig. 6

satisfies 2 of its 4 latent bonds and in doing so forms a cluster of 4 atoms having together only 8 latent bonds; whereas if the new atom attaches itself to any chain by only 1 bond, it produces a cluster of 4 atoms having together 10 latent bonds. Each latent bond is a part of the surface energy of the cluster. The square cluster has a lower surface energy and accordingly is more stable. Should a four-membered chain form, either straight, as in Fig. 6*C*, or bent, it would transform to the more stable form of Fig. 6*D* and so reduce its surface energy.

While additional condensation mechanisms may also occur, the one described shows that newly arriving atoms or clusters attach themselves in such locations as to remove the largest number of latent bonds on the arriving atom and, by the same mechanisms, to minimize the remaining number of latent bonds on the growing crystal. This leads to two features of crystals. One is a relatively high density of bonds within the crystal, a feature which is satisfied by a relatively dense pattern of atoms.

The second feature may be understood with the aid of the model already considered. In the case of the simple model of Fig. 6, the growth mechanism just discussed may be applied in the specific form that a newly arriving atom will always place itself in a position such that it bonds itself to the crystal by 2 latent bonds, as in Fig. 6*E*, unless such a location does not exist, as in Fig. 6*F*. Such a location does not exist whenever the growing layer has just been completed. The starting of each new layer therefore requires the newly arriving atom to accept some 1-bond position, after which 2-bond positions are again available until that layer is complete. This sequence requires a layer-by-layer growth of a crystal, so that the crystal pattern, in growing, tends to develop a boundary consisting, in three dimensions, of a set of planes. It is for this general reason that crystals grown in a noninhibiting environment tend to assume polyhedral shapes. The shapes are consistent with the point-group symmetry of the pattern, discussed in Chapter 2.

This simple model illustrates the basic features which characterize the addition of atoms to form a pattern. The actual case differs merely in exhibiting certain complications which do not vitiate the general scheme, specifically,

(*A*) The actual crystal exists in three dimensions.

(*B*) The composition of the growing crystal usually includes more than one kind of atomic species. This may provide several different kinds of bonds with different bond energies. The bonds with highest energies may already have permitted the formation of clusters of limited sizes in the liquid state, so that condensation may be more complicated and may occur in stages. In this event, various clusters of atoms having external latent bonds may assume the role of the individual atoms in the simple model.

(*C*) In these more complicated cases the latent bonds need not be arranged in as simple a geometrical manner as in the model discussed, and the bonds of the pattern formed by the clusters may be under some strain. Such strain is part of the internal energy of the crystal. But whatever its magnitude, it is less than the internal energy of a random jumble of atoms which would constitute the *amorphous state* of the same material.

The anisotropism of crystals

The pattern nature of crystals causes them to have some quite different properties from the unordered states of matter, namely, the gaseous, liquid, and amorphous states. All forms of matter have *scalar properties*, such as volume, mass, and density. But the patterns of crystals provide them with a sort of grain, somewhat similar to the grain of wood, and this causes certain properties to vary with direction; these are therefore *directional properties*. Unstrained glasses, liquids, and amorphous substances have no vector properties and so are termed *isotropic* (from the Greek *iso*, equal, and *tropos*, turn). Substances which have vector properties, in contrast, are called *anisotropic*, and the existence of the vector nature of the property is referred to as *anisotropism*.

In a given crystal the variation of a property with direction may be great or small, depending on the nature of the pattern; the variation may vanish because of symmetry, so that even a crystal may be isotropic with respect to some properties. Accordingly, a crystal may be anisotropic with respect to some properties yet isotropic with respect to others.

An easily understood mechanical property is tensile strength. This depends on the bonding forces between atoms of each pair across a unit area whose orientation may be specified with respect to known directions in the pattern of atoms. For unstrained amorphous materials this property is the same regardless of the orientation of the area. But the pattern nature of crystals requires this to vary with orientation. That the tensile strength must vary in this way is evident for certain crystals whose structures are called *layer structures*. In these the pattern consists of strong planar layers which are weakly bonded to each other (mica is a well-known example). A little tension across the layers easily pulls them apart, but great tension along the layers is required to tear the structure apart in that direction. In other crystals the breaking strengths may vary less extremely with direction, but most crystals have some detectable variation, and many crystals have a very obvious variation. Such crystals, when stressed arbitrarily, tend to fail by splitting along the planes of the pattern across which the bonding is weakest. This qualitative property is called *cleavage*, and the crystals displaying it are said to be *cleavable* along one or more particular planes which are specified by symbols described in Chapter 2.

Another characteristic property is the velocity of transmission of light. The refractive index of light traveling through a crystal varies, in general, with the direction of the electric vector (which crystallographers call the vibration direction of the ray). The symmetry of certain crystals—the so-called cubic crystals—requires that the variation of refractive index with direction vanish; therefore these crystals are optically isotropic. But the symmetry does not necessarily require the crystals to be isotropic with respect to other properties. For example, NaCl and CaF_2 are both cubic crystals. They are therefore optically isotropic, but they are very anisotropic with respect to mechanical properties. If crystals of these substances are crushed, the fragments of NaCl have surfaces parallel to the three pairs of opposite planes of a cube (Fig. 7); therefore the crystal is said to display cubic cleavage. The fragments of CaF_2 have surfaces parallel to the four pairs of opposite planes of an octahedron (Fig. 8); therefore the crystal is said to display octahedral cleavage.

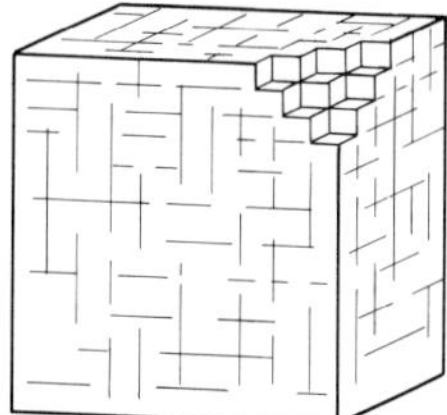

Fig. 7. A crystal displaying cubic cleavage.

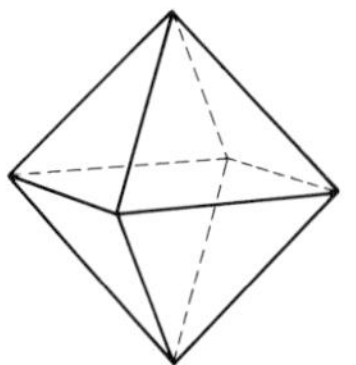

Fig. 8. An octahedron (*left*), and a crystal displaying octahedral cleavage.

These few illustrations should make it evident that the characteristic properties of a crystal depend upon its atomic pattern. Some important qualitative conclusions can be drawn about a crystal merely by knowing the symmetry of the pattern. Quantitative conclusions depend upon understanding the physics of the particular pattern; the branch of physics known as crystal physics deals with these matters. Chemistry, especially physical chemistry, is also concerned with quantitative features of the pattern, especially the energy of the pattern and the way it stores heat.

In Chapter 2, the symmetry properties of crystals are considered.

2

The geometrical nature of order in crystals

Homogeneous patterns

Significance of order. In the preceding chapter it was pointed out that the key feature of crystals is order. Order is equivalent to homogeneity of structure; that is, if every motif of the pattern has an environment which is the same as that of every other motif, the pattern is an ordered one.

Homogeneous sequences of congruent motifs. The geometrical requirements for a homogeneous series of motifs depend on whether all the motifs are congruent or not. If they are all congruent, then the most general homogeneous series can be described as a sequence of motifs equally spaced along a helical ribbon, as in Fig. 1. The geometrical operation which relates neighboring motifs is a rotation parallel to the axis of the helix combined with a translation parallel to that axis. The combination is a motion similar to that of an advancing screw, and is called a screw motion. The two components are designated

A_α, a rotation through an angle α about A, the axis of the helix, and
τ, a translation through a distance τ parallel to axis A.

The combined operation is symbolically designated as $A_{\alpha,\tau}$. This operation can be regarded as generating any motif from the one just preceding it in the helical series. More broadly, given a specific motif and specific values of α and τ, the operation $A_{\alpha,\tau}$ can be used to lay down a duplicate of the original motif; then from the duplicate it can be used to lay down a triplicate; from a triplicate a quadruplicate; etc. From this point of view the operation $A_{\alpha,\tau}$ can be regarded as generating a

Fig. 1

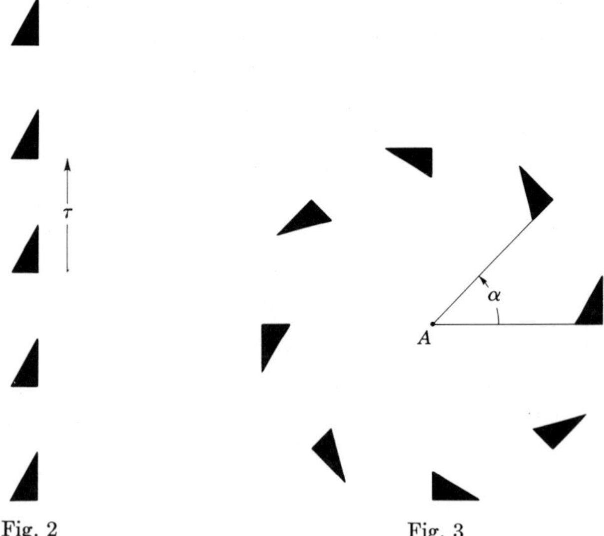

Fig. 2

Fig. 3

homogeneous pattern from a given motif. Such an operation is called the *generating operation* of the pattern.

The value of either α or τ can be zero. When $\alpha = 0$, the helical pattern degenerates to a sequence of motifs in parallel orientation arranged along a line with a common interval τ, as in Fig. 2. When $\tau = 0$, the helical pattern degenerates to a sequence of motifs arranged around a circle with a common angular interval α, the adjacent motifs differing in orientation by α, as shown in Fig. 3. In a pattern of this sort, the last of the sequence of motifs must coincide with the first motif; this requires α to be a submultiple of 360°, or

$$n\alpha = 2\pi. \qquad (1)$$

Enantiomorphic motifs. Any motif can be brought into coincidence with a congruent motif by a combination of a rotation and a translation. In the narrowest sense, two such congruent motifs are truly identical. But pairs of motifs also exist which, while not congruent, are nevertheless the same in the sense that the right and left hands are the same. The relation between the individuals of such a pair is called *enantiomorphism;* each member is said to be *enantiomorphic* with the other; and each is an *enantiomorph* of the other.

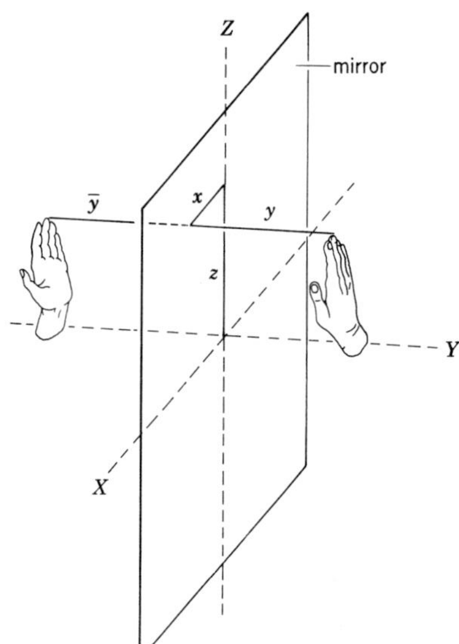

Fig. 4. Derivation of an enantiomorphic object by reflection.

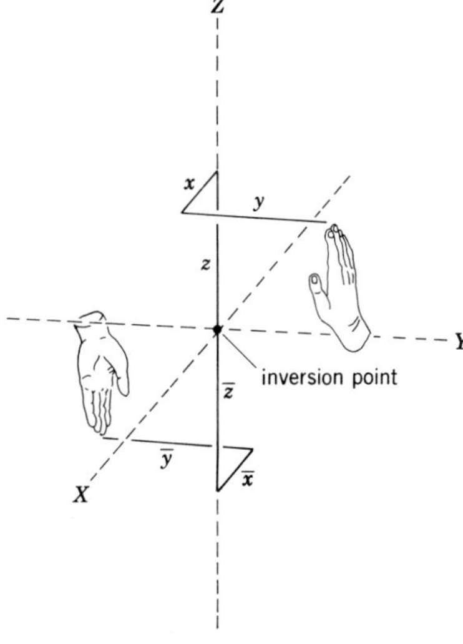

Fig. 5. Derivation of an enantiomorphic object by inversion. [*Figures 4 and 5 modified from M. J. Buerger: Encyclopedia Britannica article Crystallography* (1963) 6, 851–863.]

In three-dimensional space one enantiomorph can be derived from its mate by an operation of reflection in a mirror, as in Fig. 4, or by inversion through a point, as in Fig. 5. In Fig. 4, if an origin is taken in the mirror, with two coordinate axes in the mirror and one perpendicular to it, and in Fig. 5, if the origin is taken at the point of inversion, then it can be seen that changing the sign of the one coordinate normal to the mirror for all points on either motif generates the enantiomorph by reflection, while changing the signs of all three coordinates generates the enantiomorph by inversion. The reader can readily demonstrate to himself that changing the signs of two coordinates is the same as rotating the motif 180°, so that the motifs so related are congruent.

Homogeneous sequences of enantiomorphic motifs. A homogeneous sequence of alternating enantiomorphs cannot be generated by a screw, unless the translation τ along the screw axis is zero. The alternating sequence then degenerates to one generated by a rotation combined with a reflection (Fig. 6) or a rotation combined with an inversion (Fig. 7). Whether the motifs of the pattern appear to alternate, as in Fig. 6, or to occur in pairs, as in Fig. 7, depends on whether there is an even or odd number of repetitions per 360°.

Special cases of these two kinds of pattern appear if the circle of the pattern is increased to infinity so that α is reduced to zero. Then the alternating pattern degenerates to Fig. 8 and the paired pattern degenerates to Fig. 9. For reasons which may be appreciated later, the patterns in Figs. 6 to 8 are especially simple and can be regarded as generated by

a rotoreflection operation in Fig. 6, symbolized by $A_{\alpha,m}$,
a rotoinversion operation in Fig. 7, symbolized by $A_{\alpha,i}$, and
an operation consisting of a combination of a reflection m and translation τ in Fig. 8, symbolized by m_τ.

When $\alpha = 360°$, the pattern produced by $A_{\alpha,m}$ further degenerates into the pattern of Fig. 4, namely, an enantiomorphic pair related by a mirror; when $\alpha = 180°$, the pattern produced by $A_{\alpha,m}$ degenerates into the pattern of Fig. 5, namely, an enantiomorphic pair related by

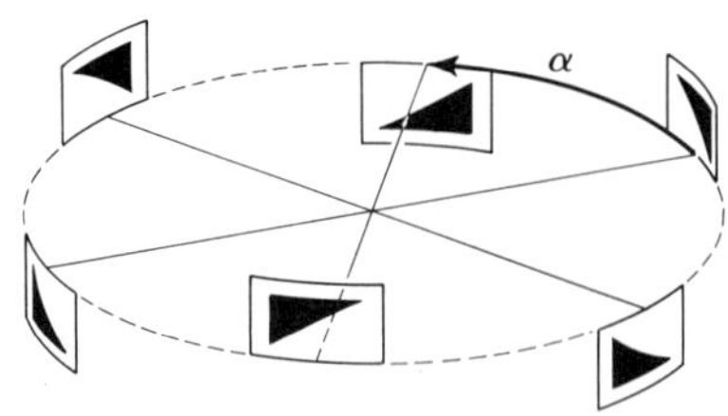

Fig. 6. $A_{\alpha,m}$.

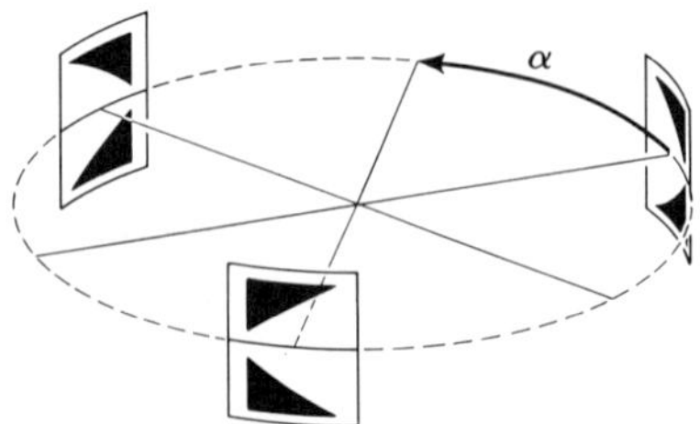

Fig. 7. $A_{\alpha,i}$.

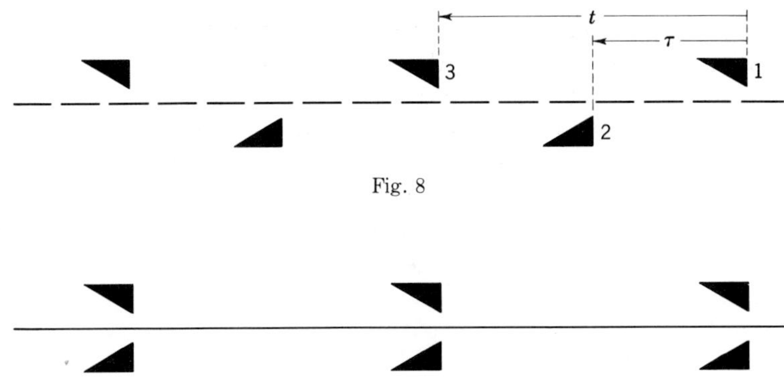

Fig. 8

Fig. 9

inversion. Similarly, $A_{\alpha,i} \to i$ when $\alpha = 360°$, and $A_{\alpha,i} \to m$ when $\alpha = 180°$.

Symmetry elements. It has just been shown that the repeated action of each of certain specific types of operation on a motif results in a simple homogeneous pattern. Table 1 presents a list of these operations and some of their characteristics. Except in the case of a pure translation, the operation leaves some specific geometrical locus unmoved. Such a locus is useful in describing the geometrical nature of the operation and also in describing the appearance of the pattern. For example, if a pattern is generated by a rotation axis whose symbol is A_α, then since (1) shows that there are $n\alpha$'s in 360°, the pattern has the same appearance from n different directions about the axis A. As a consequence the axis A is said to be an axis of n-fold symmetry; or, more exactly, A is an n-fold rotation axis. Such loci are called *symmetry elements.*

Table 1
Some features of operations whose repeated application to a motif produces homogeneous patterns

Operation			Description of corresponding symmetry element
Symbol	Description	Geometrical locus left unmoved	
$A_{\alpha,\tau}$	screw	axis A	screw axis
A_α	rotation	axis A	rotation axis
t	translation	——	——
$\{A_{\alpha,m}$	rotoreflection	axis A, plane m	rotoreflection axis $\}$
$\{A_{\alpha,i}$	rotoinversion	axis A, point i	rotoinversion axis $\}$
m_τ	translation-reflection	plane m	glide plane
m	reflection	plane m	mirror (reflection plane)
i	inversion	point i	inversion center

In Table 1 the rotoinversion and rotoreflection operations are bracketed. The significance of this is that the set of all possible rotoinversion axes is the same as the set of all rotoreflection axes, although the labeling of the α's in the two sets is, in general, different (when $n = 2\pi/\alpha$ is divisible by 4, the labels are the same). In modern crystallographic practice the labeling used is that for the set of rotoinversion axes.

Complex homogeneous patterns. So far the discussion has been limited to simple homogeneous patterns, that is, patterns which are produced by repeated application of one of the operations in Table 1 to a motif. Each of these is a pattern with one symmetry element, or else a pattern which has been produced by repeated application of one translation. More complicated patterns which are also homogeneous can be devised by properly combining these operations. Provided that the combination does not involve an inconsistency (a matter which must be investigated in each case), one homogeneous series which has been generated by one operation may act as a motif to be repeated by a second operation. A simple example is afforded by using the translation sequence of Fig. 2 as a motif for repetition by another translation. The result is the pattern shown as $p1$ in Fig. 1, Chapter 1. This is obviously a homogeneous pattern, for every triangle has the same environment as every other triangle.

Applications of homogeneous patterns to crystals

Definition of a crystal. Crystals are examples of homogeneous patterns. They are, however, restricted in two ways. First, the motif is a specific atom or group of atoms whose proportions are those of the chemical composition of the substance. Second, regardless of what possible symmetry operations may characterize its pattern, a crystal must always have three noncoplanar translations. In fact, a crystal can be defined as a pattern of atoms repeated in three dimensions by translations and possibly (but not necessarily) supplemented by symmetry.

Consistency of symmetry with translations. Since a crystal must have translation, all symmetry must be consistent with these translations. For example, in Fig. 8 two glide operations in sequence are responsible for motif 3, which is related to motif 1 by a pure translation t. Thus, the plane of every glide plane must contain a pure translation, and the translation component τ of the glide must be in the direction of the pure translation and equal in magnitude to half of it, i.e.,

$$m_\tau: \qquad \vec{\tau} = \tfrac{1}{2}\vec{t}. \tag{2}$$

Similarly, for a screw axis, the axis of $A_{\alpha,\tau}$ must be parallel to a pure translation, and a certain integral number of screw operations (containing that number of translation components τ) must be consistent with the pure translation in that direction. This requires

$$n\vec{\tau} = q\vec{t} \qquad (n,\ q,\ \text{integers};\ n = 2\pi/\alpha),$$

$$\vec{\tau} = \frac{q}{n}\vec{t}. \tag{3}$$

Finally, for A_α, $A_{\alpha,\tau}$, $A_{\alpha,m}$, and $A_{\alpha,i}$, every α must be consistent with pure translations. This restricts the possible values of α in a way illustrated in Fig. 10. Here axis A is at right angles to a pure translation $\vec{t}$, where magnitude is a, so that A', A, A'' is a sequence in which axes A act as motifs to the translation operation. But the symmetry about axis A requires translation AA'' to be repeated at an angle α, that is, along AB''. In the same way the symmetry about A requires translation AA' to be repeated as AB'. Thus AA', AA'', AB', and AB'' are all translations of magnitude a. Therefore B' and B'' are equivalent by translation, and so the interval between them, which has the same direction as AA'', must be consistent with a; this requires, in general,

$$b = pa \qquad (p \text{ an integer}). \tag{4}$$

The geometry of Fig. 10 requires that

$$\tfrac{1}{2}b = a\cos\alpha, \tag{5}$$

so

$$\cos\alpha = \frac{b}{2a} = \frac{pa}{2a} = \tfrac{1}{2}p. \tag{6}$$

This requires $\cos\alpha$ to have the values of halves of the integers, specifically, 0, $\pm\frac{1}{2}$, ± 1. The only permissible values of α are thus 90°; 60°, 120°; 360°, 180°. According to (1), these angles correspond to the following symmetry numbers n: 4; 3, 6; 1, 2. Translations therefore restrict axes to 1-, 2-, 3-, 4-, and 6-fold rotation axes, whether pure rotation, screw, rotoinversion, or rotoreflection.

Remarks on the derivation of crystal-pattern types. The types of pattern which are possible for the arrangements of atoms in crystals could be derived by establishing, first, the general type of pattern produced by the repeated action of three noncoplanar translations on an arbitrary motif. This general type could then be specialized by requiring

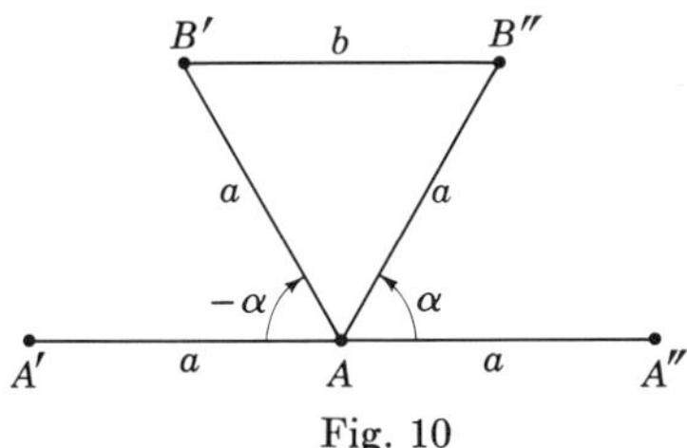

Fig. 10

it to be complicated (or specialized) in all the consistent ways by the kinds of symmetry element listed in Table 1.

This apparently complicated investigation is rendered simpler if it is tackled by a different route in comparatively easy stages. This alternative procedure, which is outlined in this chapter, has several incidental advantages: First, the individual stages are comparatively easy to understand; second, the intermediate results achieved are useful in themselves; and third, the stages present an easy introduction to the symbolism of symmetry which is used in all modern crystallographic literature.

Symmetry conforms to the mathematical *theory of groups;* as a consequence, some of the results of symmetry theory are given names suggested by this mathematical discipline. Thus, the general patterns to which crystals conform are usually referred to as *space groups.* Among the intermediate results are the crystallographic *point groups.*

In the derivation outlined here, each of the crystallographic symmetries having one symmetry axis without a translation component is first established. Then the symmetries having more than one such axis are derived from these. These axial symmetries constitute the 32 crystallographic point groups, which is a useful intermediate result.

Next, the properties of patterns generated by three noncoplanar translations are briefly investigated. This provides the properties of space lattices. The general lattice is then specialized by requiring it to be symmetrical successively with each of the 32 point groups. The intermediate result from these considerations is the set of 14 symmetrical Bravais-lattice types.

Finally, the ultimate result, the 230 space groups, is derived by making use of the intermediate results. This is made relatively easy by taking advantage of the relation of *isogony* which exists between a point group and certain space groups.

In the remainder of this chapter, this derivation is treated in outline. In this treatment rotational symmetries are limited to those which are crystallographically permissible, that is, those for which $n = 1, 2, 3, 4$, or 6.

Rotational symmetry

Proper and improper rotations. There are two general types of operations which generate patterns. The operations which generate a sequence of congruent objects are called *operations of the first sort,* while those which generate a sequence of alternating enantiomorphs are called *operations of the second sort.* These are separated by a broken line in Table 1.

The rotational operations lacking a translation component which belong to the operations of the first sort are A_α, while those which are operations of the second sort are $A_{\alpha,i}$ and $A_{\alpha,m}$. These rotational operations and the symmetry axes determined by them are specifically referred to as *proper rotations*, for A_α's and *improper rotations*, for $A_{\alpha,i}$ and $A_{\alpha,m}$.

The monaxial point groups. The pattern types corresponding to proper generating operations A_α, where $n = 2\pi/\alpha = 1, 2, 3, 4$, and 6, are shown in the upper part of Fig. 11. Each pattern is designated by a numerical symbol, namely, its n. The location of the n-fold symmetry axis is indicated by an n-sided polygon, which is omitted when $n = 1$ and is symbolized by ⬮ when $n = 2$, an equilateral triangle when $n = 3$, a square when $n = 4$, and a regular hexagon when $n = 6$.

The pattern types corresponding to the improper generating operations $A_{\alpha,i}$, where $n = 2\pi/\alpha = 1, 2, 3, 4$, and 6, are shown in the lower part of Fig. 11. These patterns are designated by the numeral corresponding to the n of the axis with a bar over the numeral, specifically, $\bar{1}$, $\bar{2}$, $\bar{3}$, $\bar{4}$, and $\bar{6}$. If these $\bar{n}$ patterns are examined, it is seen that, except for $\bar{4}$, they correspond to other entries in Table 1, or to certain combinations, specifically,

$\bar{1}$ is an inversion, listed as i in Table 1.
$\bar{2}$ is a reflection, listed as m in Table 1.
$\bar{3}$ produces a pattern the same as that produced by 3 and i.
$\bar{6}$ produces a pattern the same as that produced by 3 and a perpendicular m.

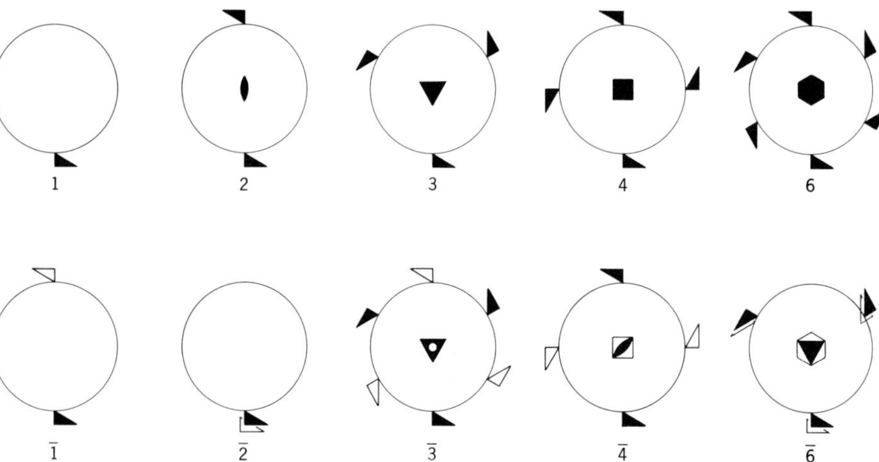

Fig. 11. The 10 types of patterns each of which is generated by a single crystallographic rotation operation. Each pattern type is designated by its (numerical) symmetry symbol, given below the pattern. The generating operation is a proper rotation for the five upper patterns; it is an improper rotation for the five lower patterns.

It is sometimes convenient to work with $\bar{n}$ as an axis, sometimes as its alternative designation, as above. Only the improper rotation $\bar{4}$ has no other designation. The locations of the $\bar{3}$, $\bar{4}$, and $\bar{6}$ axes are indicated by distinctive symbols shown in Fig. 11.

Combinations of proper rotations. When two mechanical rotations are combined, the result, in general, is the same as another rotation. An example which is readily followed is shown in Fig. 12, in which two different 3-fold axes are shown extending along two diagonals of a cube. A rotation about A through 120° brings face I of the cube to face II and brings point 1 to point 2. If this is followed by a rotation through 120° about B, face II is returned to face I, and point 2 is brought to point 3. The net result of the two rotations is to bring point 1 to point 3, which is equivalent to rotating face I about axis C through an angle of 180°.

The results of combining rotations like $A_{120°}$ and $B_{120°}$ in Fig. 12 can be investigated by means of a relation due to Euler, which has the form

$$\cos(A \wedge B) = \frac{\cos\frac{1}{2}\gamma + \cos\frac{1}{2}\alpha\cos\frac{1}{2}\beta}{\sin\frac{1}{2}\alpha\sin\frac{1}{2}\beta}, \tag{7}$$

which is based on the spherical triangle whose angles are half the angles α, β, and γ, as shown at the right of Fig. 12. Since A, B, and C can have possible n's of only 1, 2, 3, 4, or 6, the α's, β's, and γ's are limited to 360, 180, 120, 90, and 60°, respectively. Each of these possibilities is substituted into (7). Only a few combinations on the right-hand side give acceptable values for a cosine on the left-hand side. If an acceptable solution is found, corresponding formulae can be used to find the angles $B \wedge C$ and $C \wedge B$.

The six distinct nontrivial sets of solutions are shown in Fig. 13. When each of the three resulting axes A, B, and C is allowed to multiply the other, the resulting sets of axes are those shown in Fig. 14.

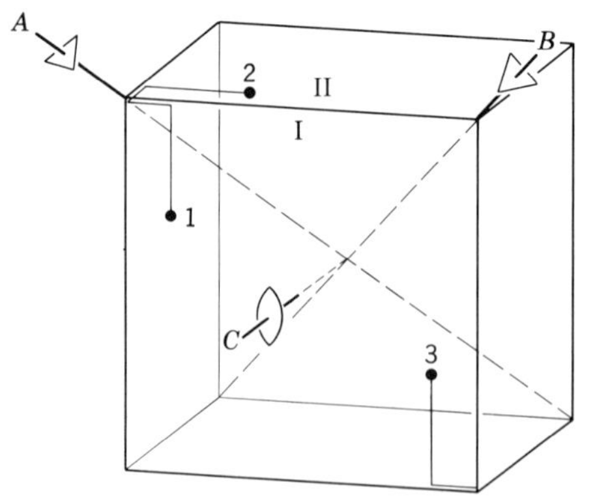

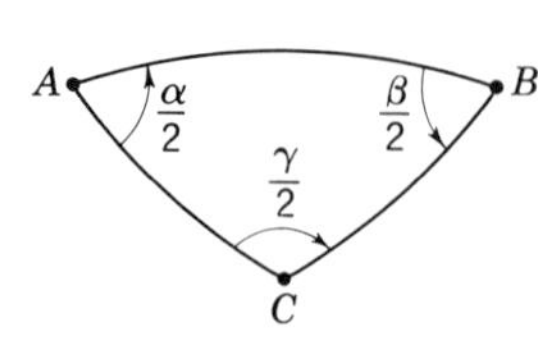

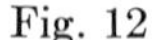
Fig. 12

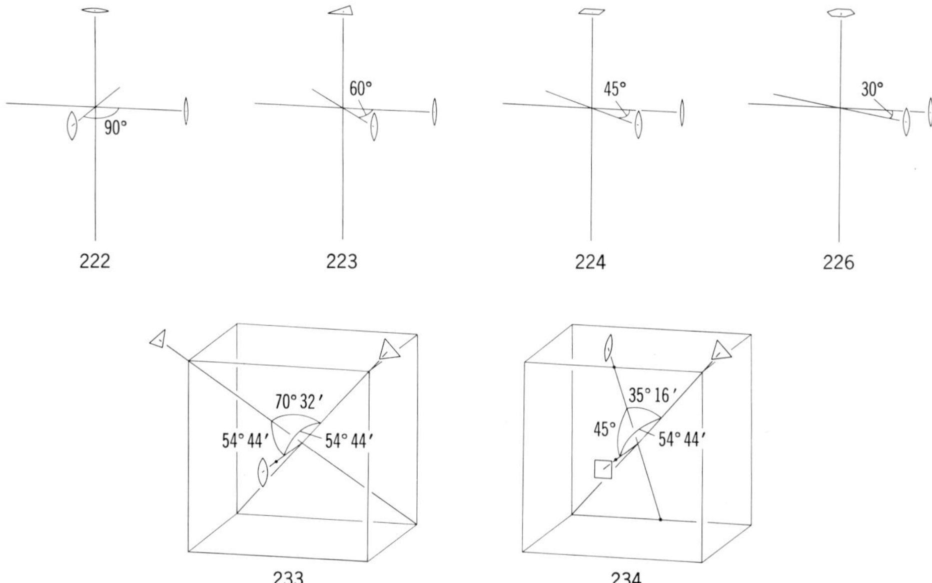

Fig. 13. The six permissible nontrivial crystallographic combinations of rotations. [*From M. J. Buerger: Elementary crystallography.* (*Wiley, New York,* 1956 *and* 1963) 43.]

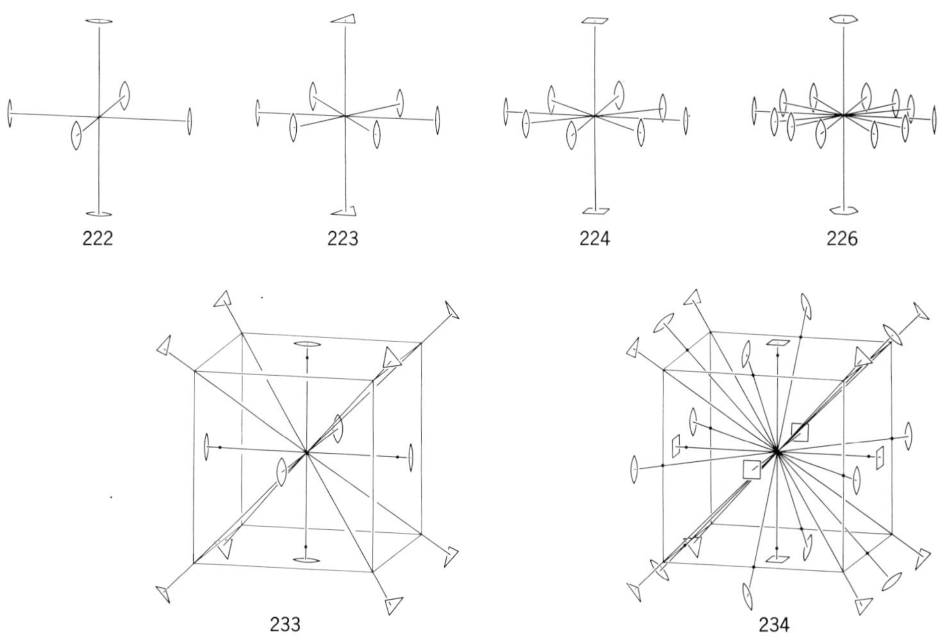

Fig. 14. The six crystallographic axial symmetries based upon the combinations in the above Fig. 13. [*From M. J. Buerger: Elementary crystallography.* (*Wiley, New York,* 1956 *and* 1963) 44.]

The total number of symmetries involving only proper rotations are the five uncombined symmetries of the upper part of Fig. 11 and the six permissible combinations of Fig. 14. These 11 symmetries are listed by their standard symbols in Table 2. The symmetries due to uncombined rotations are designated by their numerals n, as in Fig. 11, while the symmetries of the combined rotations are designated by the sequence of the three numerals of the axes combined.

Although there are certain niceties in these symbolic designations for the point groups, this use of three symbols in sequence to represent the nature of the symmetry with respect to three related rotations supplies an introduction to the modern designation of point-group symmetries. This scheme need be only slightly modified to designate space-group symmetries.

It is important to realize how closely the three rotations of a combination are tied together. Euler's relation implies that rotation A_α followed by B_β is equal to another rotation $C_{-\lambda}$. If the rotations α and β are taken in a positive direction, then the resultant γ is a negative rotation. The relation can be written

$$A_\alpha B_\beta = C_{-\lambda}. \qquad (8)$$

Here the combination of A_α with B_β is treated as a product. If both sides of (8) are multiplied on their right sides by C_λ, there results

$$A_\alpha B_\beta C_\lambda = C_{-\lambda} C_\lambda. \qquad (9)$$

The right side of (9) now represents the combination of rotations through $-\gamma$ and γ about the same axis C, which implies no resultant rotation. If this is represented by the 1 of a 1-fold rotation (as in Fig. 11), then (9) can be rewritten in the symmetrical form

$$A_\alpha B_\beta C_\lambda = 1. \qquad (10)$$

Table 2
Possible crystallographic symmetries containing only proper rotation axes

Uncombined axes	Permissible combinations
1	
2	2 2 2
3	3 2 2 (called 3 2)
4	4 2 2
6	6 2 2
	2 3 3 (called 2 3)
	4 3 2

This emphasizes the impossibility of separating these three rotations. If any two are given, the third is implied.

In order to manipulate (10), both sides must be either multiplied on the left by the same term or multiplied on the right by the same term. Derivable from (10) by this means are

$$
\begin{aligned}
&(8): & A_\alpha B_\beta &= C_{-\gamma}, & \\
& & B_\beta C_\gamma &= A_{-\alpha}, & (11)\\
& & A_\alpha C_\gamma &= B_{-\beta} & (12)\\
& & (\text{from } B_\beta &= A_{-\alpha}C_{-\gamma}). &
\end{aligned}
$$

Let P stand for a proper rotation, and I stand for an improper rotation. A first rotation P generates from a right-handed motif another right-handed motif; then a second rotation P generates a third right-handed motif from the second. Thus the combination generates from a right-handed motif another right-handed motif and so also has a proper nature P. The character of the three rotations in (10) accordingly can be represented by the sequence $P\ P\ P$.

Combinations of improper rotations. It has just been seen that the congruent character of the combination of two proper rotations can be represented by

$$P\,P = P \qquad \text{character } P\ P\ P.$$

Table 3
Point groups involving improper rotations

Uncombined axes, $\bar{n}$	Combinations conforming to				
	$\frac{n}{\bar{n}}$	$P\ I\ I$	$I\ P\ I$	$I\ I\ P$	$\frac{P}{I}\frac{P}{I}\frac{P}{I}$
$\bar{1}$	$\left[\frac{1}{\bar{1}} = \bar{1}\right]$				
$\bar{2} = m$	$\frac{2}{\bar{2}} = \frac{2}{m}$	$2\,\bar{2}\,\bar{2} = 2\,m\,m$	$[\bar{2}\,2\,\bar{2} = m\,2\,m]$	$[\bar{2}\,\bar{2}\,2 = m\,m\,2]$	$\frac{2}{\bar{2}}\frac{2}{\bar{2}}\frac{2}{\bar{2}} = \frac{2}{m}\frac{2}{m}\frac{2}{m}$
$\bar{3}$	$\left[\frac{3}{\bar{3}} = \bar{3}\right]$	$3\,\bar{2}(\bar{2}) = 3\,m$			$\frac{3}{\bar{3}}\frac{2}{\bar{2}}(\frac{2}{\bar{2}}) = \bar{3}\,\frac{2}{m}$
$\bar{4}$	$\frac{4}{\bar{4}} = \frac{4}{m}$	$4\,\bar{2}\,\bar{2} = 4\,m\,m$	$\bar{4}\,2\,\bar{2} = \bar{4}\,2\,m$	$[\bar{4}\,\bar{2}\,2 = \bar{4}\,m\,2]$	$\frac{4}{\bar{4}}\frac{2}{\bar{2}}\frac{2}{\bar{2}} = \frac{4}{m}\frac{2}{m}\frac{2}{m}$
$\bar{6}$	$\frac{6}{\bar{6}} = \frac{6}{m}$	$6\,\bar{2}\,\bar{2} = 6\,m\,m$	$\bar{6}\,2\,\bar{2} = \bar{6}\,2\,m$	$[\bar{6}\,\bar{2}\,2 = \bar{6}\,m\,2]$	$\frac{6}{\bar{6}}\frac{2}{\bar{2}}\frac{2}{\bar{2}} = \frac{6}{m}\frac{2}{m}\frac{2}{m}$
		$2\,\bar{3}(\bar{3}) = \frac{2}{m}\,\bar{3}$			$\left[\frac{2}{\bar{2}}\frac{3}{\bar{3}}(\frac{3}{\bar{3}}) = \frac{2}{m}\,\bar{3}\right]$
		$4\,\bar{3}\,\bar{2} = \frac{4}{m}\,\bar{3}\,\frac{2}{m}$	$\bar{4}\,3\,\bar{2} = \bar{4}\,3\,m$	$\left[\bar{4}\,\bar{3}\,2 = \frac{4}{m}\,\bar{3}\,\frac{2}{m}\right]$	$\left[\frac{4}{\bar{4}}\frac{3}{\bar{3}}\frac{2}{\bar{2}} = \frac{4}{m}\,\bar{3}\,\frac{2}{m}\right]$

On the other hand, every improper rotation changes right to left, or the reverse. The possible combinations of improper rotations with improper rotations and with proper rotations are therefore seen to be

$$\begin{aligned} I\,I &= P \qquad \text{character } I\ I\ P, \\ I\,P &= I \qquad \text{character } I\ P\ I, \\ P\,I &= I \qquad \text{character } P\ I\ I. \end{aligned}$$

Thus every combination must have an even number of improper rotations, specifically, either zero or two I's. Except for this requirement, *any* two rotations combine to require a third rotation exactly as if they were combinations of proper rotations.

Derivation of the 32 point groups. With this background all symmetries can be derived as if they were the 11 axial symmetries in Table 2. Each numeral in that table can be an n, as it is there, or an $\bar{n}$, provided that, in a combination, the result is $P\ P\ P$, $I\ I\ P$, $I\ P\ I$, or $P\ I\ I$. Furthermore, any position given as n in Table 2 can be occupied by both n and $\bar{n}$ along the same axis. Such a combination is represented by the fraction $n/\bar{n}$, with the improper axis in the denominator. Many combinations involving $\bar{n}$ can be transformed by making use of the certain relations noted earlier, namely,

$$\begin{aligned} \bar{2} &\equiv m, \\ \bar{3} &= 3 + \bar{1}. \end{aligned}$$

Table 4
The 32 crystallographic point groups

Crystal system	Type: n	$\bar{n}$	$\frac{n}{m}$	$n\,2\,2$	$n\,m\,m$	$\bar{n}\,2\,m$	$\frac{n}{m}\frac{2}{m}\frac{2}{m}$
Triclinic	1	$\bar{1}$					
Monoclinic	2	$\bar{2} = m$	$\frac{2}{m}$				
Orthorhombic				2 2 2	$2\,m\,m$		$\frac{2}{m}\frac{2}{m}\frac{2}{m}$
Hexagonal	3	$\bar{3}$	$\left[\frac{3}{m} = \bar{6}\right]$	3 2	$3\,m$	$\bar{3}\frac{2}{m}$	
Tetragonal	4	$\bar{4}$	$\frac{4}{m}$	4 2 2	$4\,m\,m$	$\bar{4}\,2\,m$	$\frac{4}{m}\frac{2}{m}\frac{2}{m}$
Hexagonal	6	$\bar{6}$	$\frac{6}{m}$	6 2 2	$6\,m\,m$	$\bar{6}\,2\,m$	$\frac{6}{m}\frac{2}{m}\frac{2}{m}$
Isometric				2 3			$\frac{2}{m}\bar{3}$
				4 3 2		$\bar{4}\,3\,m$	$\frac{4}{m}\bar{3}\frac{2}{m}$

Some other specific combinations help in transforming the results so obtained. Among these are:

$\bar{1}$ is an inversion.

$2 \perp m = \bar{1}$.

$2 \cdot i = m\ (\perp 2)$.

$i \cdot m = 2\ (\perp m)$.

$\frac{n}{\bar{n}} \rightarrow \frac{n}{m}$ when n is even.

$\frac{n}{\bar{n}} \rightarrow \bar{n}$ when n is odd.

The derivation of the point groups is outlined in Table 3. The symbols enclosed in brackets are duplicates of others deduced in columns to the left. A total of 32 symmetries are seen to be distinct. A useful order for these is given in Table 4. The designations in the column headed "Crystal system" will be clear later.

Space lattices

Combinations of translations. As shown in Fig. 2, a homogeneous pattern can be generated by the operations of a translation. It was noted earlier that such a pattern can act as a motif to be repeated by a translation which is noncollinear with the first; in this way a plane pattern like $p1$ of Fig. 1, Chapter 1, is generated. This pattern, in turn, can act as a motif to be repeated by a third translation which is noncoplanar with the first two. In this way a homogeneous three-dimensional pattern like Fig. 15*A* is generated.

For some purposes it is convenient to disregard the nature of the motif of the pattern by substituting for each motif unit a point, which, of course, has a geometrical significance but is not a physical object. The result of this substitution in the pattern of Fig. 15*A* is a collection of points, shown in Fig. 15*B*, called a *space lattice*, or, more simply, a *lattice*. A lattice has no physical existence, but is a purely geometrical abstraction which is useful in studying the geometrical properties that result from repetition by three noncoplanar translations.

Cells. The three translations which generate the pattern from an original motif, and which generate the lattice from an initial point, can be used to define a parallelepiped called a *cell*. This kind of cell, which has one lattice point at each vertex, is called a *primitive cell*. The three edges of a primitive cell define a lattice, but from any given lattice an infinite number of primitive cells can be selected. The one based on the three shortest noncoplanar edges is called the *reduced cell*.

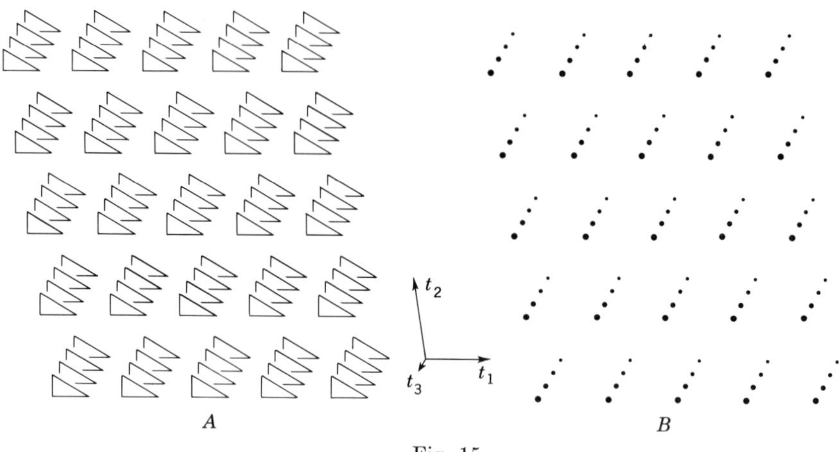

Fig. 15

To gain the advantages of symmetry, it is sometimes worthwhile to describe a crystal pattern by a cell which has one or more additional lattice points that are not at the cell vertices. Such cells are called *nonprimitive*, or *multiple*, *cells*. The multiplicity of a cell is the number of lattice points per cell, which is 1 for a primitive cell. Multiple cells are often designated by their multiplicity, for example, double cells, triple cells, and quadruple cells.

Indices. If any lattice point is taken as an origin, any other lattice point can be reached from it by a vector which is a linear sum of three vectors defined by the cell edges $\boldsymbol{t}_1$, $\boldsymbol{t}_2$, and $\boldsymbol{t}_3$, namely,

$$\boldsymbol{T} = u\boldsymbol{t}_1 + v\boldsymbol{t}_2 + w\boldsymbol{t}_3. \tag{13}$$

Here u, v, and w are integers. The term on the left is a vector, specifically, a translation from the origin point to the other lattice point. It defines a direction, called a *rational direction*, in the lattice. According to (13), this direction is characterized by the three integers u, v, w; it is symbolized by these integers placed between brackets: $[uvw]$.

A plane is determined by three points. If these points are lattice points, the plane is called a rational lattice plane, or simply a *rational plane*. Any rational plane behaves as a motif to the translations of the lattice, which produce from it a collection of parallel planes. These planes constitute a pattern which must be homogeneous. The requirement of homogeneity is that each plane have an environment which is indistinguishable from that of any other plane. This can be satisfied only if the spacings between all pairs of neighboring planes are identical. The planes therefore constitute a stack with a common spacing interval always symbolized as d.

Every lattice point must occur on a plane of the stack. Specifically, the lattice points at the cell's several vertices must occur on planes of the stack. The equally spaced planes must therefore cut each edge of the cell into an integral number of parts, as suggested in Fig. 16*A*. One of the three cell edges is normally called *a*, another *b*, and the third *c*. The integers into which these axes are cut are designated respectively by the general symbols *h*, *k*, and *l*. Specifically, the axes are cut by the planes of the stack as follows:

The *a* axis is cut into *h* parts.
The *b* axis is cut into *k* parts.
The *c* axis is cut into *l* parts.

Another aspect of these integers *h*, *k*, and *l* can be appreciated by writing the equation of the first plane from the origin. The intercept of the plane on the *a* axis is a/h; on the *b* axis it is b/k; on the *c* axis it is c/l. If the unit measures along the *a*, *b*, and *c* axes are the magnitudes of these translations *a*, *b*, and *c*, then the intercept form of the equation of the first plane (Fig. 16*B*) is

$$\frac{x}{1/h} + \frac{y}{1/k} + \frac{z}{1/l} = 1, \tag{14}$$

so that

$$hx + ky + lz = 1. \tag{15}$$

The integers *h*, *k*, and *l* are called the indices† of the plane. The symbol for a specific rational plane is this set of three integers between paren-

† Noncrystallographers commonly call these the "Miller" indices of the plane. This apparent sophistication should be discouraged because (1) no other indices of a plane are in use, and (2) Miller did not invent these, although he did popularize them.

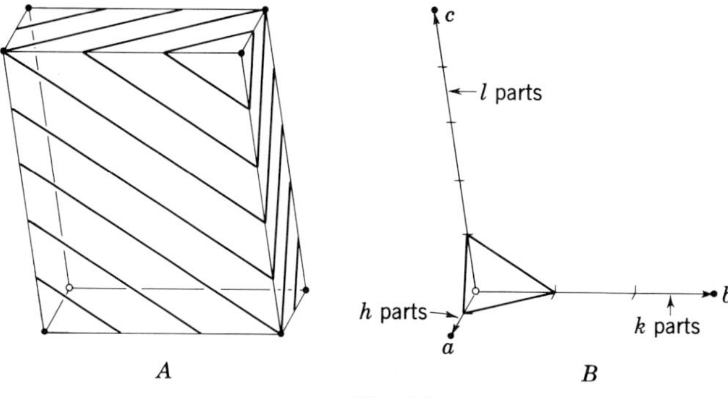

Fig. 16

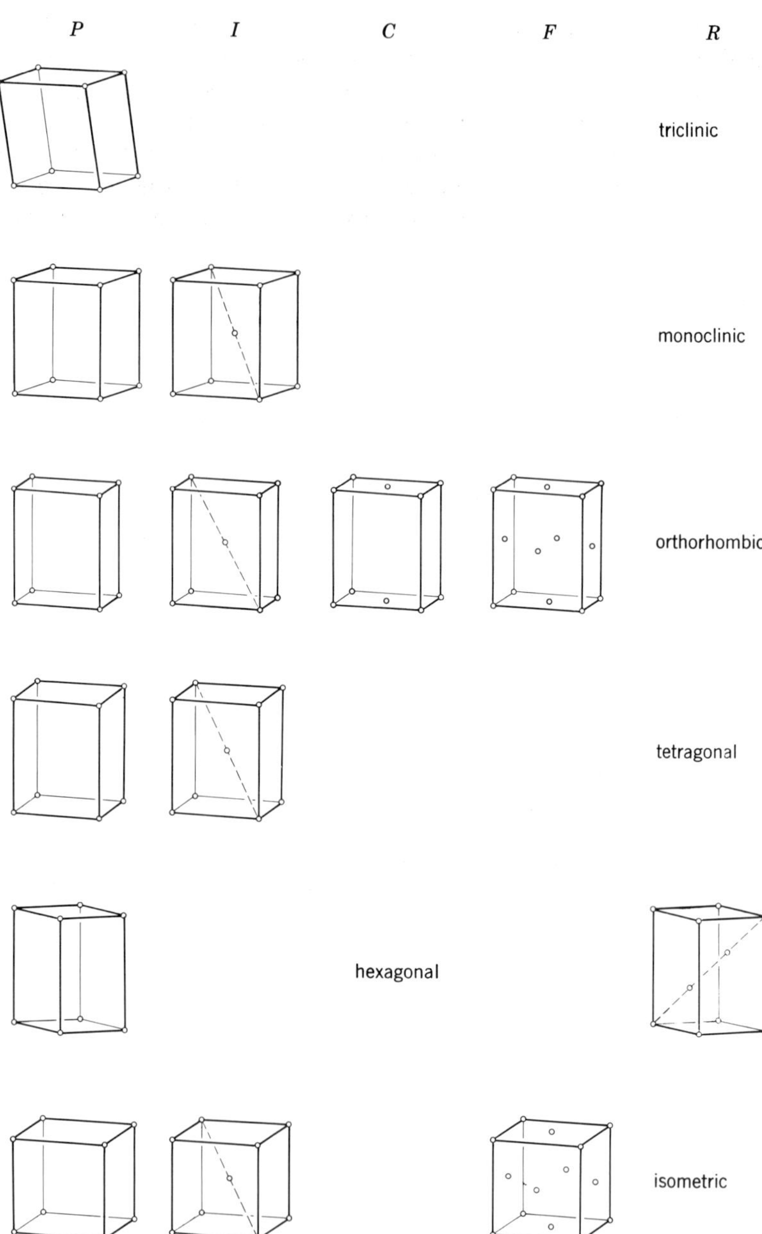

Fig. 17. The 14 symmetrical space lattices and their distribution among the crystal systems. [*From M. J. Buerger: Elementary crystallography.* (*Wiley, New York,* 1956 *and* 1963) 101.]

Table 5

The appropriate coordinate systems for the various point groups

Point groups	Space-lattice types permitted	Symmetry of space lattice	Crystal systems	Axes
$1, \bar{1}$	P	$\bar{1}$	triclinic	$a \neq b \neq c$ $\alpha \neq \beta \neq \gamma$
$2, m, \frac{2}{m}$	P, I (or A or B)	$\frac{2}{m}$	monoclinic	$a \neq b \neq c$ $\gamma \neq \alpha \equiv \beta = 90°$
$2\,2\,2, 2\,m\,m, \frac{2}{m}\frac{2}{m}\frac{2}{m}$	P, I, A (or B or C), F	$\frac{2}{m}\frac{2}{m}\frac{2}{m}$	orthorhombic	$a \neq b \neq c$ $\alpha \equiv \beta \equiv \gamma = 90°$
$4, \bar{4}, \frac{4}{m}$; $4\,2\,2, 4\,m\,m, \bar{4}\,2\,m, \frac{4}{m}\frac{2}{m}\frac{2}{m}$	P, I	$\frac{4}{m}\frac{2}{m}\frac{2}{m}$	tetragonal	$a \equiv b \neq c$ $\alpha \equiv \beta \equiv \gamma = 90°$
$3, \bar{3}$; $3\,2, 3\,m, \bar{3}\frac{2}{m}$ $6, \bar{6}, \frac{6}{m}$; $6\,2\,2, 6\,m\,m, \bar{6}\,2\,m, \frac{6}{m}\frac{2}{m}\frac{2}{m}$	$\begin{cases} R \\ P \end{cases}$ P	$\left.\begin{matrix} \bar{3}\frac{2}{m} \\ \bar{3}\frac{2}{m} \end{matrix}\right\}$ $\frac{6}{m}\frac{2}{m}\frac{2}{m}$	hexagonal	$a \equiv b \neq c$ $90° = \alpha \equiv \beta \neq \gamma = 120°$
$2\,3, \frac{2}{m}\bar{3}$; $4\,3\,2, \bar{4}\,3\,m, \frac{4}{m}\bar{3}\frac{2}{m}$	P, I, F	$\frac{4}{m}\bar{3}\frac{2}{m}$	isometric	$a \equiv b \equiv c$ $\alpha \equiv \beta \equiv \gamma = 90°$

theses: (*hkl*), for example, (312), (100), and (201). The three integers of a rational plane cannot have a common factor. If there were a common factor p, the stack would consist of the rational stack whose indices are these with the common factor removed, plus $p - 1$ additional planes equally spaced between the rational ones. The additional planes do not contain lattice points, and so are not rational.

Symmetrical lattices. Since a crystal must have a pattern which is referrable to a space lattice, and since a crystal may also have one of the 32 symmetries of Table 4, it follows that its space lattice must conform to that symmetry. These symmetrical lattices have been known since the early 1800s. They can be deduced readily by elementary methods, but this somewhat lengthy process is not reproduced here. It turns out that, except for the most general symmetries 1 and $\bar{1}$, there are at least two possible lattice types which conform to each symmetry. These, together with information still to be discussed, are listed in Table 5 and illustrated in Fig. 17; some explanations of the symbolism are given in Table 6.

The distribution of these general types among the different symmetries is shown in Table 5. This tabulation calls for the following remarks. In the second line, I, A, and B are not distinct but merely represent alternative permissible ways of selecting a cell from the lattice points. In the third line, A, B, and C are not distinct but merely represent the orientation selected for the centered cell. Finally, it will be observed that P for the 3-fold case and the P lattice for the 6-fold case have the

Table 6
The general kinds (or "modes") of space lattices

Multiplicity of cell	Label	Lattice requirements
Single	P	Primitive cell must have symmetry of lattice.
	I	Body-centered cell must display symmetry of lattice.
Double	A	Cell centered on A or (100) face must display symmetry of lattice.
	B	Cell centered on B or (010) face must display symmetry of lattice.
	C	Cell centered on C or (001) face must display symmetry of lattice.
Quadruple	F	Cell centered on all faces A, B, and C must have symmetry of lattice.
Single (Triple)	R	Primitive cell must have shape of rhombohedron. (This lattice is more conveniently referred to a cell with $a = b \neq c$, $\beta = 120°$, but having two internal points spaced $\frac{1}{3}$ and $\frac{2}{3}$ along the long body diagonal.)

same symmetry, as noted in the third column; therefore these two cases give the same lattice type. If these duplications are taken into account, there are 14 distinct specialized space-lattice types. A unit cell of each lattice type is shown in Fig. 17.

The crystallographic coordinate systems

The edges of the unit cell of the lattice provide the units of the natural coordinate system of the crystal. It should be noted that in Table 5 the P lattices for 3- and 6-fold axes are the same, and that in Table 6 the R lattice can be referred to a cell of the same external shape as P for 3- and 6-fold axes, but having two additional internal lattice points. If these two lattices are referred to the same cell edges, it can be seen that only six symmetrically distinct coordinate systems are required for crystals. These are called the six *crystal systems.* (Some authors distinguish, in different ways, a subdivision of the hexagonal system, calling it "trigonal" or "rhombohedral." Such practices lead to serious inconsistencies and should be avoided.)

The 32 symmetries permitted to crystals and the six coordinate systems to which crystals can be referred both constitute ways of classifying crystals; the most fundamental classification is that based upon the 32 symmetries. The significance of the crystal systems can be overestimated, and the restrictions they place upon crystals are frequently misunderstood by beginners. Basically, the six crystal systems provide six convenient coordinate systems to which to refer the geometry of crystals. Each crystal system, shown in the fourth column of Table 5, embraces several symmetries, as shown in the first column of that table.

Axes of systems. Crystallographers designate the cell edges as a, b, and c. These correspond to the units along the X, Y, and Z axes to be used as the coordinate systems for the crystals. The angles between the a, b, and c axes are labeled thus:

$$b \wedge c = \alpha,$$
$$c \wedge a = \beta,$$
$$a \wedge b = \gamma.$$

Symmetry causes the lengths of a, b, and c and the angles between them to be specialized. This specialization is listed in the last column of Table 5. The information contained in that column is so frequently subject to misinterpretation that an attempt is made here to avoid this. The sign $\equiv$ is the mathematical one for "is identically equal to," and implies that this is true under all circumstances. In this use, the equality is required by the symmetry (noted in the first column). The sign $\not\equiv$ means "is not generally equal to." In the present use, it does not prevent the two quantities connected by the sign from being fortuitously equal on occasion. Thus, in the orthorhombic case, the last column of Table 5 describes the relation between the lengths of the axes as $a \neq b \neq c$. This means that, in general, a, b, and c are not required by symmetry to be identical. It should be emphasized that, although symmetry does not require identity, nothing prevents them from being equal within the precision of measurement. If, on measurement of the cell edges of a certain crystal, it turned out that a and b were, by some chance, equal within the precision of measurement, this would imply $a = b$, but not $a \equiv b$. The finding that $a = b$ dimensionally, therefore, does not place the crystal in the tetragonal system, for which the last column of Table 5 gives $a \equiv b \neq c$. More generally, it should be clearly understood that the descriptions of axial dimensions in the last column of Table 5 are not definitions of the crystal systems: they are merely statements of dimensional restrictions in the crystal systems. On the contrary, the first column of Table 5 shows that a crystal is tetragonal if and only if its symmetry contains one axis with $n = 4$. This requires that $a \equiv b$. (If the symmetry contained more than one axis with $n = 4$, the $a \equiv b$ requirement would place the crystal in the isometric system, where the symmetry requires $a \equiv b \equiv c$.)

The specialization of axes required by symmetry, such as $a \equiv b$, implies that these axes cannot be distinguished. Under these circumstances they are called a_1 and a_2, instead of a and b, etc.

By common agreement, crystals are ordinarily drawn in a conventional orientation. The convention requires, first, that c be oriented in a vertical position with $+c$ upward. Second, b is oriented approximately left and right (if $\alpha = 90°$, b becomes exactly horizontal) with $+b$ to the observer's right. This leaves a oriented approximately toward and away from the observer; it is taken as + toward the observer. The resulting axial system is right-handed.

Special features of the several systems. There follow some remarks about conventions and other special features of the several crystal systems.

Isometric system. The three cell edges are orthogonal and equal, and, since they are indistinguishable, they are labeled a_1, a_2, and a_3. The axes therefore constitute the ordinary cartesian-coordinate system.

Tetragonal system. The three cell edges are orthogonal, but two are indistinguishable and are called a_1 and a_2. The third (tetragonal) axis is unique and is labeled c. It is customarily oriented vertically.

Hexagonal system. One axis corresponds to the symmetry axis with $n = 3$ (or 6). It is unique, labeled c, and customarily oriented vertically. The other two axes are in a plane normal to c; they make an angle $\gamma = 120°$ with each other, are indistinguishable, and are called a_1 and

a_2. Neither a_1 nor a_2 can be distinguished from a third axis with which they form a set of three related by this 3-fold axis c. This unnecessary but inescapable axis is called a_3. The hexagonal system is customarily referred to four axes: $a_1 \equiv a_2 \equiv a_3 \neq c$. The indices of a plane with respect to these are, respectively, h, k, i, and l. Because a_1, a_2, and a_3 are linearly dependent in their plane, the integers h, k, and i are also dependent, as follows:

$$h + k + i = 0. \tag{16}$$

The indices of a plane in the hexagonal system are written fully as $(hkil)$, which, because of the linear dependence in (16), is often abbreviated $(hk \cdot l)$.

Orthorhombic system. The three axes a, b, and c are distinct but orthogonal. By universal agreement they are labeled so that $a < b$. An earlier generation of crystallographers placed no restriction on the length of c; however, it is desirable to describe newly measured crystals with the convention $a < b < c$. (Some crystallographers prefer $c < a < b$, which is much like defining 1, 2, and 3 so that $3 < 1 < 2$. No logic prevents this, but $a < b < c$ requires a minimum of arbitrary ruling.)

Monoclinic system. If the unique axis with $n = 2$ is called c (as in the tetragonal system with unique axis $n = 4$, and the hexagonal system with unique axis $n = 3$ or 6), the monoclinic system is said to be described in the *first setting*. In this setting the 2 or $\bar{2}$ axis is called c and taken vertically, while the plane of a and b is orthogonal to c. The angle $\gamma = a \wedge b$ is taken as obtuse. (If the system is described in the *second setting*, b is taken as the 2 or $\bar{2}$ axis and set horizontally, and the plane of a and c is taken orthogonal to b. The second setting was usual with a past generation; however, it does not fit in with the principle of a minimum of arbitrary rulings, and therefore should be dropped.)

Triclinic system. The three cell edges are entirely unspecialized in lengths and interaxial angles. Since there is no symmetrical specialization, there are an infinite number of ways in which the cell can be chosen from the lattice. By convention that cell is chosen which has for edges the three shortest noncoplanar edges. This cell is called the reduced cell. Such cells occur in two types: one can be described by a set of three edges surrounding a corner with three acute angles, the other by a set of three edges surrounding a corner with three obtuse angles. These are called all-acute cells, or type I cells, and all-obtuse cells, or type II cells, respectively.

All crystallographers agree with the convention of labeling the cell edges to conform with $a < b$. For reasons mentioned in the discussion of orthorhombic crystals above, it is desirable to describe newly measured crystals with the additional restriction $a < b < c$.

Space groups

General features of derivation. The derivation of the space groups is not difficult, but it is tedious. The derivation of the point groups was based largely on understanding how the results of several rotations can be combined. To derive space groups it is necessary to know not only how these combinations can be made but also how rotations, screws, reflections, inversions, glides, and lattice translations can be combined with each other. In this book, no attempt is made to develop these combinations, but the general ways that the results of these combinations are restricted are pointed out.

Permissible operations. Table 1 shows the symmetry elements which give rise to homogeneous patterns. The ones corresponding to operations with translation components τ have not yet been considered.

In particular, screws $A_{\alpha,\tau}$ applicable to crystallographic patterns must have $n = 1, 2, 3, 4$, or 6, so that $\alpha = 0, 180, 120, 90$, or $60°$. Furthermore, as pointed out in (3), τ must be a rational fraction of t smaller than unity. Each such screw is designated by the symbol n_q, where $n = 1, 2, 3, 4$, or 6, and q is an integer smaller than n. When $q = 0$, $\tau = (0/n)t$, and so the screw degenerates to a pure rotation. The crystallographic screws are illustrated in Fig. 18.

As pointed out in (2), a glide plane m_τ must have its translation component parallel to a lattice translation and equal to half of it. A glide plane receives a symbol which indicates to what lattice translation its τ is parallel. The symbols and what they convey are listed in Table 7.

Table 7
Some characteristics of glide planes

Translation component	Type of glide plane	Symbol
$\frac{a}{2}$	axial-glide plane	a
$\frac{b}{2}$		b
$\frac{c}{2}$		c
$\frac{a}{2}+\frac{b}{2}$, or $\frac{a}{2}+\frac{c}{2}$, or $\frac{b}{2}+\frac{c}{2}$	diagonal-glide plane	n
$\frac{a}{4}+\frac{b}{4}$, or $\frac{a}{4}+\frac{c}{4}$, or $\frac{b}{4}+\frac{c}{4}$, or $\frac{a}{4}+\frac{b}{4}+\frac{c}{4}$	"diamond"-glide plane	d
Zero	mirror	m

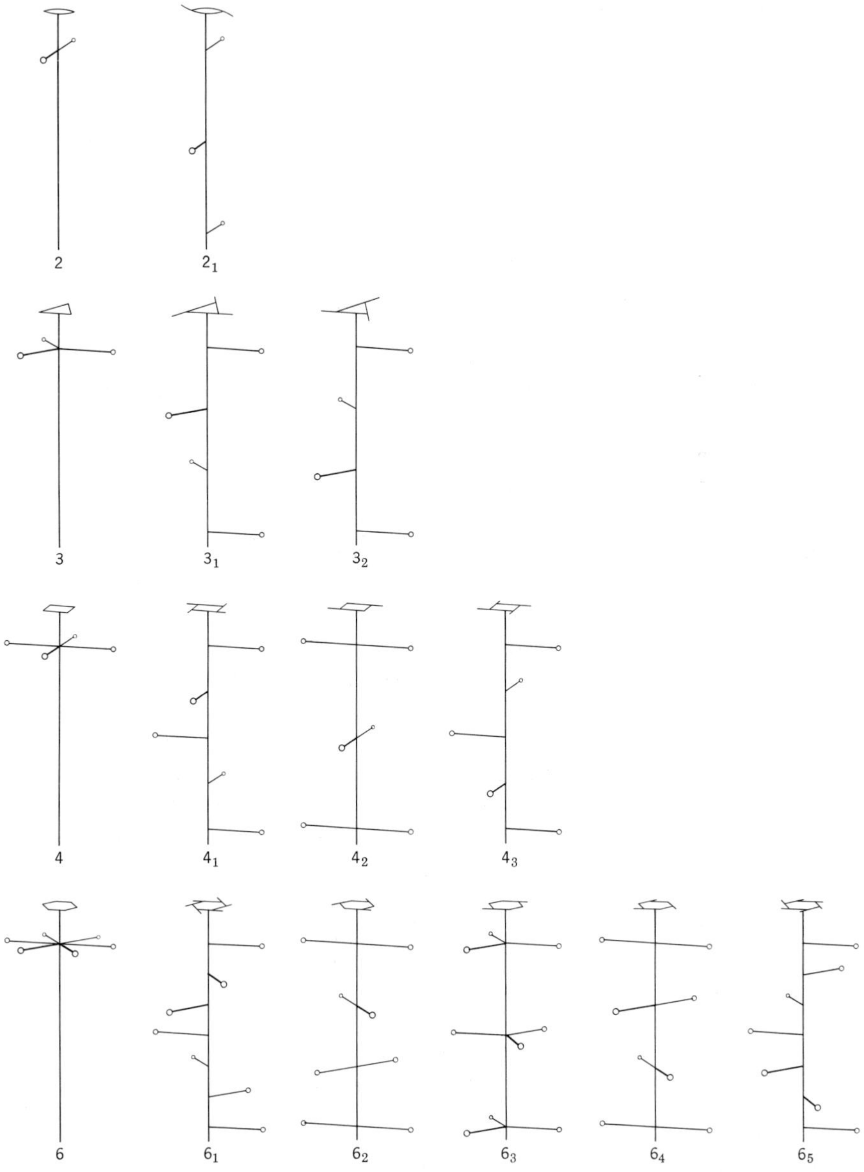

Fig. 18. The 15 crystallographic screw axes. Each axis appears as a vertical line capped by the geometrical symbol of the axis. The written symbol of the axis is given just below the bottom of the line. The kind of symmetry which each axis requires is demonstrated by the way an arbitrary point is repeated about it. [*From M. J. Buerger: Elementary crystallography.* (*Wiley, New York,* 1956 *and* 1963) 205.]

Isogonal relations. Rotations are isogonal if they have the same rotational component α. Thus the axes 6, 6_1, 6_2, 6_3, 6_4, and 6_5 all have $\alpha = 60°$, and so are mutually isogonal.

If two rotational operations also have translation components, these translations cannot affect the rotational components since the motions are orthogonal. It follows that if two pure rotations A_α and B_β can be combined at a specific angle $A \wedge B$ with a result $C_{-\gamma}$, possible translation components cannot affect the angular aspects of the combination. Accordingly, A_{α,t_1} combined with B_{β,t_2} must produce $C_{-\gamma,t_3}$, in which the α, β, and γ are the same as for the isogonal set, and $A \wedge B$, $B \wedge C$, and $C \wedge A$ are the same as for the isogonal set. That is, the combination of two screws which are isogonal with two pure rotations is a screw which is isogonal with the combination of the pure rotations; the inclinations of the axes within one set are isogonal with the inclinations in the isogonal set.

From this it follows that to every point group there exists a set of one to several isogonal space groups. The point group and the set of space groups are isogonal with respect to their respective rotations and the angles which their axes make with one another. Note also that a reflection plane m in a point group is isogonal with a mirror m or a glide m_τ in the isogonal space groups.

Examples of isogonal space groups. No attempt is made here to deduce or to enumerate the results of combining lattice translations with symmetry operations, although in each case such a combination results in a new symmetry operation; or, more crudely, a lattice and a symmetry element combine to produce a new symmetry element usually at a different location. Although these results are not treated, examples of the space groups isogonal with two point groups are illustrated in Fig. 19.

Space-group symbols. Space groups are characterized by

(1) the general lattice type (that is, P, A, B, C, I, F, or R),

(2) the several particular symmetry axes in the space group which are isogonal with the corresponding symmetry axes in the isogonal point group, and

(3) the several reflection planes or glide planes in the space group which are isogonal with the corresponding mirrors in the isogonal point group.

The point-group symbol consists of a sequence of one, two, or three items, any one of which may be a fraction. If a fraction, the item has the symbol for the proper rotation in the numerator and a mirror in the denominator, for example,

$$4 \qquad \frac{4}{m} \qquad 2\,3 \qquad 4\,m\,m.$$

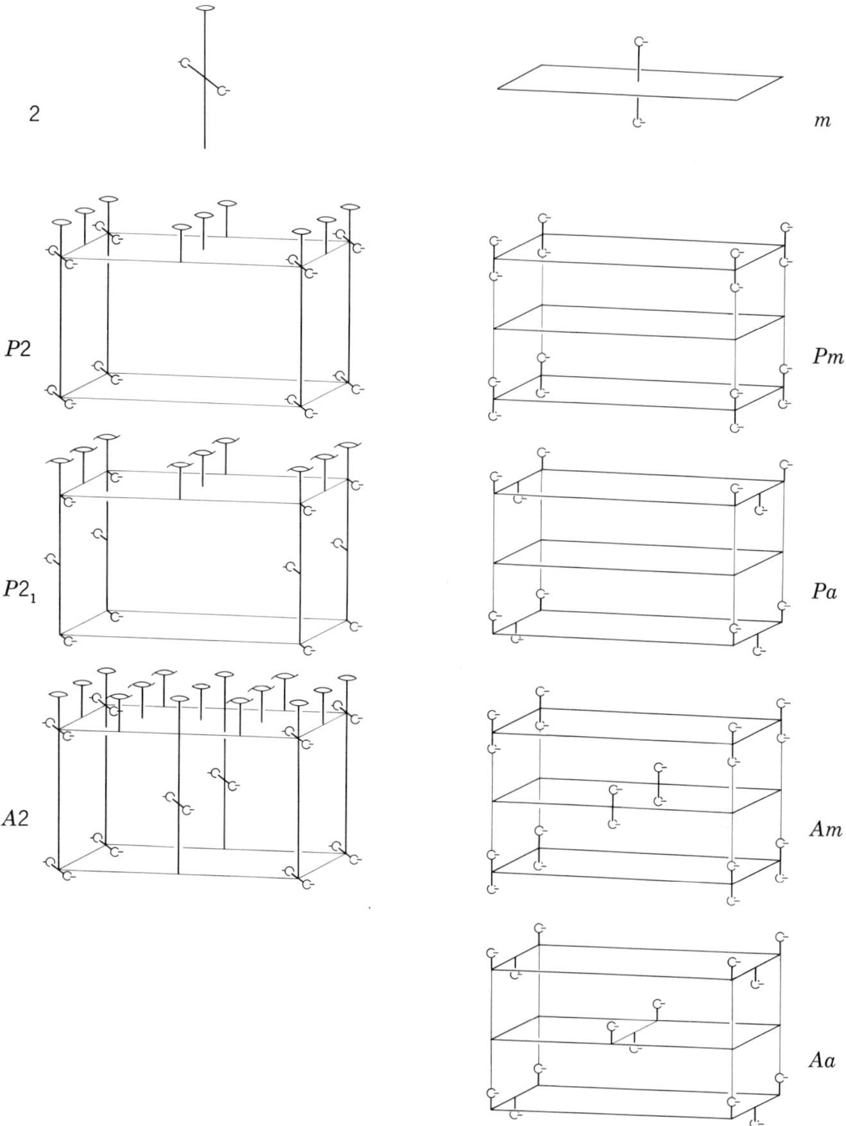

Fig. 19. The upper pair of drawings show the symmetry elements of two monoclinic point groups, 2 and *m*, and the way they require an arbitrary point to be repeated. Below each point group is its several isogonal space groups and the repetition of an arbitrary point by each space group.

Table 8
Correspondence between symbols of point groups and isogonal space groups

Point group	Isogonal space groups		
4	$P\,4$ $P\,4_1$ $P\,4_3$ $P\,4_2$	$I\,4$ $I\,4_1$	
$\frac{4}{m}$	$P\frac{4}{m}$ $P\frac{4}{n}$ $P\frac{4_2}{m}$ $P\frac{4_2}{n}$	$I\frac{4}{m}$ $I\frac{4_1}{a}$	
$2\,3$	$P\,2\,3$ $P\,2_1\,3$	$I\,2\,3$ $I\,2_1\,3$	$F\,2\,3$
$4\,m\,m$	$P\,4\,m\,m$ $P\,4\,c\,c$ $P\,4\,b\,m$ $P\,4\,n\,c$ $P\,4_2\,m\,c$ $P\,4_2\,c\,m$ $P\,4_2\,b\,c$ $P\,4_2\,n\,m$	$I\,4\,m\,m$ $I\,4\,c\,m$ $I\,4_1\,m\,d$ $I\,4_1\,c\,d$	

Any space groups which are isogonal with these point groups have a sequence of isogonal items plus a prefixed capital letter indicating the lattice type. For example, the space groups isogonal with the four point groups just noted have symbols with items as shown in Table 8.

Notes on history

The development of space-group theory was a nineteenth-century affair. During the first half of that century the symmetrical lattices had been conceived, first by Frankenheim in 1842 and later by Bravais in 1850. Each of these lattices has the property that all its lattice points have the same symmetry with respect to their surrounding points. It was first supposed that, in a crystal structure, each lattice point was occupied by a molecule having the symmetry of the point. But this had the disadvantage that the possible symmetries of the lattice points can only be one of the 7 so-called holohedral crystal classes, so that this theory of internal symmetry cannot account for the other 25 classes. Bravais, who held this view, later modified it by permitting the molecule at the lattice point to have a symmetry less than that of the lattice point. By this relaxation the theory accounted for the 32 crystal classes by attributing the symmetry of a crystal essentially to the shape of the molecule.

In an attempt to define the general nature of the structural state which gives a crystal its unique properties, such as planes with rational intercepts, cleavage along rational planes, and anisotropic optical properties, the structure was considered to be a homogeneous arrangement of molecules. In terms of Bravais' theory, this homogeneity, defined in geometrical terms, amounted to the requirement that every molecule in a crystal be related to every other molecule by lattice translations. Thus, "homogeneity" was thought to be identical with translation equivalence. But about 1870 a less stringent view was independently suggested by three men: Wiener, Sohncke, and Jordan. They believed that a structure composed of points could be regarded as homogeneous if the appearance of the structure from every point was the same.

Sohncke based an entirely new theory of the internal arrangement and symmetry in crystals on this view. His interpretation of the geometrical requirement of homogeneity can be illustrated by considering a pattern of molecules generated by repeating one molecule by periodic application of the same translation. The pattern has the obvious property that the entire structure appears the same from every molecule. If this pattern is wound around a circular cylinder in such a way that the line of translations becomes a helix, the structure still appears the same from every molecule. Thus, while homogeneity can be produced by the geometry of translation, it can also be achieved by a screw motion. Sohncke's theory of homogeneous structures is based on such screw motions. If the structure also has a lattice, then the points on one helix, and also points on parallel helices, must be related by the translations of that lattice. Sohncke used geometrical means essentially the same as employed in this chapter to develop this subject. For example, he showed that the combination of screw motions about two nonparallel axes is equivalent to a screw motion about a third axis.

By geometrical means Sohncke shows that there are 65 "regular, infinite point systems." These have the symmetries of what are now known as the 65 axial space groups. Sohncke's contribution to the theory of crystal symmetry constituted a true breakthrough in the science of crystallography, for it constituted the firm basis of both method and result upon which all further progress was founded. His work, unfortunately, is not usually accorded the credit it deserves, because few were interested in the subject until about 1915, and by then others had added to Sohncke's foundation.

Sohncke's point systems did not account for any symmetries of the second sort. It remained to scientists of three nationalities and from diverse fields to add this feature later. About 1890 this added feature came independently and almost simultaneously from Schoenflies (a German mathematician), Barlow (an English chemist), and Fedorow (a Russian crystallographer). Schoenflies added glide planes and $\bar{4}$ axes to the 65 point systems of Sohncke to produce the additional 165 space groups. Barlow and Fedorow used different methods to achieve the same results. Their theories were complete well before the opening of the twentieth century, but they aroused little interest until 1919, when Niggli showed how the space groups of crystals could be determined by x-ray studies. Chapter 5 of this book is concerned with this matter.

Additional reading

Harold Hilton. *Mathematical crystallography and the theory of groups of movements.* (Clarendon Press, Oxford, 1903) 262 pages.

Paul Niggli. *Geometrische Kristallographie des Diskontinuums.* (Gebrüder Borntraeger, Leipzig, 1919) 576 pages.

Artur Schoenflies. *Theorie der Kristallstruktur, ein Lehrbuch.* (Gebrüder Borntraeger, Berlin, 1923) 555 pages.

Johann Jakob Burckhardt. *Die Bewegungsgruppen der Kristallographie.* (Birkhäuser, Basel, 1947) 186 pages.

M. J. Buerger. *Elementary crystallography. An introduction to the fundamental geometrical features of crystal.* (Wiley, New York, 1963) 528 pages.

Significant literature

Chr. Wiener. *Die Grundzüge der Weltordnung (Atomlehre).* (Leipzig and Heidelberg, 1863, 1869) 82.

L. Sohncke. *Die Gruppirung der Moleküle in den Krystallen.* Poggendorffs Ann. **132** (1867) 75–106.

C. Jordan. *Mémoire sur les groupes de mouvements.* Ann. Mat. F. Brioschi e L. Cremona [Milan] **2** (1868, 1869) 167–215, 322–345.

L. Sohncke. *Die unbegräntzen regelmässigen Punktsysteme als Grundlage einer Theorie der Krystallstructur.* (Braun, Karlsruhe, 1876).

Leonhard Sohncke. *Entwickelung einer Theorie der Kristallstruktur.* (Leipzig, 1879) 247 pages.

Arthur Schoenflies. *Krystallsysteme und Krystallstruktur.* (Teubner, Leipzig, 1891) 638 pages.

William Barlow. *Ueber die geometrischen Eigenschaften homogener starrer Structuren und ihre Anwendung auf Krystalle.* Z. Kristallogr. **23** (1894) 1–63.

E. von Fedorow. *Symmetrie der regelmässigen Systeme der Figuren* [in Russian, 1890]. *Theorie der Krystallstructur.* Z. Kristallogr. **24** (1895) 209–252; **25** (1896) 113–224.

3

The diffraction of x-rays by crystals

Prediffraction background

The study of crystals has made use of whatever experimental tools have been available. In 1669 Nicolaus Steno showed, by cutting sections across quartz crystals, that regardless of the sizes of the faces on the crystals, they were always inclined to one another at constant dihedral angles. This was the beginning of the law of constancy of angles between corresponding faces. Rome de l'Isle put this on a firm footing in 1772 by many measurements made with a contact goniometer, which is, essentially a protractor with an attached bar. The reflecting goniometer, invented by Wollaston in 1809, permitted measurement of interfacial angles on a much more precise basis and led to the design of the single-circle goniometer, with which interfacial angles could be measured to about a minute of arc. This instrument was especially suited to establishing the angular relations between the several faces in a zone. The two-circle goniometer, with which the normals to various faces could be located in terms of spherical coordinates similar to latitude and longitude, was independently invented by Miller in 1874, Fedorov in 1889, and Victor Goldschmidt in 1893. This tool became the popular one for characterizing crystals in North America in the early part of this century, but its use has declined as the last generation trained to employ it gave way to the first generation having x-ray diffraction available as an experimental tool.

The study of the optical properties of crystals might be said to have begun in 1828 when William Nicol, a teacher of physics in Edinburgh, invented his polarizing prism. Eventually, this became part of the polarizing microscope, which was, and still is, extensively used by mineralogists and petrologists. The immersion method was used by Maschke in 1871 and by Schroeder von der Kolk in 1892. With this method the refractive indices of crystal fragments are compared with a number of liquid standards in which they are successively immersed. The comparison is made as the fragments are viewed through the polarizing microscope.

Until 1912, the optical goniometer and the polarizing microscope were the only tools available to the crystallographer for routine study and characterization of crystals. The goniometer permitted the study of the crystal surface or morphology, and the data so obtained could be used to establish the axial ratio and point-group symmetry. The polarizing microscope yielded information from which the crystal could be assigned to one of several broad symmetry categories, and also usually supplied the refractive indices of the crystal. But none of the tools available allowed the crystal pattern itself to be studied. X-ray diffraction eventually supplied the tool for studying crystal patterns and thus opened the door to an intelligent understanding of crystals.

Although x-ray diffraction was discovered in 1912, the First World War prevented its extensive use until the post-war period. Between the two world wars x-ray diffraction underwent extensive development in both theory and technique and became the popular means for studying crystals. The expanded use of x-ray diffraction methods was accompanied by such a decline in the use of goniometric and optical methods that these older methods are all but unknown to many present-day crystallographers, especially those drawn from the ranks of physics and chemistry. On the other hand, crystallographers coming from a background of mineralogy are ordinarily trained in optical and goniometric methods and so are at home in classical methods as well as in the newer ones.

This chapter initiates a sequence whose purpose is to introduce the reader to an outline of the theory and experimental means of studying crystal patterns by means of x-ray diffraction.

Diffraction

When matter is in the path of a train of waves, it may, under appropriate circumstances, scatter the waves. If the scale of detail of the matter is approximately that of the wavelength, the waves scattered in the same direction by various features of this detail combine with one another in such a way that in some directions the scattered waves reinforce one another and in others they annul one another; this general phenomenon is called diffraction of the waves by the object. For crystals the scale of detail is atomic, and radiation having wavelengths of this order of magnitude can be expected to be diffracted by the atoms of crystals. In particular, x-rays constitute electromagnetic waves, and atoms compose electrical systems capable of being disturbed by these waves. The fluctuation of the electric field due to the electromagnetic wave disturbs the electrons of the atoms, which absorb the x-rays and reemit them at the same frequency; in this specific way, atoms are said to *scatter* the x-rays which impinge upon them.

The nature of diffraction is dependent upon the pattern which produces it. This can be formulated in several ways, for example, by an equation, one side of which contains a function representing the details of the patterns, the other the manner in which the diffraction effects depend upon it. If the pattern is known, its diffraction can be predicted. If full data on the diffraction are available, the equation can also be solved for the pattern. This possibility is examined in more detail in Chapters 12, 13, and 14. In this chapter the relation between the geometry of the diffraction and the lattice of the crystal structure is outlined. Some preliminary attention is also given to the manner in which the motif of the pattern affects the diffraction.

Diffraction by an atom

With each atom are associated a number of electrons, each of which scatters the impinging x-rays. Since the effective cross section of the atom has a scale comparable to x-ray wavelengths, waves scattered in a

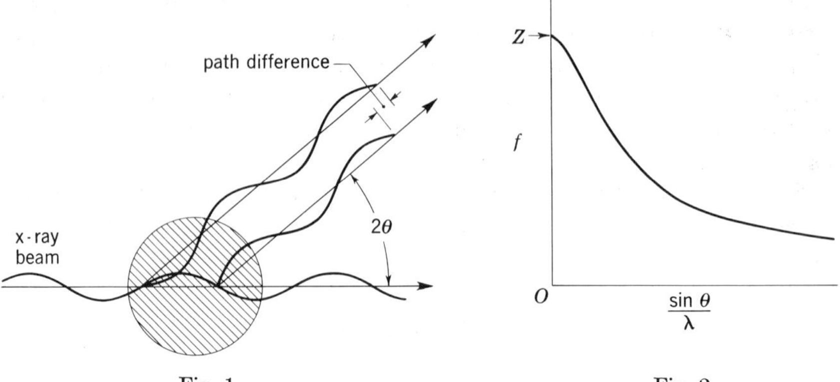

Fig. 1

Fig. 2

particular direction by the electron within the atom are, in general, not in phase with one another, as suggested in Fig. 1. The path difference is zero for scattering in the direction of the incident ray, but increases with the angle which the direction of scattering makes with the direction of the direct beam. As a consequence, if the amplitude of the resultant wave scattered by the atom is plotted against this angle, the amplitude is found to decline as the angle increases, as seen in Fig. 2. The amplitude is customarily measured in terms of the scattering power of one electron as a unit. In the direction of the direct beam all the electrons of the atoms scatter in phase; therefore the amplitude is equal to that of Z electrons, where Z is the atomic number of the atom. With increasing angle 2θ (Fig. 1) the amplitude falls off, and this is usually plotted as a function of $(\sin \theta)/\lambda$, where λ is the wavelength of the x-rays; $(\sin \theta)/\lambda$ is a convenient quantity which increases with θ and which is invariant with certain features of the experiment. It should be observed in Fig. 2 that, although the scattering declines with θ, it never becomes zero but is always some appreciable quantity.

Diffraction by crystals

The scattering by all the atoms in a crystal can be demonstrated in a number of different ways. A convenient way is first to study how a set of translation-equivalent atoms scatters and then to combine the scattering from as many such sets as are contained in the unit cell. As an easy preliminary, the scattering by a single row of translation-equivalent atoms is considered first.

Scattering by a row of atoms. When an x-ray wavefront impinges on a row of translation-equivalent atoms, as suggested in Fig. 3A, each atom scatters the incoming wave as a spherical wavelet centered on the atom. If wavefronts are regarded as succeeding one another at intervals equal to the wavelength λ, each atom scatters these fronts as they arrive, and so becomes surrounded by a series of expanding spherical wavelets whose successive radii differ by λ. According to Huygens' principle, any tangent common to the wavelets about neighboring atoms marks out a new wavefront moving in the direction of normal to that front. Many such normals can be formed, only a few of which are illustrated in Fig. 3*A* by arrows. Each constitutes the direction of a wave which is said to be diffracted by the row of atoms. When the path difference for pairs of rays from an incoming wavefront to a pair of adjacent atoms, and on to the new wavefront, is zero, the wave is said to constitute the zero-order diffraction maximum; when the paths differ by one wavelength, the wave is said to constitute the first-order diffraction maximum; when the paths differ by two wavelengths, the wave constitutes the second-order diffraction maximum; etc.

Some important geometrical aspects of the diffraction can be derived with the aid of Fig. 3*A* and *B*. Let $\bar{\mu}$ be the angle the incoming ray makes with the row of atoms, and let $\bar{\nu}$ be the angle the diffracted ray makes with the row. (The convention adopted here is that both angles are measured about the same atom, and $\bar{\mu}$ is taken as 90° or less.) For the arrangement shown in Fig. 3*B*, the total path difference is composed of two parts: the additional path from a wavefront to the atom it reaches last, and the additional path from that atom to the new diffracted wavefront. In order for the wavefront to be a diffraction maximum, the sum of these two parts must restore the original phase relation at the wavefront; this requires the path difference to be an integral number of wavelengths. The condition for a diffraction maximum is therefore

$$a \cos \bar{\mu} + a \cos \bar{\nu} = m\lambda, \tag{1}$$

where m is an integer. A minor rearrangement gives the relation

$$\cos \bar{\mu} + \cos \bar{\nu} = \frac{m\lambda}{a}. \tag{2}$$

This is known as the *Laue condition* for scattering in phase by all the members of a row of translation-equivalent point scatterers.

The convention of angles used here differs a little from those used originally by Laue. Both $\bar{\mu}$ and $\bar{\nu}$ are measured about the same atom. The angle to the incoming ray, $\bar{\mu}$, can always be taken as acute; but the angle of the diffracted ray, $\bar{\nu}$, can then be either acute, as seen in Fig. 3*B*, or obtuse, as seen in Fig. 3*D*. In the first case $\cos \bar{\nu}$ is positive; in the second, negative. When these are equal and opposite (Fig. 3*C*), the left side of (2) vanishes; since both a and λ must be positive, this requires

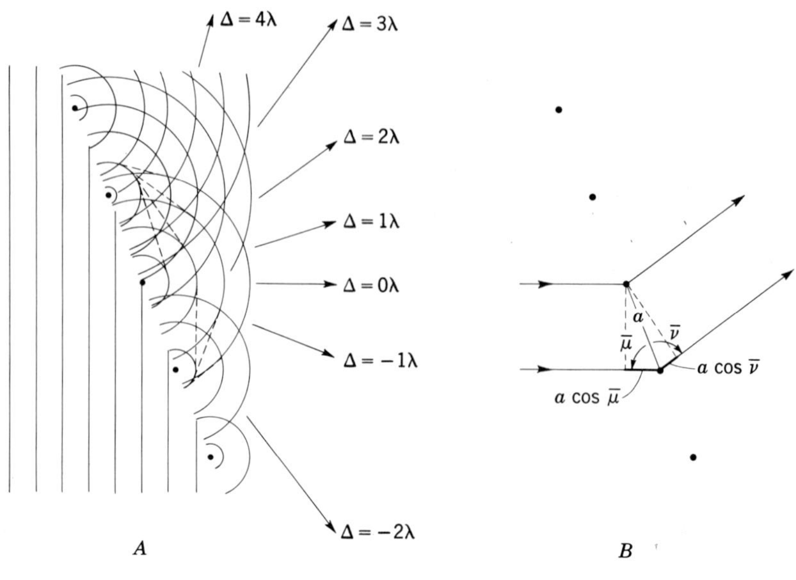

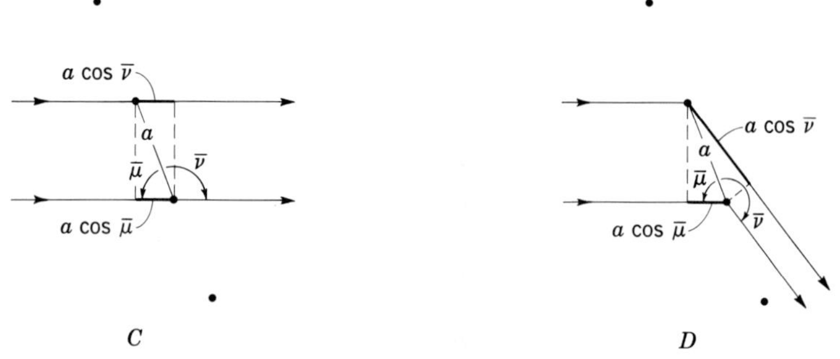

Fig. 3

order m to be zero. When the left side of (2) is negative, the order m must be negative also.

Another arrangement of (2), useful for predicting $\bar{\nu}$ when λ, a, m, and $\bar{\mu}$ are given, is

$$\bar{\nu} = \cos^{-1}\left(\frac{m\lambda}{a} - \cos \bar{\mu}\right). \tag{3}$$

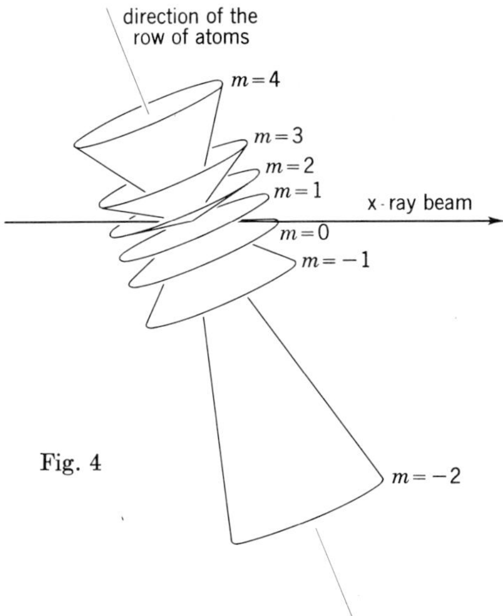

Fig. 4

The integer m may assume only such values that the term in parentheses lies within the permissible range of the cosine, namely, $+1$ to -1.

The locus of directions making a particular angle $\bar{\nu}$ with a given line lies along the generators of a cone coaxial with the line and having half-opening angle $\bar{\nu}$. Accordingly, the directions corresponding to the permissible solutions of (3) for various values of the integer m lie along the generators of a nest of cones coaxial with the row, one cone for each permissible value of m in (3). An example is illustrated in Fig. 4. These cones are called the *Laue cones* corresponding to the rational direction of the row. It will be seen subsequently that to every rational direction of the lattice there corresponds such a nest of Laue cones.

Scattering by a lattice array of atoms. The simplest kind of crystal is one in which every atom is equivalent to all others by the translation of the lattice.† Such a set of atoms is not a lattice (since a lattice is an imaginary set of points) but rather a *lattice array* of atoms. The scattering by such an array can be understood in terms of the scattering by a row of atoms (which is the same as a one-dimensional lattice array of atoms).

When an x-ray beam strikes a lattice array of atoms, all atoms scatter the x-rays, but the scattered wavelets do not reinforce one another to produce maxima unless certain conditions are satisfied. Under ordinary circumstances any arbitrary row scatters wavelets which are in phase at various angles $\bar{\nu}$ given by (3) and illustrated by Figs. 3 and 4. But all

† This simple kind of pattern occurs in some (but not all) of the elements, for example, iron, titanium, copper, gold, and argon.

the atoms of the lattice array do not scatter in phase unless conditions like (2) are satisfied simultaneously, for some particular direction of diffraction, by three noncoplanar rows, specifically,

$$\begin{aligned}\cos \bar{\mu}_X + \cos \bar{\nu}_X &= \frac{m\lambda}{a},\\ \cos \bar{\mu}_Y + \cos \bar{\nu}_Y &= \frac{n\lambda}{b},\\ \cos \bar{\mu}_Z + \cos \bar{\nu}_Z &= \frac{p\lambda}{c}.\end{aligned} \tag{4}$$

Here a, b, and c are the translation intervals along any rows OX, OY, and OZ; while $\bar{\mu}_X$, $\bar{\mu}_Y$, and $\bar{\mu}_Z$ are the angles which the incoming beam makes with these rows, and $\bar{\nu}_X$, $\bar{\nu}_Y$, and $\bar{\nu}_Z$ are the angles which the diffracted beam makes with these rows. Relations (4) are the Laue conditions for diffraction by a three-dimensional lattice array. The conditions can be understood by reference to Fig. 4, which is a graphical representation of the Laue condition for one row alone. Three such rows scatter in phase only if three cones (one from each nest) intersect in a common generator and so define a common diffraction direction. Such an intersection for an arbitrary direction of the x-ray beam and an arbitrary orientation of the lattice would be fortuitous, and in general does not occur.

To understand the conditions to be satisfied for all atoms to scatter in phase in the same direction, the conditions can be simplified by considering the corresponding situation in two dimensions, which brings out the essential features. In Fig. 5, let row OX scatter in the mth order and row OY scatter in the nth order. This means neighboring points along OX scatter with a path difference of $m\lambda$, while neighboring points along OY scatter with a path difference of $n\lambda$. The path from the incoming wavefront to an arbitrary point P (whose coordinates with respect to the lattice origin O are $u\ v$) and on to the diffracted wavefront exceeds the corresponding path to and from the origin point O by $(um + vn)\lambda$. Since u, v, m, and n are all integers, the arbitrary point P scatters with a path difference with respect to the origin which is an integral number of wavelengths. Provided that the scattered wavelets constitute a common wavefront proceeding in the same direction, then, *all points of the lattice array scatter in phase when any one arbitrary row* (OX in Fig. 5) *scatters in the mth order and any other nonparallel row* (OY in Fig. 5) *scatters in the nth order.* This rule can be readily generalized to three dimensions.

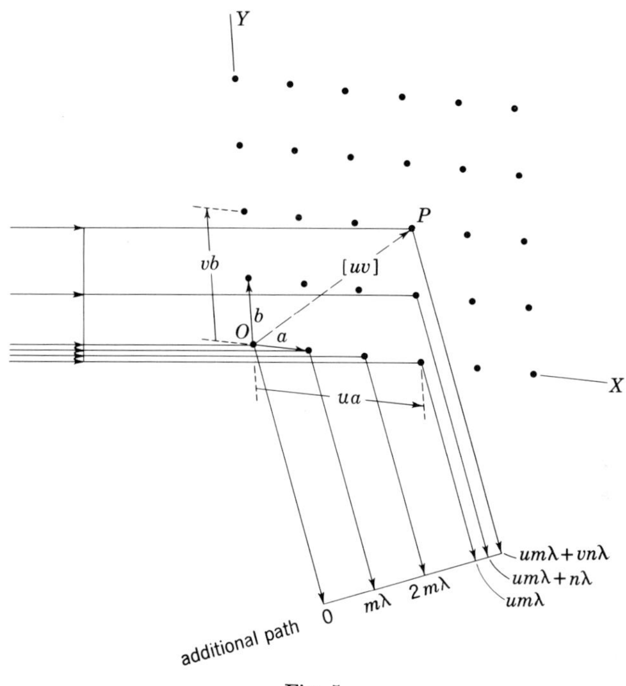

Fig. 5

There is a simple interpretation of the Laue conditions for three noncoplanar rows. This is again illustrated in simplified two-dimensional form. In Fig. 6, if the row OX scatters in the mth order and the row OY scatters in the nth order, then the Ath atom along the row OX scatters with a path difference $Am\lambda$ with respect to the origin atom O, and the Bth atom in row OY scatters with a path difference $Bn\lambda$ with respect to the origin atom O. Both of the path differences $Am\lambda$ and $Bn\lambda$ are equal to $mn\lambda$ when $A = n$ and $B = m$. In other words, the nth atom along OX and the mth atom along OY both scatter with the same path difference $mn\lambda$ with respect to the origin atom, so that the paths from the incoming wavefront to each atom A and B to the diffracted wavefront are the same. This is also the condition for the optical reflection of the rays by the plane connecting A and B (Fig. 6). This plane has intercepts $A = n$ and $B = m$, which define the crystallographic plane represented by (mn).

This result can be readily generalized to three dimensions along the following outline: If a three-dimensional lattice array scatters in phase so that the diffraction can be described in terms of the Laue orders

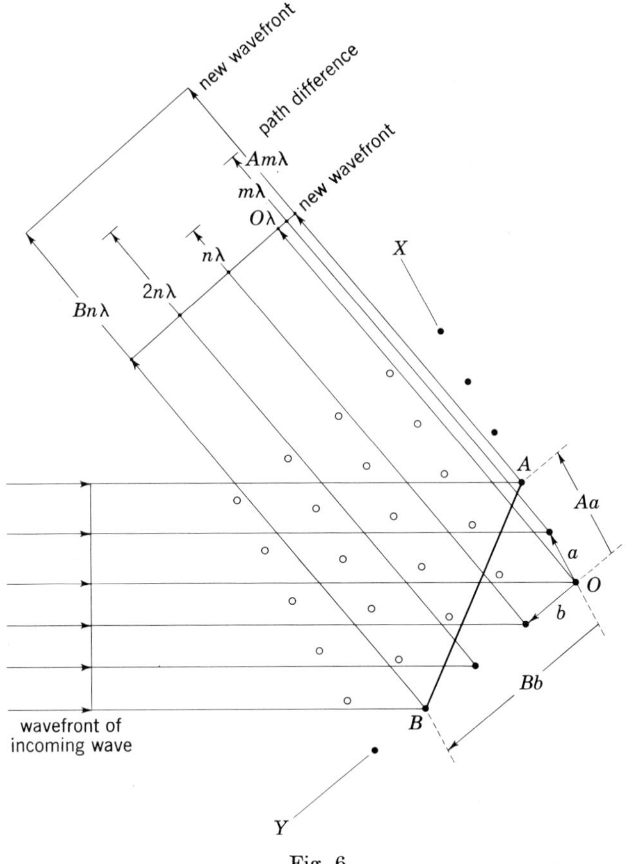

Fig. 6

m, n, and p for three noncoplanar rows (which can be taken as axes OX, OY, and OZ), this is equivalent to reflection by a plane whose intercepts are np, pm, and mn, respectively. The equation of this plane is

$$\frac{x}{np} + \frac{y}{pm} + \frac{z}{mn} = 1, \tag{5a}$$

which can be readily transformed into

$$mx + ny + pz = mnp. \tag{5b}$$

This is the equation of the mnpth plane from the origin. The first plane is represented by the equation

$$mx + ny + pz = 1. \tag{6}$$

This is a plane whose crystallographic indices (hkl) are (mnp).

Theorem: If a lattice array of atoms scatters in the Laue orders m, n, p, this requires all points on the crystallographic plane (mnp) to scatter in the same phase, which is geometrically equivalent to reflection of the x-rays by atoms in the crystallographic plane (mnp).

Bragg's law. The theorem just stated shows that there is an alternative way of considering x-ray diffraction by a lattice array of atoms. Such an array can also be regarded as a distribution of atoms on the crystallographic planes of a stack (hkl). The atoms in each plane diffract as if they were optically reflecting the rays. The question arises as to what conditions must be satisfied so that each plane of the stack reflects in phase with the other planes. Let the common spacing between planes be d_{hkl}. Then the condition that the reflections from neighboring planes be in phase is that the path difference of rays reflected by the neighboring planes be an integral number of wavelengths; Fig. 7 shows that this path difference is $2d_{hkl} \sin \theta$, and so the condition is

$$n\lambda = 2d_{hkl} \sin \theta_{hkl}. \tag{7}$$

This relation is known as *Bragg's law*, after Sir Lawrence Bragg, who first stated it in English.†

† The Russians attribute this to Wulf. See the personal reminiscence of A. V. Shubnikov in P. P. Ewald's *Fifty years of x-ray diffraction.* (N. V. A. Oosthoek's Uitgeversmaatschappij, Utrecht, 1962) 648.

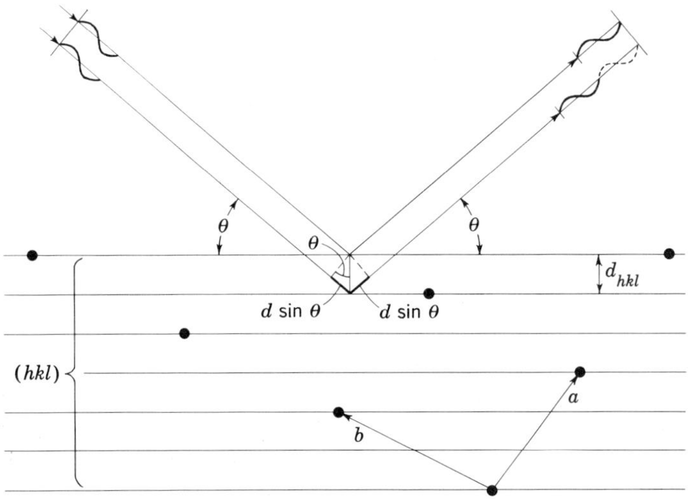

Fig. 7

It is useful to rearrange (7) in the following way:

$$\theta_{hkl} = \sin^{-1}\left(\frac{n\lambda}{2d_{hkl}}\right). \tag{8}$$

Here n is an integer which gives the number of wavelengths of path difference for the rays reflected by neighboring planes of the stack. Relation (8) shows that this kind of reflection by the planes of the stack (hkl), unlike ordinary optical reflections, takes place only if the crystal is turned so that a particular plane makes an angle θ, which is given by (8), with the incoming x-ray beam. This angle has discrete values because only the discrete values of the integers are allowable for n in (8). Furthermore, only a limited number of such integers n are permissible, namely, only those for which the term in parentheses has an absolute value less than or equal to unity.

The order of reflection n in Bragg's law is not commonly retained, but instead is merged with the indices hkl in the following way: Only those planes of a lattice are rational (and so have a real existence in this sense) for which the three integers h, k, and l have no common factor. Nevertheless, consider the geometry of a stack of planes which would be specified by indices that do have a common factor n, namely, the planes (nh,nk,nl). These planes cut the axes a, b, and c into n times as many parts as do the planes of the stack (hkl). They therefore have a spacing $1/n$th as large as do the planes (hkl). Therefore

$$d_{nh\,nk\,nl} = \frac{1}{n}\,d_{hkl}. \tag{9}$$

If for n/d_{hkl} in (8) there is substituted the equivalent from the left side of (9), then (8) can be rewritten as

$$\theta_{hkl} = \sin^{-1}\left(\frac{\lambda}{2d_{nh\,nk\,nl}}\right). \tag{10}$$

In this form of Bragg's equation the order n has now appeared as a possible common factor in the indices.

But $nh\ nk\ nl$ no longer represents a rational crystal plane. It now represents both the plane which is reflecting and the order in which it is reflecting; in short, it now characterizes the *reflection*. To distinguish these two kinds of indices, the set of indices representing a plane is written, as usual, in parentheses, while the set of indices representing a reflection is written without parentheses. For example, (123) represents a rational crystallographic plane. It can give rise to first-, second-, and third-order reflections; these are designated reflections 123, 246, and 369, respectively.

Diffraction by all the atoms of a crystal. Only a very few crystals are so simple in structure that they consist of a single lattice array. In general, crystal structures are characterized by several (often many) atoms per unit cell. The entire cell is repeated by the translations of the lattice, so that each of the atoms in the cell is repeated by the same translations to become a separate lattice array. The crystal structure can therefore be thought of as n intermeshed lattice arrays, one for each of the n atoms in the cell. Each such lattice array diffracts as described in the last section. The question arises as to how their diffraction effects interact.

For the sake of clarity, suppose there are only two different atoms in a unit cell, as illustrated in Fig. 8. For convenience, let these be referred to as white atoms and black atoms. Suppose that the two lattice arrays are reflecting in the first order; then the path difference for rays reflected by two successive planes of the white lattice array is λ. The path difference for rays reflected by two successive planes of the black lattice array is also λ. But the path difference for corresponding planes of the two lattice arrays is a smaller amount, $\Delta\lambda$. In exactly the same way that (7) was derived from Fig. 7, the following relation for the path difference $\Delta\lambda$ in terms of the separation Δd of the two lattice arrays can be derived:

$$\Delta\lambda = 2\Delta d_{hkl} \sin \theta_{hkl}. \tag{11}$$

If this is compared with (7), in which n is taken as unity, it is seen that

$$\frac{\Delta\lambda}{\lambda} = \frac{\Delta d_{hkl}}{d_{hkl}}. \tag{12}$$

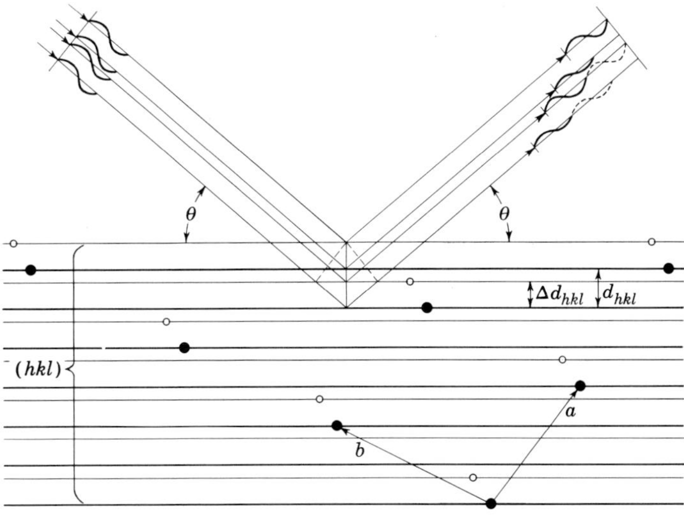

Fig. 8

It is convenient to express path difference in terms of the phase difference of the wave. Since a path difference of λ corresponds to one cycle, or an angular phase difference of 2π, it is evident that the phase difference in the waves scattered by corresponding planes of the two lattice arrays can be found from the proportion

$$\frac{\Delta\phi}{2\pi} = \frac{\Delta\lambda}{\lambda}, \tag{13}$$

so that the phase difference is

$$\Delta\phi = \frac{\Delta\lambda}{\lambda} 2\pi = \frac{\Delta d}{d} 2\pi. \tag{14}$$

A qualitative answer can now be given to the question raised at the beginning of this section. A crystal composed of n lattice arrays of atoms reflects x-rays in exactly the same geometric way as a single lattice array of atoms having the same lattice geometry. But whereas a single lattice array reflects x-rays so that all atoms scatter in phase, and so gives rise to a maximum scattering amplitude for all reflections, this is not true for the amplitudes scattered by several lattice arrays, because, in general, there are phase differences among the waves scattered by them. This reduces the amplitudes of the combined scattering to less than the maximum; and, as will be seen, the amount of reduction depends on the indices h, k, and l and on the coordinates x, y, and z of the atoms doing the scattering. This gives rise to the following simple theorem:

Theorem: The geometry of x-ray diffraction depends *only* on the cell geometry, but the amplitudes of the various diffraction maxima depend on the locations of the atoms in the cell.

The second part of the theorem is probably obvious from the qualitative discussion just given. The magnitudes of the amplitudes as a function of the coordinates of the atoms are derived in the next section.

Diffraction amplitude. It is a simple matter to put the phase relations mentioned in the last section on a basis suitable for computation. Figure 9 shows a two-dimensional unit cell in which there is an atom having fractional coordinates x (a fraction of the a axis) and y (a fraction of the b axis). In two dimensions the stack of reflecting planes corresponds to a series of lines represented by the symbol (hk). These lines cut the a axis in h parts and the b axis in k parts. Accordingly, the distance between lines as measured along the a axis is a/h; along the b axis, b/k. Neighboring lines scatter with a phase shift of 2π, so that

a shift of a/h parallel to a corresponds to 2π, and
a shift of b/k parallel to b corresponds to 2π.

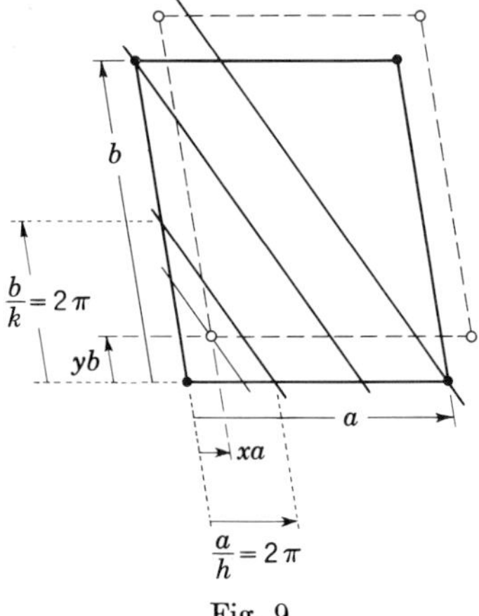

Fig. 9

If the scattering atom is first regarded as at the origin and then displaced by a distance xa in the direction of the a axis, it produces a phase shift which is given by the proportion

$$\frac{\Delta\phi_a}{2\pi} = \frac{xa}{a/h} = hx; \tag{15}$$

and if it is displaced by a distance yb in the direction of the b axis, it produces a phase shift provided by

$$\frac{\Delta\phi_b}{2\pi} = \frac{yb}{b/k} = ky. \tag{16}$$

Accordingly, if an atom is displaced from the origin to coordinates $x\,y$ the entire phase shift is

$$\Delta\phi = \Delta\phi_a + \Delta\phi_b = 2\pi(hx + ky). \tag{17}$$

If the zero value of phase is taken as that of a wave scattered by a point at the origin, then $\Delta\phi$ for an atom simply becomes ϕ, and the phase of a wave scattered by an atom whose coordinates are $x\,y$ is merely

$$\phi = 2\pi(hx + ky). \tag{18}$$

In three dimensions the corresponding expression is

$$\phi = 2\pi(hx + ky + lz). \tag{19}$$

It is convenient to represent the properties of a wave graphically on a circular diagram, as in Fig. 10. The amplitude of the wave is a radial vector, and the phase is represented by the angle which the vector makes with the horizontal axis. The superposition of two wave disturbances is

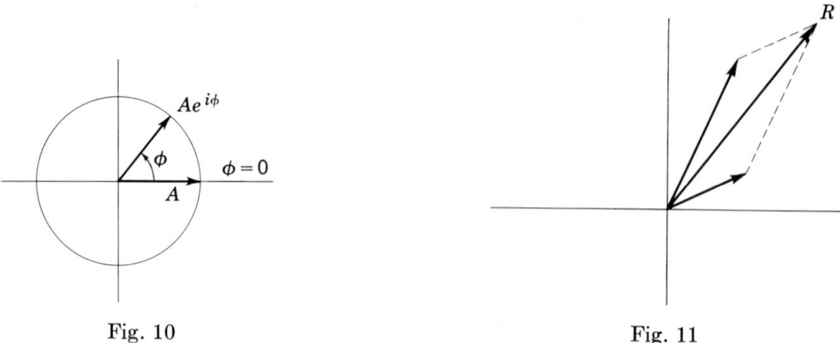

Fig. 10

Fig. 11

then the sum of the two vectors representing their relative amplitudes and phases (Fig. 11). The algebra of complex numbers can be neatly applied to the geometry of this wave diagram: The radial vector terminates in a point which is a complex number; that is, it has a real component on the horizontal axis, and it has an imaginary component on the vertical axis. The unit on the imaginary axis is $i = \sqrt{-1}$. The algebra of the complex plane provides an interesting and useful role for the term $e^{i\phi}$. When any complex number is multiplied by this term, $e^{i\phi}$ behaves as an operator which rotates the vector through an angle $+\phi$. Thus, if F represents a wave having an amplitude $|F|$, then $Fe^{i\phi}$ represents a similar wave whose phase is greater by angle ϕ than that of F.

Figure 12 shows a diagram in the complex plane, called the *Argand diagram*, on which two waves are represented, one having an amplitude whose magnitude is f_1 and whose phase is ϕ_1, and another whose amplitude has magnitude f_2 and phase ϕ_2. These waves are expressed by $f_1e^{i\phi_1}$ and $f_2e^{i\phi_2}$. In Fig. 12 their resultant is labeled F. This graphical representation can be written analytically as

$$F = f_1e^{i\phi_1} + f_2e^{i\phi_2}. \tag{20}$$

The wave resulting from scattering by two atoms in a unit cell can be expressed in the same way. If these atoms have coordinates $x_1y_1z_1$ and $x_2y_2z_2$, and their scattering powers are f_1 and f_2, respectively, then if (19) and (20) are used, the wave scattered by the two atoms can be written

$$F_{hkl} = f_1e^{i2\pi(hx_1+ky_1+lz_1)} + f_2e^{i2\pi(hx_2+ky_2+lz_2)}. \tag{21}$$

More generally, if there are j atoms, their resultant can be expressed by the summation

$$F_{hkl} = \Sigma f_je^{i2\pi(hx_j+ky_j+lz_j)}. \tag{22}$$

This expression is commonly called the *structure factor* for reflection hkl. In general, it is a complex quantity; that is, it has both magnitude and phase.

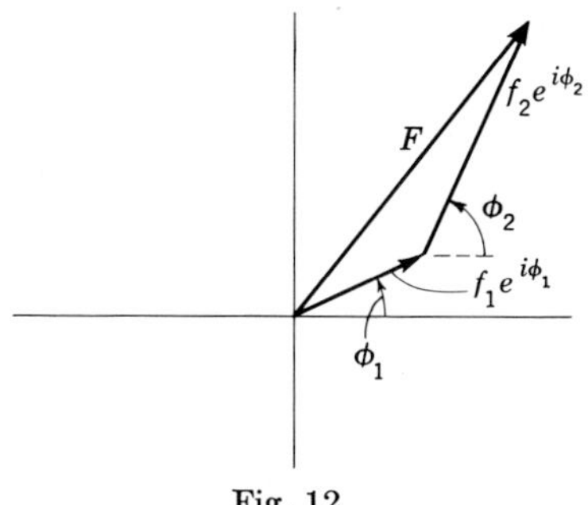

Fig. 12

Summary

A three-dimensional periodic pattern of matter can diffract a train of waves whose wavelength is of the order of the pattern periods. The characteristics of the diffraction can be used to deduce information about the pattern which produced it.

In particular, an atom scatters x-rays which impinge upon it. A periodic row of atoms scatters in such a way that there are maxima in the directions of the generators of a set of cones, called *Laue cones,* which are coaxial with the row. The directions are controlled by an equation known as the *Laue condition.* For a three-dimensional lattice array of atoms a Laue condition must be satisfied simultaneously for each of three such rows. This can occur only for certain specific orientations of the lattice array with respect to the x-ray beam.

The geometrical conditions for the occurrence of diffraction maxima in terms of the Laue conditions are not easy to visualize. Bragg reformulated the situation in terms of reflections by the planes of a stack whose indices are (hkl). The formulation is neatly condensed in *Bragg's law:* $n\lambda = 2d \sin \theta$, where n is an integer, λ is the wavelength of the x-rays, d is the spacing of the planes, and θ, known as the *glancing angle,* is equal to the complement of the angle of incidence or reflection of geometrical optics. It turns out that the indices of the reflecting plane are the same as the orders of diffraction corresponding to three axial Laue cones, provided indices with a common factor are permitted. These three indices hkl, written without parentheses, are called the *indices of the reflection.*

An actual crystal, except in some very simple cases such as the native metals, ordinarily is a composite of several parallel, displaced lattice arrays. The geometry of reflections from such a composite lattice array is exactly the same as that from one of the component lattice arrays, but the amplitudes of the reflections are functions of the relative displacements of the arrays. Another statement of this, which is more useful, is that the directions of the diffraction maxima depend on the geometry of the cell, while their amplitudes depend on the arrangement of the atoms in the cell. The amplitude, in general, is a complex quantity, commonly known as the *structure factor*.

Notes on history

The discovery of the diffraction of x-rays by crystals, about May, 1912, constituted the beginning of one of the most remarkable developments in science. Before that time the nature of x-rays was not known. Some physicists thought they were streams of corpuscles; others, that they were wave motion akin to light, and evidence from attempts to observe diffraction by a slit suggested that if they were indeed wave motion they would have wavelengths of the order of 4×10^{-9} cm. Cathode rays were known to be caused by charged particles called electrons, but the nature of atoms, if they existed, was not understood. Bohr's quantum theory of the atom would not be presented until 1913. On the other hand, the theory of both the external and the internal symmetry of crystals had been established for nearly a quarter of a century.

Paul Ewald, then a doctoral candidate at the University of Munich, was writing his thesis under Professor Arnold Sommerfeld on the behavior of a light wave in an anisotropic lattice array of dipoles. He discussed some of the results he was obtaining for his thesis with Privatdozent Max Laue. The latter's interest in diffraction prompted him to raise the question of what would occur if the light had a wavelength near that of the separation of dipoles in the lattice array. Laue became so interested in the possibility of diffraction by the lattice array that he influenced two assistants, W. Friedrich and P. Knipping, to set up an experiment in which a pencil of x-rays fell on a copper sulfate crystal. The hoped-for diffracted beams were photographically recorded in their second experiment, and the results were communicated to the Bavarian Academy of Sciences in June and July, 1912, and also presented to several other meetings of physicists about the same time.

This unusual development was of tremendous interest to William Lawrence Bragg, a young student at the University of Cambridge, England. Noting the geometrical shapes of the spots in Friedrich and Knipping's photographs, Bragg believed that this kind of diffraction could be regarded as cooperative reflections by the internal planes of the crystal, and accordingly reformulated diffraction by a crystal in these terms. This led to what we now know as Bragg's law; this was based upon a simpler phenomenon to visualize than the simultaneous satisfying of three Laue conditions, although these were, as shown in this chapter, equivalent.

As early as 1883, William Barlow had published the arrangements of atoms to be expected in certain symmetrical crystals such as the alkali halides and zinc sulfides on the assumption that the atoms would behave as spheres. The Professor of Chemistry in Cambridge, Pope, persuaded Bragg to test these predicted structures in explaining the x-ray diffraction results. This Bragg did, with successful results, and so began a new era in which the structure of crystals could be analyzed by x-ray diffraction. Bragg was able to set forth the structures of NaCl, KCl, KBr, KI, diamond, CaF_2, Cu_2O, ZnS, pyrite, $NaNO_3$, and the $CaCO_3$ family of minerals before the First World War intervened. But this beginning set the stage for an increase in the detailed knowledge of the structure of matter which is unique in the history of science.

Additional reading

P. P. Ewald. *Fifty years of x-ray diffraction.* (N. V. A. Oosthoek's Uitgeversmaatschappij, Utrecht, 1962) 31–73.

Significant literature

William Barlow. *Probable nature of the internal symmetry of crystals.* Nature **29** (1883) 186–188, 205–207.

W. Friedrich, P. Knipping, and M. Laue. *Interferenz-Erscheinungen bei Röntgenstrahlen.* Sitzungsber. math.-phys. Klasse der Königlich Bayer. Akad. Wiss. München (1912) 303–322; reprinted in Naturwiss. (1952) 361–368.

M. Laue. *Eine quantitative Prüfung der Theorie für die Interferenz-Erscheinungen bei Röntgenstrahlen.* Sitzungsber. math.-phys. Klasse der Königlich Bayer. Akad. Wiss. München (1912) 363–373; reprinted in Naturwiss. **39** (1952) 368–372.

M. v. Laue. *Röntgenstrahlinterferenzen.* Phys. Z. **14** (1913) 1075–1079.

W. H. Bragg and W. L. Bragg. *The reflection of x-rays by crystals.* Proc. Roy. Soc. (London) (A) **88** (1913) 428–438.

W. L. Bragg. *The structure of some crystals as indicated by their diffraction of x-rays.* Proc. Roy. Soc. (London) (A) **89** (1913) 248–277.

W. H. Bragg and W. L. Bragg. *The structure of the diamond.* Proc. Roy. Soc. (London) (A) **89** (1913) 277–291.

W. Lawrence Bragg. *The analysis of crystals by the x-ray spectrometer.* Proc. Roy. Soc. (London) (A) **89** (1914) 468–489.

W. H. Bragg and W. L. Bragg. *X rays and crystal structure.* (G. Bell, London, 1915) especially 8–21.

W. H. Bragg and W. L. Bragg. *The discovery of x-ray diffraction.* Current Sci. (India), special number on "Laue Diagrams," (1937) 9–10.

Lawrence Bragg. *The history of x-ray analysis.* (Address given before First Conference on X-ray Analysis in Industry, Institute of Physics, Cambridge, England, 1942. Issued by the British Council as part of a pamphlet series entitled "Science in Britain") (Longmans, London, 1943) especially 8.

4

The reciprocal lattice

Geometrical interpretation of Bragg's law

While Bragg's law permits a simple interpretation of the diffraction of radiation by a triperiodic array, it is not easy to visualize at the same time all stacks of planes of a crystal with their various orientations. Bragg's law can be reformulated in an interesting way which permits an easy visualization of the situation in terms of a geometrical device known as the *reciprocal lattice.*

The standard form of Bragg's law, given in (7) of Chapter 3, is

$$n\lambda = 2d_{hkl} \sin \theta_{hkl}. \tag{1}$$

If, now, only first-order maxima ($n = 1$) are considered, a convenient rearrangement of (1) is

$$\sin \theta_{hkl} = \frac{\lambda(1/d_{hkl})}{2}. \tag{2}$$

A simple geometrical interpretation of this is shown in Fig. 1: If a circle of unit radius is drawn, then $\sin \theta$ is half the length of the chord subtended by the angle θ, one of whose sides is the diameter of the circle; the other side is then a line representing the slope of the reflecting plane. If some of these features are entered onto the diagram, as in Fig. 2, then it has the property of showing

(*a*) the direction of the direct beam,
(*b*) the direction of the diffracted beam, and
(*c*) the normal to the crystal plane.

The last item is the direction of the line segment whose length is $\lambda(1/d)$. Can any meaning be given to this vector?

It can indeed, for this is the vector from one point of a lattice, known as the *reciprocal lattice*, to another point of that lattice. The reciprocal lattice is a geometrical device discovered by the French mining engineer Bravais. It was merely an elegant relation between two lattices when he published his discovery in a memoir in 1850. It was not until 1912 that its use in connection with x-ray diffraction was discovered, and only about 1927 did it become an indispensable tool for understanding diffraction by crystals. The existence and some properties of this geometrical tool will be discussed next.

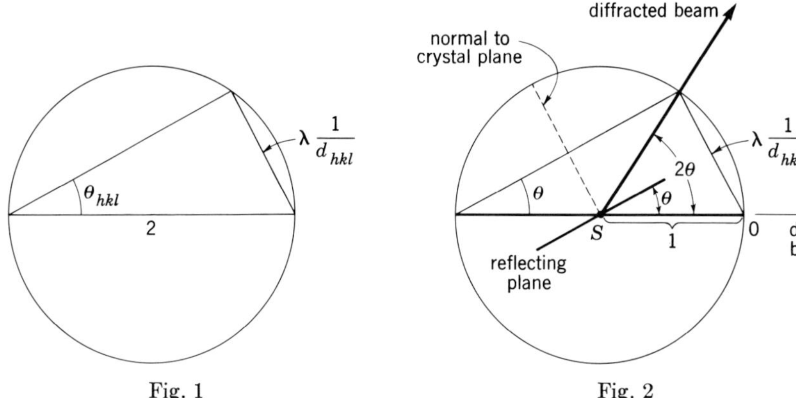

Fig. 1

Fig. 2

Theory of reciprocal lattices

It is an interesting fact that with every lattice can be associated another imaginary lattice called the reciprocal lattice. This new lattice has properties which are useful in describing certain properties of the original lattice. The relations between these two lattices will first be discussed in two dimensions, since this will provide the key to the geometrical relations without bringing in the complications of the third dimension.

The reciprocal lattice in two dimensions. Figure 3i shows a region of a plane lattice with lines, corresponding to the planes (hkl) in three dimensions, drawn through lattice points. In two dimensions these lines are represented by the symbol (hk). For any of the lines of the set (hk), a segment L_{hk} of the line which is delimited by neighboring lattice points, is the base of a parallelogram which is a possible primitive cell of the lattice. The altitude of the parallelogram is the spacing d_{hk} between neighboring lines. Now, the areas of all possible primitive cells of a lattice are equal; and, in terms of the parallelogram whose base is the segment of (hk), the area is given by

$$\text{Area} = L_{hk}d_{hk}. \tag{3}$$

At the beginning of this chapter it was seen that $1/d_{hk}$ is an important quantity in diffraction. Its magnitude can be found from (3), its specific

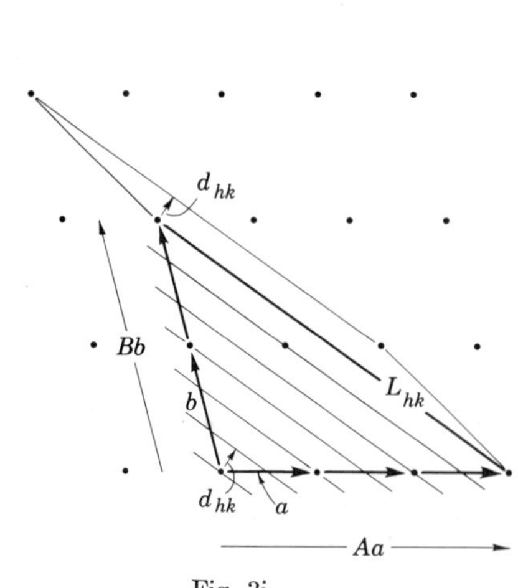

Fig. 3i

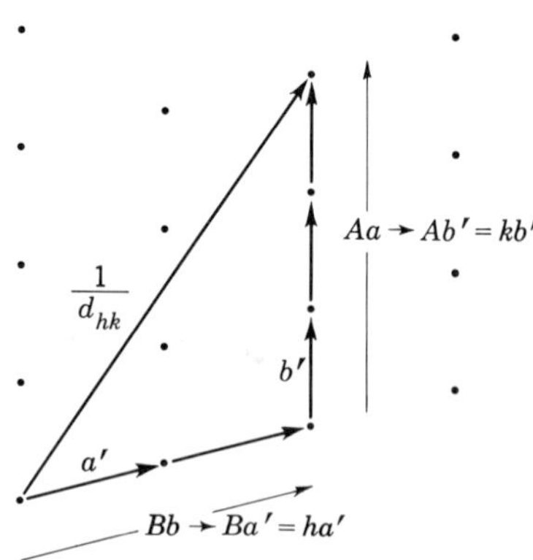

Fig. 3ii

value being

$$\frac{1}{d_{hk}} = \frac{1}{\text{area}} L_{hk}. \tag{4}$$

Thus the magnitude of the reciprocal of d_{hk} is always proportional to the magnitude of the line segment L_{hk}. But the directions of d_{hk} and L_{hk} are orthogonal, and this holds for every (hk). If L_{hk} were turned through 90°, it would become the desired $1/d_{hk}$ in both magnitude and direction, provided the proportionality constant 1/area is regarded as unity. All possible $1/d$'s can evidently be found by merely rotating the original lattice in Fig. 3i through 90° so that it becomes Fig. 3ii; every L_{hk} is then transformed into $1/d_{hk}$. Now in Fig. 3i, the line segment L_{hk} had indices h and k because it had intercepts A and B, for in two dimensions the relations between intercepts and indices are given in the sequence of equations which begins with the equation of a rational-intercept plane, as follows:

$$\frac{x}{A} + \frac{y}{B} = 1,$$

$$\frac{AB}{A} x + \frac{AB}{B} y = AB,$$

$$Bx + Ay = AB.$$

Let $$h = B$$

and $$k = A;$$

then $$hx + ky = AB.$$

The relations between the pair h, k and the pair A, B are as follows: In the original lattice (Fig. 3i), the line segment L_{hk} was determined by its intercepts $h = B$ and $k = A$, so that it might also be expressed as L_{BA}. In the rotated lattice, the integers B, A become the integers u, v of the components $u\mathbf{a}' + v\mathbf{b}'$ of the translation $1/d_{hk}$.

The important features of the relations just discussed may be restated in the following way: For any given plane lattice there exists a related plane lattice, known as its reciprocal lattice, which has certain useful features to be noted below. Provided that the area of a primitive cell is regarded as unity, in two dimensions the reciprocal lattice can be derived simply by rotating† the original lattice through 90°. (If the area of the primitive cell is not regarded as unity, the lattice must be not only rotated but also scaled by a factor equal to the reciprocal of its area.)

The fundamental property of the reciprocal lattice is that for every

† This simple rotation relation is replaced in three dimensions by a more general and complicated relation.

spacing vector between lines in the direct lattice there is a corresponding parallel vector in the reciprocal lattice

(*a*) which is a rational translation,

(*b*) whose components u and v with respect to the cell edges of the reciprocal lattice are the same integers as the indices hk of the lines of the original lattice, and

(*c*) whose magnitude is the reciprocal of the magnitude of the original spacing vector.

Thus, to every spacing d corresponding to indices hk in the original lattice there corresponds a translation t' whose indices are also hk in the reciprocal lattice. These two magnitudes are related by

$$t'_{hk} = \text{area}_{\text{primitive cell}} \frac{1}{d_{hk}}. \tag{5}$$

This can be transformed into a neat, symmetrical form by replacing the quantity t'_{hk} by area $\times\ t^*_{hk}$. Then (5) can be rewritten in the very symmetrical form

$$t^*_{hk} d_{hk} = 1. \tag{6}$$

In this and other similar expressions, the asterisk (ordinarily pronounced "star") is appended to the label of any geometrical feature which occurs in the reciprocal lattice.

If special values are substituted in (6), the axes of the lattice may be determined, as illustrated in Fig. 4. The axes are given by

$$\begin{aligned} t^*_{10} d_{10} &= a^* d_{10} = 1, \\ t^*_{01} d_{01} &= b^* d_{01} = 1. \end{aligned} \tag{7}$$

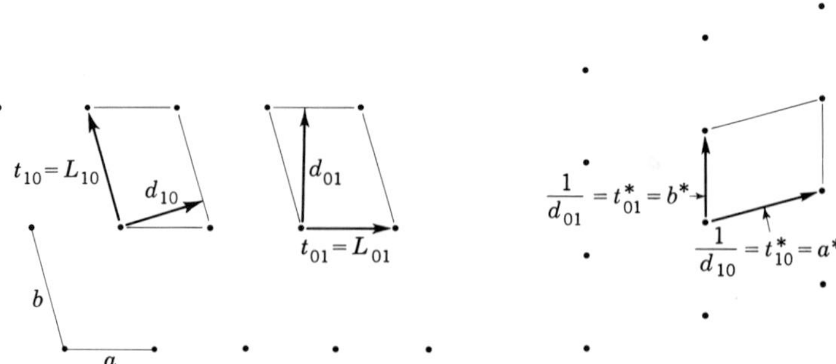

Fig. 4

These relations are fundamental, and they hold for any plane lattice whatever. Accordingly, they must hold not only for the original lattice but also for the reciprocal lattice itself. Thus, an equally valid relation, corresponding to (6) but in which the reciprocal lattice is regarded as the original lattice, is

$$t_{hk}d^*_{hk} = 1. \tag{8}$$

The axes of the new lattice are provided by relations similar to (6), namely,

$$\begin{aligned} ad^*_{10} &= 1, \\ bd^*_{01} &= 1. \end{aligned} \tag{9}$$

The chief relation of the original lattice and its reciprocal is that to every spacing in one there corresponds in the other a parallel translation having a magnitude which is the reciprocal of the spacing magnitude. Since, by its direction and magnitude, the spacing characterizes a line, and since a translation vector characterizes the point at its end, it follows that for every set of parallel rational lines in one lattice there corresponds a specific lattice point in the other. This permits properties of all the sets of parallel lines of a lattice to be visualized in terms of positions of all the points of its reciprocal lattice.

The reciprocal lattice in three dimensions. The derivation of the reciprocal lattice for a three-dimensional lattice is more complicated than that for a two-dimensional lattice, and the results have a more general form. Nevertheless, the discussion follows a similar pattern, with allowances made for the complication due to three dimensions. The argument starts with the constancy of the volume of any primitive cell, and continues by expressing this in terms of the area of the primitive plane cell of any rational plane, times its spacing. In examining the consequences of this, the most direct route is to express the geometrical features involved in terms of vectors and then to manipulate the resulting expressions with the rules of vector algebra.†

The volumes of all primitive cells of a given lattice are the same and are equal, for example, to the volume of the parallelepiped based upon the conjugate axes used to describe the lattice. In general, a primitive cell is the parallelepiped that is included between any two neighboring rational planes and whose base is a primitive parallelogram-shaped cell in these planes. Let such a parallelogram be chosen so that its vertex lattice points are on extensions of the three cell axes which are referred to some lattice point O taken as origin, as shown in Fig. 5. If the conjugate translations used to describe the lattice are the vectors **a**, **b**, and **c**, then the rational plane has rational intercepts closest to the origin

† If the reader is unfamiliar with vector algebra, he can probably follow the argument until results similar to those found for two-dimensional lattices are attained.

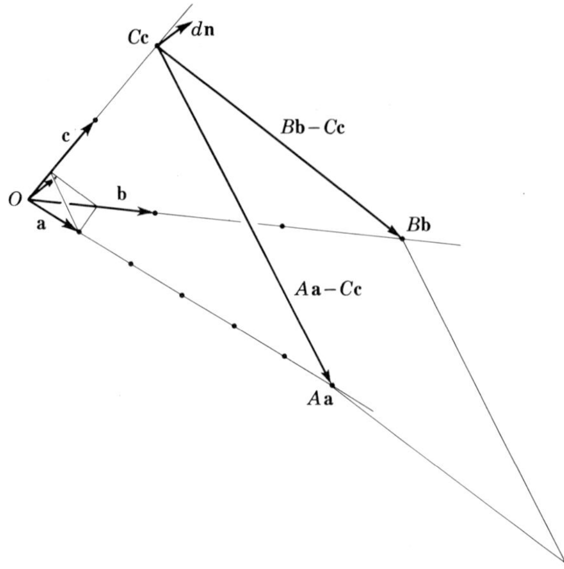

Fig. 5

which are integral multiples of these primitive vectors, specifically, $A\mathbf{a}$, $B\mathbf{b}$, and $C\mathbf{c}$, where A, B, and C are integers. Two adjacent sides of the primitive plane cell in the rational plane can be specified by the vector differences $A\mathbf{a} - C\mathbf{c}$ and $B\mathbf{b} - C\mathbf{c}$. The area of this primitive parallelogram-shaped cell is the cross product of these, namely, $(A\mathbf{a} - C\mathbf{c}) \times (B\mathbf{b} - C\mathbf{c})$. This area times the spacing vector $d\mathbf{n}$, where $\mathbf{n}$ is a unit vector normal to the base, is the volume V of the primitive three-dimensional cell, and this is constant for all such primitive cells. Generally, then,

$$V = (A\mathbf{a} - C\mathbf{c}) \times (B\mathbf{b} - C\mathbf{c}) \cdot d_{hkl}\mathbf{n} \tag{10}$$

$$= (AB\mathbf{a} \times \mathbf{b} - AC\mathbf{a} \times \mathbf{c} - CB\mathbf{c} \times \mathbf{b} + CC\mathbf{c} \times \mathbf{c}) \cdot d_{hkl}\mathbf{n}$$

$$= (AB\mathbf{a} \times \mathbf{b} + CA\mathbf{c} \times \mathbf{a} + BC\mathbf{b} \times \mathbf{c} + 0) \cdot d_{hkl}\mathbf{n}. \tag{11}$$

Let the pairs AB, CA, and BC be replaced by

$$BC = h, \quad CA = k, \quad AB = l. \tag{12}$$

When these replacements are made in (11), and after the sequence of terms in the parentheses is rearranged to give the sequence h, k, l, (11) becomes

$$V = (h\mathbf{b} \times \mathbf{c} + k\mathbf{c} \times \mathbf{a} + l\mathbf{a} \times \mathbf{b}) \cdot d_{hkl}\mathbf{n}. \tag{13}$$

This has a form which corresponds to the two-dimensional form of (3). The term in parentheses is the sum of three terms, each a cross product of two of the three vectors which were chosen as the conjugate translations of the cell for purposes of describing the lattice, each product scaled by h, k, or l. Each such cross product is a vector normal to the plane of the two vectors of the product. It is convenient to define the new vectors, such as $\mathbf{b} \times \mathbf{c}$, in such a way as to eliminate the term V in (13). This can be done by defining

$$\begin{aligned} \frac{\mathbf{b} \times \mathbf{c}}{V} &= \mathbf{a}^*, \\ \frac{\mathbf{c} \times \mathbf{a}}{V} &= \mathbf{b}^*, \\ \frac{\mathbf{a} \times \mathbf{b}}{V} &= \mathbf{c}^*. \end{aligned} \tag{14}$$

On substituting these, (13) takes the simpler form

$$1 = (h\mathbf{a}^* + k\mathbf{b}^* + l\mathbf{c}^*) \cdot d_{hkl}\mathbf{n}. \tag{15}$$

This is merely a reduced and relabeled equivalent of (10). The cross product in (10) is a vector whose direction is normal to the plane of the parallelogram in Fig. 5 whose area it represents. This vector is parallel to the spacing vector $d\mathbf{n}$. The dot product in (10) and (15) is therefore between two vectors having the same direction, so that their magnitudes are given by

$$1 = |h\mathbf{a}^* + k\mathbf{b}^* + l\mathbf{c}^*|\, d_{hkl}, \tag{16}$$

and this can be rearranged to give

$$\frac{1}{d_{hkl}} = |h\mathbf{a}^* + k\mathbf{b}^* + l\mathbf{c}^*|. \tag{17}$$

Both sides are the magnitudes of vectors having the same direction; therefore the relation can be written again as an equality between these two vectors, namely,

$$\frac{1}{d_{hkl}}\mathbf{n} = h\mathbf{a}^* + k\mathbf{b}^* + l\mathbf{c}^*. \tag{18}$$

Relations (17) and (18) show that the reciprocal to any interplanar spacing can be expressed in terms of the magnitude of a vector composed of a linear sum of certain vectors $\mathbf{a}^*$, $\mathbf{b}^*$, and $\mathbf{c}^*$. These were defined by (14), but they can also be evaluated by assigning certain special values to h, k, and l in (18) and then setting down the result. These special cases of (18) are:

$$\frac{1}{d_{100}}\mathbf{n} = \mathbf{a}^* \qquad a^*d_{100} = 1;$$
$$\frac{1}{d_{010}}\mathbf{n} = \mathbf{b}^* \qquad b^*d_{010} = 1; \tag{19}$$
$$\frac{1}{d_{001}}\mathbf{n} = \mathbf{c}^* \qquad c^*d_{001} = 1.$$

These relations are similar to (7) in two dimensions.

The collection of vectors parallel to the spacings of rational planes of a lattice and having magnitudes equal to the reciprocals of the spacings are shown by (18) to terminate at points at the ends of vectors $h\mathbf{a}^* + k\mathbf{b}^* + l\mathbf{c}^*$. But these points constitute another lattice whose conjugate translations are $\mathbf{a}^*$, $\mathbf{b}^*$, and $\mathbf{c}^*$. According to (14) these translations are normal to pairs of translations in the original lattice, and their magnitudes are provided by both (14) and (19).

As in the case of two-dimensional lattices, the second lattice is called the reciprocal lattice, and features in it are distinguished by appending an asterisk to the symbol of the feature. But since a reciprocal lattice can be constructed for any lattice, it can be constructed for a reciprocal lattice, so that the asterisks can be transferred to labels without asterisks. Thus the relations which have been discussed are mutually reciprocal.

Some further relations between reciprocal lattices. The fundamental relations between a lattice and its reciprocal lattice are that any spacing vector in one is parallel to a translation in the other and that the magnitudes of these are reciprocals, specifically,

$$t^*_{hkl}d_{hkl} = 1, \qquad t_{uvw}d^*_{uvw} = 1. \tag{20}$$

The two lattices are referred to sets of axes such that each axis of one lattice is perpendicular to pairs of axes in the other, a condition which is expressed in vector-algebraic symbolism as

$$\mathbf{a}^* = \frac{\mathbf{b} \times \mathbf{c}}{V} \qquad \mathbf{a} = \frac{\mathbf{b}^* \times \mathbf{c}^*}{V^*};$$
$$\mathbf{b}^* = \frac{\mathbf{c} \times \mathbf{a}}{V} \qquad \mathbf{b} = \frac{\mathbf{c}^* \times \mathbf{a}^*}{V^*}; \tag{21}$$
$$\mathbf{c}^* = \frac{\mathbf{a} \times \mathbf{b}}{V} \qquad \mathbf{c} = \frac{\mathbf{a}^* \times \mathbf{b}^*}{V^*}.$$

These two sets of axial vectors constitute an example of two systems of reciprocal vectors of vector algebra.

Another relation between direct and reciprocal-cell edges can be derived by taking the scalar product of the vector representing an axis with the corresponding reciprocal axis, beginning with the basic definition, for example, (14):

$$\mathbf{a}^* = \frac{\mathbf{b} \times \mathbf{c}}{V};$$

$$\mathbf{a} \cdot \mathbf{a}^* = \frac{\mathbf{a} \cdot \mathbf{b} \times \mathbf{c}}{V} = \frac{V}{V} = 1.$$

Similarly, (22)

$$\mathbf{b} \cdot \mathbf{b}^* = 1$$

and

$$\mathbf{c} \cdot \mathbf{c}^* = 1.$$

The volumes of the two cells defined by these reciprocal sets are also reciprocals, as can be seen by starting with the fundamental volume relation and substituting from (21):

$$V^* = \mathbf{a}^* \cdot \mathbf{b}^* \times \mathbf{c}^* \tag{23}$$

$$= \frac{\mathbf{b} \times \mathbf{c}}{V} \cdot \frac{\mathbf{c} \times \mathbf{a}}{V} \times \frac{\mathbf{a} \times \mathbf{b}}{V}$$

$$= \frac{(\mathbf{b} \times \mathbf{c}) \cdot ([\mathbf{cab}]\mathbf{a} - [\mathbf{caa}]\mathbf{b})}{V^3}$$

$$= \frac{(\mathbf{b} \times \mathbf{c}) \cdot (V\mathbf{a} - 0)}{V^3}$$

$$= \frac{\mathbf{b} \times \mathbf{c} \cdot \mathbf{a}V}{V^3}$$

$$= \frac{V^2}{V^3} = \frac{1}{V}. \tag{24}$$

Thus,

$$V^*V = 1. \tag{25}$$

Each axis of the reciprocal cell is perpendicular to one face of the cell of the original lattice; consequently, the interaxial angles of the two cells are related to each other and do not depend on the length of any of the axial vectors. This mutual relation is suggested in Fig. 6, in which the axes of one set meet at the acute apex at the left, while the axes of the reciprocal set meet at the obtuse apex at the right of the figure. In each of the quadrilaterals meeting at the right-hand apex, each of the two inner sides is orthogonal to its adjacent outer side so that, for example, $C + 90° + 90° + \gamma^* = 360°$, which requires $C = 180° - \gamma^*$. If the several angular relationships are plotted on a spherical triangle (Fig. 7), and the law of sines is applied, there results

$$\frac{\sin A}{\sin \alpha} = \frac{\sin B}{\sin \beta} = \frac{\sin C}{\sin \gamma}, \tag{26}$$

and from the angular relations such as $C = 180° - \gamma^*$ within each

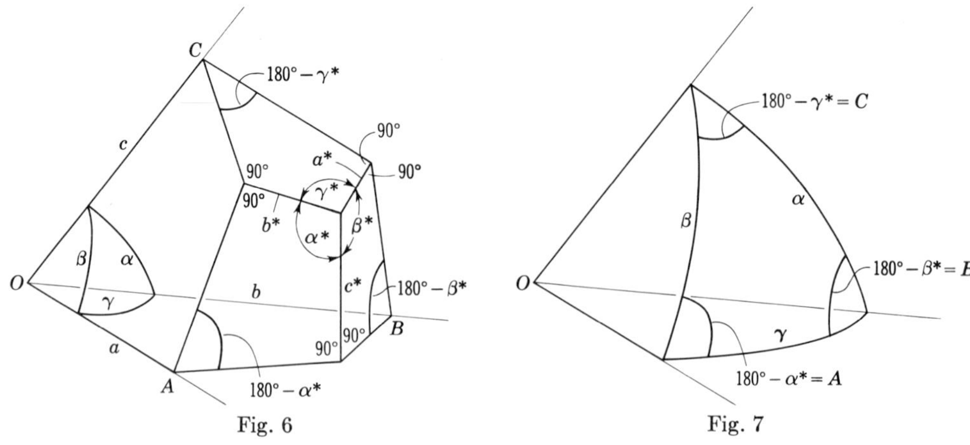

Fig. 6 Fig. 7

Table 1A

$$\frac{\sin \alpha^*}{\sin \alpha} = \frac{\sin \beta^*}{\sin \beta} = \frac{\sin \gamma^*}{\sin \gamma}$$

$$\cos \alpha^* = \frac{\cos \beta \cos \gamma - \cos \alpha}{\sin \beta \sin \gamma}$$

$$\cos \beta^* = \frac{\cos \gamma \cos \alpha - \cos \beta}{\sin \gamma \sin \alpha}$$

$$\cos \gamma^* = \frac{\cos \alpha \cos \beta - \cos \gamma}{\sin \alpha \sin \beta}$$

$$a^* = \frac{bc \sin \alpha}{V}$$

$$b^* = \frac{ca \sin \beta}{V}$$

$$c^* = \frac{ab \sin \gamma}{V}$$

$$V^* = \frac{1}{V}$$

$$V^{*2} = \begin{vmatrix} a^{*2} & a^*b^* \cos \gamma^* & a^*c^* \cos \beta^* \\ b^*a^* \cos \gamma^* & b^{*2} & b^*c^* \cos \alpha^* \\ c^*a^* \cos \beta^* & c^*b^* \cos \alpha^* & c^{*2} \end{vmatrix}$$

$$V^* = a^*b^*c^* \sqrt{1 - \cos^2 \alpha^* - \cos^2 \beta^* - \cos^2 \gamma^* + 2 \cos \alpha^* \cos \beta^* \cos \gamma^*}$$

$$\begin{aligned} V^* &= a^*b^*c^* \sin \alpha^* \sin \beta^* \sin \gamma \\ &= a^*b^*c^* \sin \alpha^* \sin \beta \;\; \sin \gamma^* \\ &= a^*b^*c^* \sin \alpha \;\; \sin \beta^* \sin \gamma^* \end{aligned}$$

quadrilateral, it follows that

$$\frac{\sin \alpha^*}{\sin \alpha} = \frac{\sin \beta^*}{\sin \beta} = \frac{\sin \gamma^*}{\sin \gamma}. \tag{27}$$

Another relation between interaxial angles can be derived by applying the law of cosines:

$$\cos \alpha = \cos \beta \cos \gamma + \sin \beta \sin \gamma \cos A. \tag{28}$$

The angle A is shown in Figs. 6 and 7; it is the supplement of α^*. If this is substituted in (28) and rearranged, α^* is seen to be given by

$$\cos \alpha^* = \frac{\cos \beta \cos \gamma - \cos \alpha}{\sin \beta \sin \gamma}. \tag{29}$$

Similar relations hold for the other interaxial angles as well as for the reciprocal interaxial angles.

The basic formulae useful in transforming from the direct lattice to the reciprocal lattice are assembled in Tables 1A and 1B.

Table 1*B*

$$\frac{\sin \alpha}{\sin \alpha^*} = \frac{\sin \beta}{\sin \beta^*} = \frac{\sin \gamma}{\sin \gamma^*}$$

$$\cos \alpha = \frac{\cos \beta^* \cos \gamma^* - \cos \alpha^*}{\sin \beta^* \sin \gamma^*}$$

$$\cos \beta = \frac{\cos \gamma^* \cos \alpha^* - \cos \beta^*}{\sin \gamma^* \sin \alpha^*}$$

$$\cos \gamma = \frac{\cos \alpha^* \cos \beta^* - \cos \gamma^*}{\sin \alpha^* \sin \beta^*}$$

$$a = \frac{b^* c^* \sin \alpha^*}{V}$$

$$b = \frac{c^* a^* \sin \beta^*}{V}$$

$$c = \frac{a^* b^* \sin \gamma^*}{V}$$

$$V = \frac{1}{V^*}$$

$$V^2 = \begin{vmatrix} a^2 & ab \cos \gamma & ac \cos \beta \\ ba \cos \gamma & b^2 & bc \cos \alpha \\ ca \cos \beta & cb \cos \alpha & c^2 \end{vmatrix}$$

$$V = abc \sqrt{1 - \cos^2 \alpha - \cos^2 \beta - \cos^2 \gamma + 2 \cos \alpha \cos \beta \cos \gamma}$$

$$V = abc \sin \alpha \sin \beta \sin \gamma^*$$

$$V = abc \sin \alpha \sin \beta^* \sin \gamma$$

$$V = abc \sin \alpha^* \sin \beta \sin \gamma$$

Application of the reciprocal lattice to diffraction

At the beginning of this chapter it was pointed out that the reciprocal lattice could be utilized to provide a neat geometrical picture of x-ray diffraction by a crystal. On the one hand this yields a simple device for explaining the geometry of the x-ray diffraction effects produced by a crystal, and the record they produce; and on the other it provides a means of deducing the geometry of the reciprocal lattice of any specific crystal from the record of its diffraction. The mathematical relations given in the last section permit transforming a set of dimensions measured in the reciprocal lattice to a corresponding set in the crystal lattice. In this way the dimensions of the cell of a crystal can be determined. Many x-ray diffraction methods have been developed, each having its own technique for accomplishing this; the most commonly used are discussed briefly in Chapters 7 to 9. Of these the neatest is the precession method, in which the reciprocal lattice is photographed in undistorted form, level by level.

In applying the reciprocal lattice to problems of x-ray diffraction the reader is warned that two conventions of dimensions are in use by scientists who apply the reciprocal lattice to somewhat different problems. Relation (2) is satisfied by the dimensions given in either drawing of Fig. 8. With the Ewald convention, which is the older, the reciprocal lattice remains constant but the radius of the sphere changes according to the wavelength or wavelengths of the experiment. This is a convenient convention when the wavelength is a variable, as in the Laue method. With the alternative convention the sphere is regarded as having unit radius, and the reciprocal lattice accordingly is scaled by the constant λ. This convention is commonly followed by x-ray crystallographers who routinely use a method which employs monochromatic x-radiation.

Whichever convention is used, the crystal is thought of as having its center, or origin, at the center of the sphere, while its reciprocal lattice

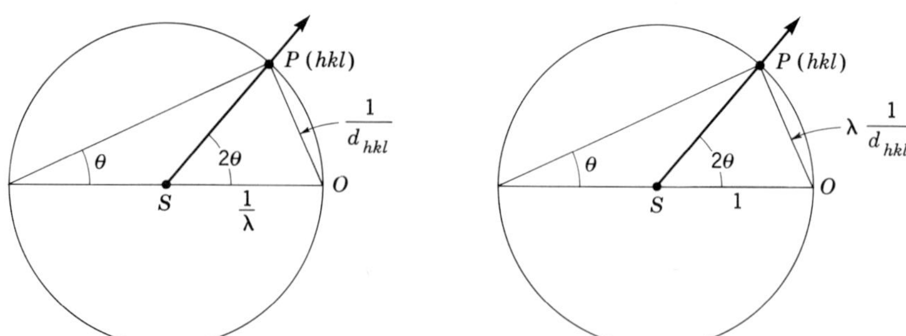

Fig. 8

has its origin (that is, its lattice point whose coordinates are 000) at the point where the x-ray beam leaves the sphere. The x-ray beam is always a diameter of the sphere and may be thought of as having the sphere as a partial envelope; the crystal is at the center of the sphere, and the reciprocal-lattice origin is located at the point of exit of the x-ray beam.

From the derivation of the reciprocal lattice it follows that if the crystal is turned, the reciprocal lattice turns with it; that is, the reciprocal lattice and crystal always retain parallel orientation. The crystal may now be forgotten, except that it must be remembered that the diffracted rays all emanate from its location at the center of the sphere. In an arbitrary orientation, in general, no reciprocal-lattice point lies on the surface of the sphere, so that the condition represented in Figs. 1, 2, and 8 is not achieved; this corresponds to an orientation in Fig. 7, Chapter 3, such that the path difference is not an integral number of wavelengths, and so (7) and (8) of Chapter 3 are not fulfilled. Under these circumstances no x-ray reflection results. But whenever (7) and (8) of Chapter 3 are fulfilled, some point P, whose index is hkl, lies on the surface of the sphere, and a reflection in direction SP takes place.

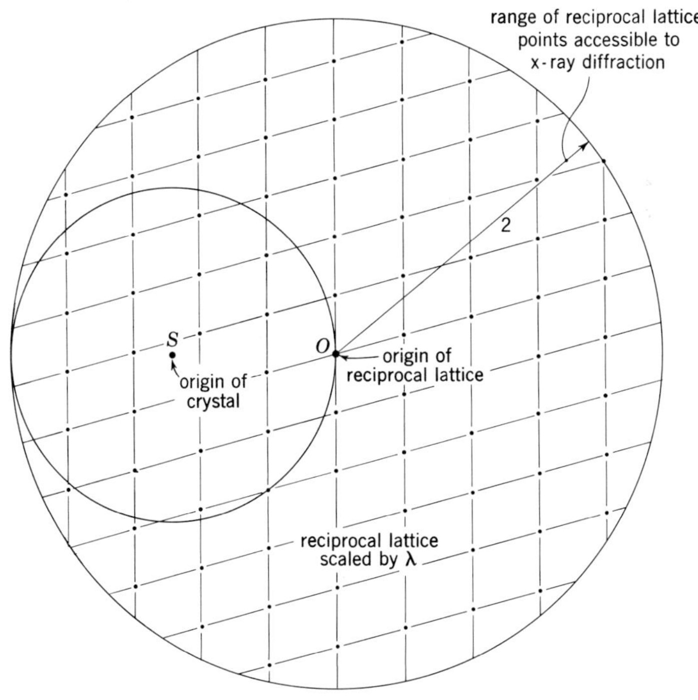

Fig. 9

To find the directions of possible reflections, the entire reciprocal lattice is imagined to exist in the neighborhood of the sphere, as shown in Fig. 9. The reciprocal lattice is then turned according to some scheme, for example, by a steady rotation about an axis through the origin O. Whenever any reciprocal-lattice point P (Fig. 8) makes contact with the sphere, a reflection is generated in the direction SP. These various reflections may be recorded on a photographic film or detected by a photon counter. Understanding each kind of diffraction method requires an inquiry into the geometry of these reflections as a function of the crystal motion and the geometry of the recording device, for example, shape and the placement of the film.

The most natural coordinate system for the reciprocal lattice is the set of the three translations adopted to describe the lattice. Occasionally it is convenient to refer the lattice points to the cylindrical-coordinate system shown in Fig. 10. In this case the axis of the cylinder is a rational translation of the crystal (generally $[uvw]$, but usually a cell axis), while the base of the cylinder is the reciprocal-lattice plane which is perpendicular to this crystal translation. This is always rational, generally the plane $(uvw)^*$, but usually an axial plane. The cylindrical coordinates of the reciprocal-lattice point are ϕ, ξ, and ζ, where ϕ is the angle in the plane $(uvw)^*$ between the ξ, ζ plane and a selected direction, such as a reciprocal-cell axis. The distance of the reciprocal-lattice point from the origin is

$$r^* = \sqrt{\xi^2 + \zeta^2}. \tag{30}$$

Figure 10 also shows the relation of the cylindrical coordinates to the spherical coordinates commonly used in two-circle goniometry; in that

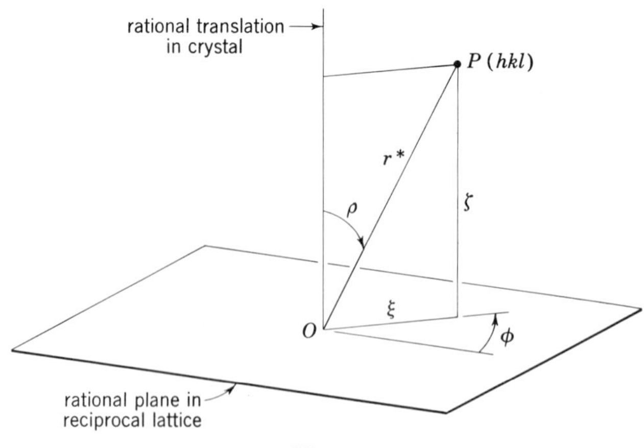

Fig. 10

system ϕ is the longitude and ρ is the colatitude of the normal r^* to a crystal face.

Summary

Bragg's law can be given a neat geometrical interpretation in terms of a lattice known as the *reciprocal lattice*, which bears a simple relation to the crystal lattice. In the ordinary lattice of the crystal, each stack of planes has a spacing which is characterized by a vector whose direction is normal to the planes and whose magnitude is indicated by the symbol d_{hkl}. If each such vector is replaced by one having the same direction but a length which is the reciprocal of the original, namely, $1/d_{hkl}$, the points at the ends of this collection of vectors also constitute a lattice which is called the reciprocal of the original lattice. A lattice and its reciprocal are mutually reciprocal. The simplest law connecting them is that the magnitude of any spacing in one is the reciprocal of the magnitude of the parallel translation in the other. This can be formulated as $dt^* = 1$ and $d^*t = 1$, where the asterisk (pronounced "star") indicates the feature in the reciprocal lattice. Other simple mathematical relations permit transforming the dimensions of the cell of any lattice to the dimensions of the cell of its reciprocal, or the reverse.

If Bragg's law is written $\sin\theta = [\lambda \cdot 1/d_{hkl}]/2$, it is seen that the term in brackets is the same as the reciprocal-lattice vector t^*_{hkl} scaled by the wavelength λ. Thus Bragg's law in reciprocal-lattice terms is $\sin\theta = \lambda t^*/2$. If the diameter of a circle of unit radius is drawn, θ is the angle subtended at one end of a diameter by a vector of length λt^*, whose origin is the other end of the diameter and which terminates on the circumference. This can be interpreted in the following way: Let a unit sphere be constructed with the x-ray beam as diameter; place the origin of the reciprocal lattice at the point where the x-ray beam leaves the sphere. Then, if the orientation of the reciprocal lattice is adjusted in any way which brings one of its points to the surface of the sphere, Bragg's law, $\sin\theta = \lambda t^*/2$, is satisfied. The direction of the reflected beam is the line from the center of the sphere to the reciprocal-lattice point on the sphere's surface, because this direction makes an angle 2θ with the x-ray beam. This geometrical interpretation of Bragg's law is most useful in understanding the production and recording of reflections by the various x-ray diffraction methods.

Notes on history

The geometrical features of essentially what is now known as the reciprocal lattice were discovered by the French mining engineer Bravais about 1849. In his famous 1850 monograph he not only derived the 14 symmetrical space lattices but also devoted a score of pages to what he called the *polar lattice*. Little use was made of this new concept until Ewald rediscovered it while considering Laue's theory and the results of the Friedrich-Knipping experiment in connection with his own doctoral thesis. In his 1913 paper Ewald pointed out that x-ray diffraction could be explained in terms of the points of a "reciprocal lattice" whose translations were the reciprocals of the translations of the crystal lattice (for his orthogonal case), and a sphere whose radius was equal to the wave number of the radiation, namely, $K = 1/\lambda$. In a later paper, published in 1921 in the Zeitschrift für Kristallographie, Ewald recognized the relation of his reciprocal lattice to Bravais' polar lattice, and reformulated it with the aid of Gibbs' vector algebra. It turned out, therefore, that the polar lattice, the reciprocal lattice, and sets of reciprocal vectors had a closely related geometry.

In Ewald's 1921 paper the reciprocal lattice was expressed in a form which would appeal to crystallographers, but it was really popularized by a paper published by Bernal in 1926 in the Proceedings of the Royal Society of London. Bernal showed that Ewald's reciprocal lattice and sphere of reflection provided a simple means for indexing photographs made by the rotating-crystal method and the oscillating-crystal method (discussed in Chapter 7). These gradually became the standard tools for the interpretation of such x-ray photographs, and eventually also of Weissenberg (Chapter 8) and precession photographs (Chapter 9). Indeed, precession photographs are really undistorted photographs of the reciprocal lattice itself, scaled by a factor λ times an instrumental constant M.

Additional reading

M. J. Buerger. *X-ray crystallography.* (Wiley, New York, 1942) 107–132.

Significant literature

M. A. Bravais. *Mémoire sur les systèmes formé par des points distribués régulièrement sur un plan ou dans l'espace.* J. École Polytech., Cahier 33, Tome XIX, (Paris, 1850) 1–128. [Translation by Amos J. Shaler. *On the systems formed by points regularly distributed on a plane or in space.* American Crystallographic Association Monograph 3 (1949) especially 92–111.]

J. Willard Gibbs and Edwin Bidwell Wilson. *Vector analysis.* (Yale, New Haven, Conn., 1901) especially 81–87.

P. P. Ewald. *Zur Theorie der Interferenzen der Röntgenstrahlen in Kristallen.* Phys. Z. **14** (1913) 465–472.

P. P. Ewald. *Das "reziproke Gitter" in der Strukturtheorie.* Z. Kristallogr. **56** (1921) 129–156.

J. D. Bernal. *On the interpretation of x-ray, single crystal, rotation photographs.* Proc. Roy. Soc. (London) (A) **113** (1926) especially 118–123.

5

Routine determination of symmetry by x-ray diffraction

Pattern characteristics to be established

In Chapter 1 it was pointed out that three features characterize a crystal pattern; these features can be rephrased in the following language:

(*a*) its symmetry,
(*b*) its cell dimensions, and
(*c*) the location of the atoms in the cell.

The first two of these can be established by relatively simple experimental methods. The general aspects of symmetry determination are dealt with in this chapter. In Chapters 7 to 9 an outline is given of the application of three common methods of x-ray diffraction to finding the symmetry and cell of a crystal. The general ways that the atoms can be located with regard to the symmetry elements of the cell are discussed in Chapter 10. But the problem of finding the specific locations of the atoms in the cell is a more difficult one, which requires a more sophisticated approach. Discussion of the theory and its application to solving this problem is reserved for later chapters.

The matters to be discussed in this chapter, then, concern the general features of establishing the space group, or pattern type. This is a qualitative aspect of the pattern, and no measurements are required to establish it. It follows that if suitable experimental means are employed, whatever can be found out about the symmetry should be obvious on inspection. It will become apparent from the discussion in Chapters 7 to 9 that the precession method provides an ideal means for determining symmetry features generally, although the same information can be obtained in less simple form from the Weissenberg method, much information can be obtained by using the Laue method, some can be obtained from a form of the rotating-crystal method, but none whatever can be gleaned by using the powder method (which is not discussed in detail in this book).

Reciprocal space

The points of the reciprocal lattice are related to the points of the direct lattice in the elegant way demonstrated in Chapter 4. It is convenient to speak of the crystal, including its structure, lattice, and symmetry, as occurring in *crystal space*, and to speak of its reciprocal lattice as occurring in another space called *reciprocal space*. So far, the reciprocal lattice has been put forward merely as a convenient device for visualizing the geometry of x-ray reflections. The concept of reciprocity, some important features of which will now be mentioned without mathematical formalities, is capable of considerable generalization.

Every x-ray reflection is characterized by a unique index hkl, and to each such reflection there also corresponds a unique point in the reciprocal lattice whose coordinates (measured in units of a^*, b^*, and c^*, respectively) are also h, k, and l. Thus the points of the reciprocal lattice offer a convenient device for filing information about any reflection hkl.

Toward the end of Chapter 3 it was seen that each x-ray reflection has the properties of a wave, with a characteristic amplitude and phase. This information is represented by the structure factor F_{hkl}, which, in general, is a complex quantity. If to each reciprocal-lattice point there is attributed not only the coordinates h, k, l but also F_{hkl}, the points of the reciprocal lattice take on added meaning. But when this is done, the points which were formerly thought of as pure lattice points are no longer translation-equivalent, and the collection of points can therefore no longer be described as a lattice, although the point locations are *on* a lattice. The new collection of weighted points is commonly designated as a *weighted reciprocal lattice*, but this is still incorrect since it still implies translation equivalence of the points. It will develop later that this collection of points is the Fourier transform of the crystal structure; and since this kind of transform is a generalization of the reciprocal property of lattices, a more appropriate designation for the reciprocal lattice weighted by F_{hkl}'s would be *reciprocal structure*.

In the next section the relation between the symmetry of the crystal and the symmetry of its representation in reciprocal space will be examined.

Point-group symmetry of the reciprocal structure

Actual symmetry. The reciprocal representation for a crystal structure consists of a collection of points each of which has a specific location, magnitude, and phase. It will now be shown that the symmetry of this set of characteristics is related to the point-group symmetry of the crystal in a simple way.

In the first place, each reciprocal-lattice point is located on the normal to the plane to which it corresponds. The locations of the normals obviously have the same symmetry as the locations of the planes. Furthermore, the distance of the reciprocal-lattice point along the normal is the same for planes whose spacings are the same, which is true of symmetrically equivalent planes. These two statements make it certain that the locations of the reciprocal-lattice points representing two symmetrically equivalent planes are related by at least the same symmetry as the planes themselves. But the point locations may have additional symmetry not present in the point-group symmetry, because the point locations constitute a pure lattice, and a lattice can have only one of the seven symmetries listed in the third column of Table 5, Chapter 2. These are sometimes called the holohedral crystal classes.

When the magnitude and phase of the reflection are taken into account, this holohedral location symmetry of the points is, in general, reduced. Consider some possible structures having, respectively, only an n-fold axis (Fig. 1), a mirror (Fig. 2), and an inversion center (Fig. 3). In each case the amplitude and phase of the reflections from symmetrically equivalent stacks of planes have the same symmetry as that relating the planes, and no more, so that the holohedral symmetry characteristic of the point locations alone is reduced unless holohedral symmetry also characterizes the symmetry relations between the planes themselves. Accordingly, the symmetry relation between reciprocal structure and crystal structure can be stated in the following terms:

Theorem: The symmetry of the reciprocal structure conforms to the point-group symmetry of the crystal.

But the reciprocal structure has a curious further symmetry, which is illustrated in Figs. 4 and 5. Figure 4 shows a small fragment of a simple structure deliberately drawn without symmetry to illustrate the relation between reflections from the opposite sides of the same stack of planes, that is, to illustrate the relation between the reflections hkl and $\bar{h}\bar{k}\bar{l}$. It can be seen that waves reflected from these two sides are composed of the same contributions, that these contributions have the same phase differences, but that these phase differences have opposite senses. Accordingly, these two waves are equal in magnitude but opposite in phase. On the Argand diagram representing these waves (Fig. 5), such waves are represented by equal vectors making equal but opposite angles with the origin direction. In this particular diagram the zero for phase measurement was arbitrarily taken to be the phase scattered by the black atoms. Then the phase scattered by the white atoms represents a phase lag for reflection hkl, but a phase advance for reflection $\bar{h}\bar{k}\bar{l}$.

Two vectors on the Argand diagram which are equal but have opposite phases are said to be *complex conjugates* (Latin: yoked, married). This relation can be represented by using the notation of complex numbers. In this instance the vector representing the magnitude and phase of the wave of the hkl reflection is F_{hkl}. Its magnitude is the absolute value of this, namely, $|F_{hkl}|$. It has just been demonstrated that the magnitudes for reflections hkl and $\bar{h}\bar{k}\bar{l}$ are equal; this can be expressed as

$$|F_{hkl}| = |F_{\bar{h}\bar{k}\bar{l}}|. \tag{1}$$

If this magnitude is attributed to a vector along the real axis, then since $e^{i\phi}$ behaves as an operator which rotates any vector through angle ϕ, the

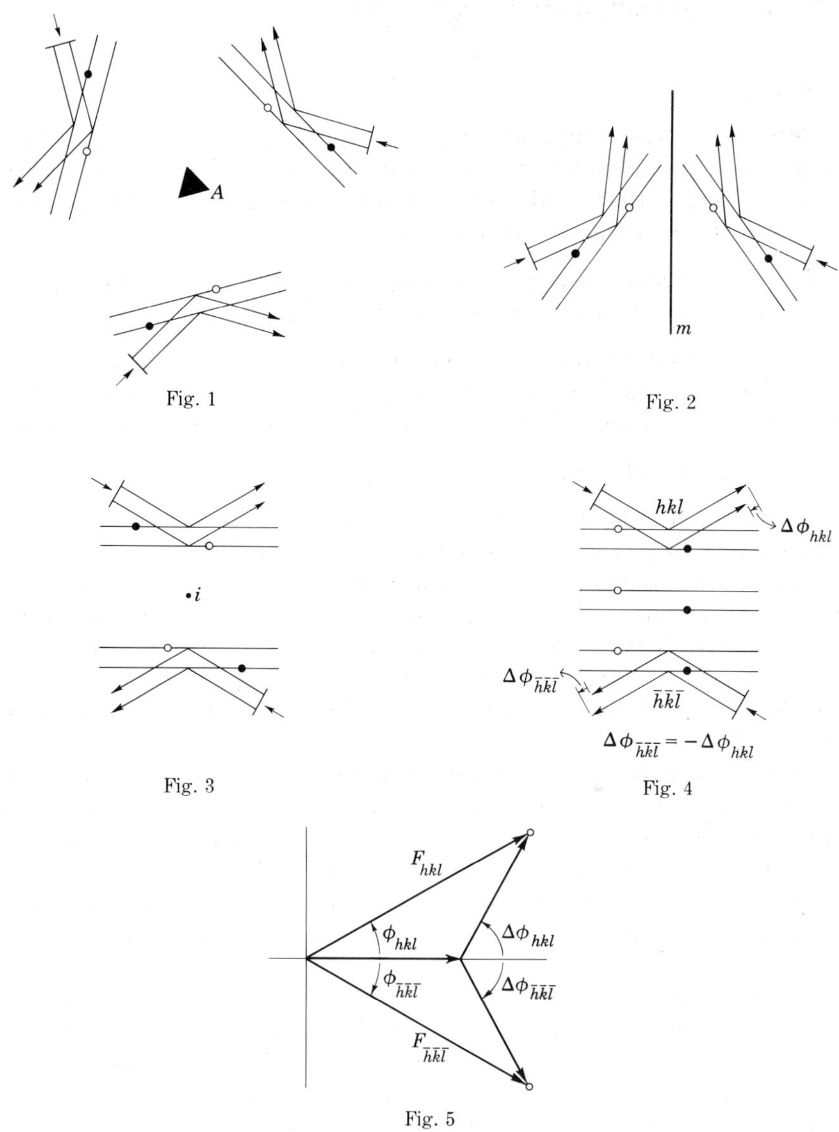

Fig. 1

Fig. 2

Fig. 3

Fig. 4

Fig. 5

relation between these complex conjugates can be written

$$F_{hkl} = |F_{hkl}|e^{i\phi}, \tag{2}$$
$$F_{\bar{h}\bar{k}\bar{l}} = |F_{hkl}|e^{-i\phi}. \tag{3}$$

The complex conjugate of F is written $\widetilde{F}$, so that the relation between the left sides of (2) and (3) can be expressed as

$$F_{\bar{h}\bar{k}\bar{l}} = \tilde{F}_{hkl},$$

as well as $$F_{hkl} = \tilde{F}_{\bar{h}\bar{k}\bar{l}}. \tag{4}$$

Because of the complex-conjugate relation between F_{hkl} and $F_{\bar{h}\bar{k}\bar{l}}$, these two waves are said to have *conjugate symmetry*, which is a symmetry over and above that of the point-group symmetry. Accordingly, the foregoing theorem may be improved as follows:

Theorem: The symmetry of the reciprocal structure is the point-group symmetry of the crystal supplemented by conjugate symmetry.

There is an interesting and important consequence of conjugate symmetry for crystals which are centrosymmetrical. This symmetry requires the reflections hkl and $\bar{h}\bar{k}\bar{l}$ to be the same in both amplitude and phase:

Centrosymmetrical crystal: $$F_{hkl} = F_{\bar{h}\bar{k}\bar{l}}. \tag{5}$$

But (4) must also hold, so that for centrosymmetrical crystals the complex conjugates must be equal (that is, equal in both magnitude and phase). But this can be true only if both coincide along the real (horizontal) axis, that is, if F_{hkl} is real for every hkl. Thus for centrosymmetrical crystals the phase of every reflection can assume only the value 0 or π, which is equivalent to saying that the phase of every F_{hkl} can be represented by placing a + or − sign before the number indicating its magnitude.

Friedel symmetry. Methods for detecting x-rays which have been devised up to the present time are able to record a measure of the intensity (and thus provide the amplitude) of a reflection but do not give any indication of its phase. Because of this, detection devices cannot distinguish between (5), the condition for a centrosymmetrical crystal, and (4), the more general condition for a noncentrosymmetrical crystal.

Friedel came to the same conclusion on different and less compelling grounds, namely, that x-rays behave like light rays and should be expected to give rise to similar phenomena if the travel direction in a ray is reversed. On this basis Friedel announced a result, now commonly called *Friedel's law*, in essentially the following form:

An x-ray photograph cannot reveal the absence of a center of symmetry.

Some small deviations of Friedel's law occur because of anomalous scattering which takes place when the x-rays have a wavelength in the neighborhood of the absorption edge of any of the atoms of the crystal structure. Such small deviations are not ordinarily noticed and do not disturb the conclusions regarding the point-group symmetry of the

Table 1
The Friedel symmetries and their distribution among the point groups

Crystal		Symmetry of diffraction effect
System	Point group	
Triclinic	1 $\bar{1}$	$\bar{1}$
Monoclinic	2 m $\frac{2}{m}$	$\frac{2}{m}$
Orthorhombic	$2\,2\,2$ $m\,m\,2$ $\frac{2}{m}\frac{2}{m}\frac{2}{m}$	$\frac{2}{m}\frac{2}{m}\frac{2}{m}$
Tetragonal	4 $\bar{4}$ $\frac{4}{m}$	$\frac{4}{m}$
	$4\,2\,2$ $4\,m\,m$ $\bar{4}\,2\,m$ $\frac{4}{m}\frac{2}{m}\frac{2}{m}$	$\frac{4}{m}\frac{2}{m}\frac{2}{m}$
Hexagonal	3 $\bar{3}$	$\bar{3}$
	$3\,2$ $3\,m$ $\bar{3}\,\frac{2}{m}$	$\bar{3}\,\frac{2}{m}$
	6 $\bar{6}$ $\frac{6}{m}$	$\frac{6}{m}$
	$6\,2\,2$ $6\,m\,m$ $\bar{6}\,m\,2$ $\frac{6}{m}\frac{2}{m}\frac{2}{m}$	$\frac{6}{m}\frac{2}{m}\frac{2}{m}$
Isometric	$2\,3$ $\frac{2}{m}\bar{3}$	$\frac{2}{m}\bar{3}$
	$4\,3\,2$ $\bar{4}\,3\,m$ $\frac{4}{m}\bar{3}\frac{2}{m}$	$\frac{4}{m}\bar{3}\frac{2}{m}$

crystal. Deviations can also be useful, especially when deliberately caused in exaggerated amounts by selection of appropriate wavelengths. (They may then aid in establishing the locations of the atoms which cause the anomalous scattering, or in establishing the absolute configuration of a particular enantiomorph.)

Except as just noted, Friedel's law prevents distinguishing a point-group symmetry lacking an inversion center from the corresponding point-group symmetry containing an inversion center.† Another statement of this is that detectable x-ray diffraction effects can have only the 11 centrosymmetrical point symmetries, so that if a crystal does not possess a center it nevertheless gives diffraction effects as if it had one. The symmetries of the diffraction effects given by crystals of the 32 point groups are listed in Table 1. It is convenient to call the 11 centrosymmetrical diffraction symmetries *Friedel symmetries* after the scientist who first recognized and tabulated them. (The symmetries are commonly called *Laue symmetries*, probably because Laue photographs once constituted the only means of recording diffraction symmetries. Since 1924 it has been common to determine diffraction symmetry by the Weissenberg method, and since about 1944 by the precession method; therefore the term "Laue symmetry" is not generally applicable.)

Space-group determination

Criteria for the lattice types. Niggli made many contributions to crystallography. One of the most useful was to point out how space-group information could be derived from x-ray diffraction results. This was included in his first book, *Geometrische Kristallographie des Diskontinuums*, which was written during the First World War.

Niggli's system, which is still used whenever possible in the early stage of each routine crystal-structure analysis, is based upon the fact that the different space groups isogonal with a particular point group differ in respect to one or more of the following features:

(*a*) the presence of lattice translations terminating elsewhere than at cell vertices,
(*b*) the presence of glide planes, and
(*c*) the presence of screw axes.

Each of these repeating devices contains translations which are submultiples of the cell-edge translations, and this feature causes them to repeat a normal stack of planes (hkl) in such a way that for certain classes of indices a new set is interleaved with the original set. The specific

† But in Chapter 13 it will be shown that the data can be used to compute a Patterson function, which can be used to circumvent this difficulty.

way this condition comes about for the several cases, and its effect on the x-ray reflections, is outlined below.

Lattice type. Whenever a cell is chosen which contains lattice points anywhere but at its corners (in other words, any kind of multiple cell such as a centered cell), then the same translations which relate a centered point to a corner point also translate a plane at the origin to one through the centered point. For certain classes of indices this is one of a new set of planes which are therefore interleaved with the usual set.

End-centered cells. The way this comes about is illustrated in Fig. 6, which is a view normal to the *ab* face of a cell. The planes of a stack with indices (hkl) divide the a axis into h parts and the b axis into k parts; the intersections of the planes of this stack with the *ab* face of the cell are the oblique lines of Fig. 6. There are $h + k$ such intersections, counting along the face diagonal $\mathbf{a} + \mathbf{b}$. Now if the *ab* face is centered, the centering lattice point is located at $\frac{1}{2}\mathbf{a} + \frac{1}{2}\mathbf{b}$. The middle plane of the (hkl) stack contains this point anyway, provided that there is an even number of planes along the face diagonal, that is, provided $h + k = 2n$, where n is an integer. In this case there is nothing unusual about the stack of planes (hkl), and it is the same as the set occurring in a primitive cell, as illustrated in Fig. 7. But if the number of planes is odd, that is, $h + k = 2n + 1$, then none of the planes of the stack (hkl) contains the centering point (Fig. 8), in which case the translation $\frac{1}{2}\mathbf{a} + \frac{1}{2}\mathbf{b}$, which generates the centering point from the corner point, also generates a new stack whose planes are interleaved halfway between the planes of the old set, as shown at the right of Fig. 8. This contrasts with the situation shown at the right of Fig. 7, which illustrates $h + k$ even. When planes are interleaved halfway between the planes of a stack, the spacing d of the stack is halved. Under these circumstances the planes which, for the primitive cell are (hkl), $h + k$ odd, simply do not exist in the C-centered cell, and so cannot give rise to a reflection.† Thus the only reflections which occur for a C-centered lattice are hkl with $h + k = 2n$, where n is an integer (i.e., where $h + k$ is even).

These considerations give rise to the following criteria for determining the lattice type from the observed reflections.

Conditions for P: If the general reflections hkl present are without restriction, the lattice is P.

Conditions for A: If the general reflections hkl present are only those for which $k + l = 2n$, the lattice is A.

Conditions for B: If the general reflections hkl present are only those for which $h + l = 2n$, the lattice is B.

† An alternative description, due to Niggli and used by early writers (but no longer in current use), is that the spacing of the planes (hkl) is halved when $h + k$ is odd. The abnormally halved spacing is easily detected by the x-ray diffraction experiment.

Conditions for C: If the general reflections hkl present are only those for which $h + k = 2n$, the lattice is C.

Face-centered cells. If the indexing of the planes of a crystal is made with respect to a face-centered cell, then these planes must conform to the conditions just given for A-centered, B-centered, and C-centered cells

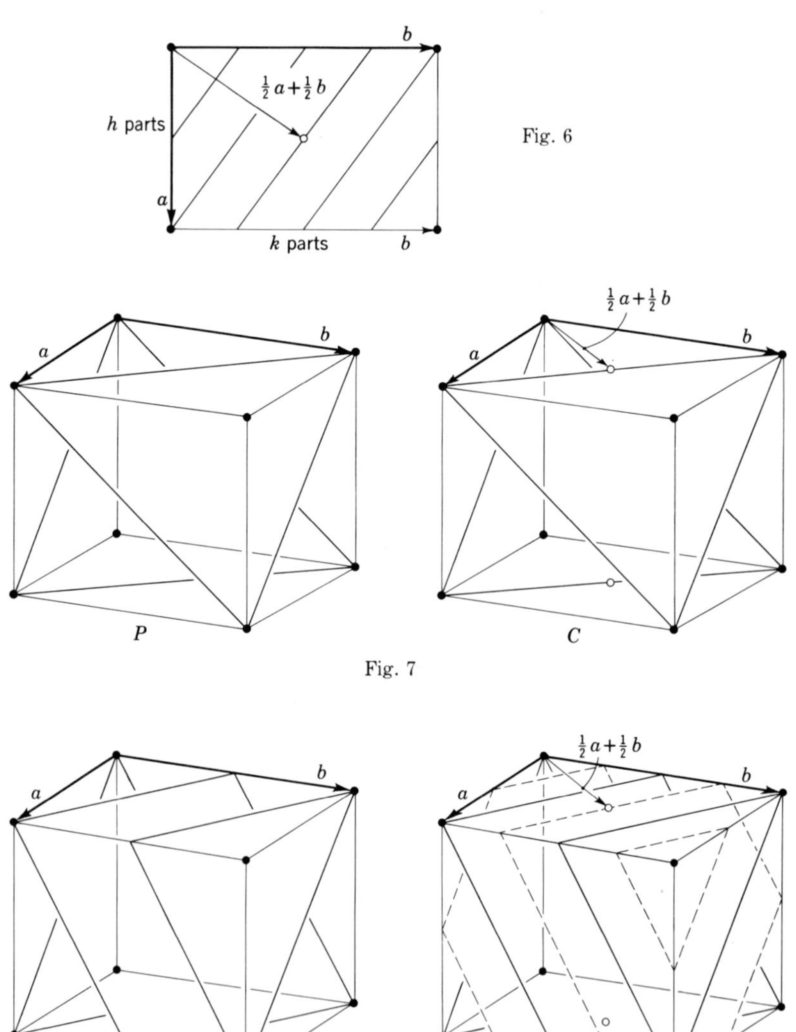

simultaneously, and the x-ray reflections must do the same, specifically,

Conditions for F: If the general reflections hkl present are only those for which

$$\begin{aligned} h + k &= 2n, \\ k + l &= 2n, \\ h + l &= 2n, \end{aligned}$$

then the lattice is F.

These three conditions are equivalent to a single condition that the indices h, k, and l are either all even or all odd numbers, which is succinctly stated by saying that the criterion for an F lattice is that the indices are unmixed.

Body-centered cells. Figure 9 shows that the number of planes (hkl), counted along the length of the body diagonal $\mathbf{a} + \mathbf{b} + \mathbf{c}$, is the sum $h + k + l$. In a body-centered cell the centering point is related to the origin point by the translation $\frac{1}{2}\mathbf{a} + \frac{1}{2}\mathbf{b} + \frac{1}{2}\mathbf{c}$. It follows that the middle plane of the stack (hkl) contains this point when the sum $h + k + l$ is even, but no plane of the stack contains it when $h + k + l$ is odd. In the latter case the same translation which relates the centering point to the origin point translates the stack of planes so that the origin plane occurs also at the centering point. This causes the planes of the translated stack to be interleaved halfway between the planes of the original stack. Accordingly, when the sum $h + k + l$ is odd, a stack whose planes cut the axes into h, k, and l parts cannot occur, and so the planes (hkl) do not exist. This gives rise to the following criterion.

Condition for I: If the general reflections hkl present are only those for which $h + k + l = 2n$, the lattice is I.

Rhombohedron-centered cells. The crystal classes 3, $\bar{3}$, 3 2, 3 m, and $\bar{3}$ $2/m$ can have either a rhombohedral lattice or a hexagonal lattice. In

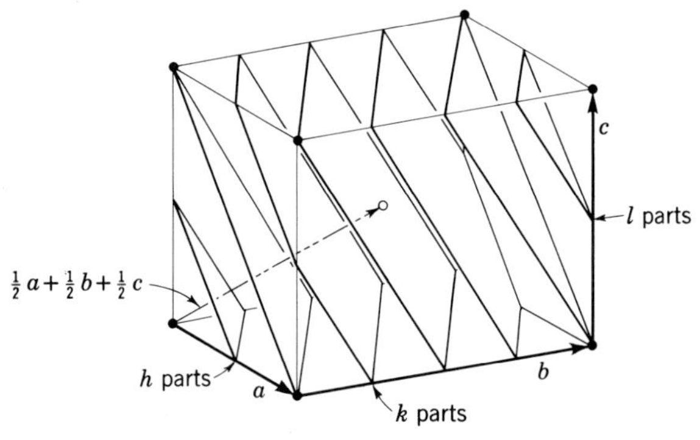

Fig. 9

advance of determining which alternative occurs for a crystal being investigated, it is customary to refer the crystal to a hexagonal cell and then to apply a criterion to the reflections present to determine whether additional lattice points do or do not occur within the interior of the cell. Even when the lattice turns out to be rhombohedral, it is customary to continue to refer the crystal to the hexagonal cell because the orthogonality of the c axis to the plane of a_1 and a_2 gives rise to a simpler geometry. When this is done, the hexagonal cell is no longer primitive, but triple. This treatment of rhombohedral-lattice points by referring them to a rhombohedron-centered hexagonal cell is analogous to referring all isometric crystals to a cubic cell and then, if the lattice turns out to be I or F, regarding the cubic cell as a double (body-centered) or quadruple (face-centered) cubic cell, respectively.

The rhombohedral cell can be referred to a hexagonal cell in two ways:

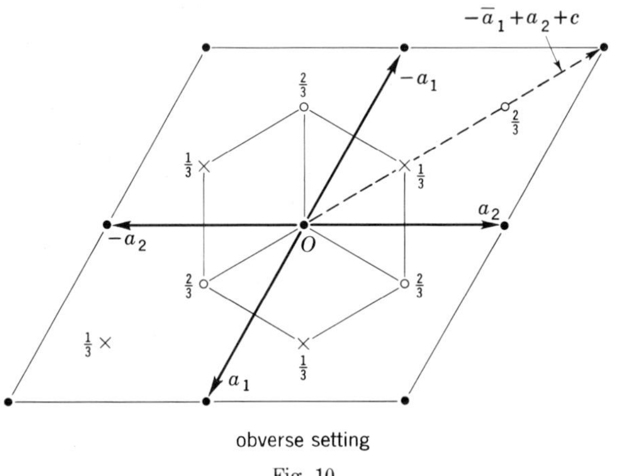

obverse setting

Fig. 10

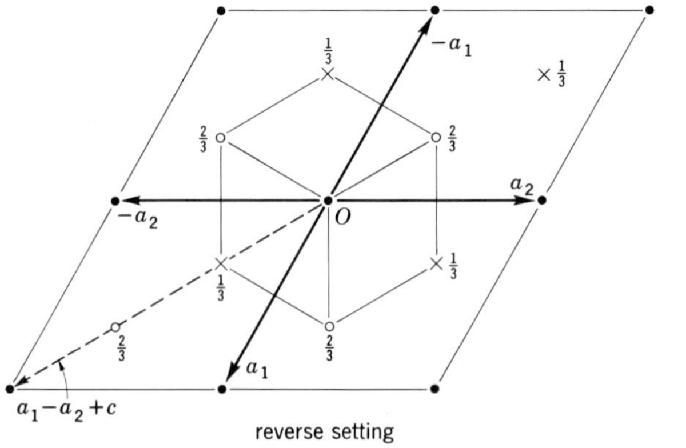

reverse setting

Fig. 11

the "obverse" setting (Fig. 10), or the "reverse" setting (Fig. 11). For definiteness, consider the obverse case first. Here the rhombohedral-lattice points occur at coordinates (referred to the hexagonal cell) of $\frac{2}{3}\frac{1}{3}\frac{1}{3}$ and $\frac{1}{3}\frac{2}{3}\frac{2}{3}$. The lattice points can be described alternatively as occurring at distances of $\frac{1}{3}$ and $\frac{2}{3}$ along the long diagonal $-\mathbf{a}_1 + \mathbf{a}_2 + \mathbf{c}$ of the cell. If the reasoning used for the I cell (discussed in the foregoing section) is followed, the planes of the stack (hkl) divide the long diagonal of a cell, referred to a set of hexagonal axes $-a_1$, a_2, c, into $h + k + l$ parts; the same planes in the same cell referred to a set of axes a_1, a_2, c (that is, the sense of a_1 is reversed) become $(\bar{h}kl)$, so that these planes divide this cell into $-h + k + l$ parts. When this sum is divisible by 3, two of the planes of the stack contain the rhombohedron-centering points. When this sum is not divisible by 3, the same translation which relates the origin point to these two additional points also translates the planes of the stack (hkl) to these two points and thus interleaves the planes of the original stack with the planes of two new stacks; the resulting composite stack has a spacing one-third that of (hkl).

These considerations, and a corresponding set for the reverse setting, give rise to the following criteria.

Conditions for P and R: For a hexagonal crystal referred to a hexagonal cell, if the general reflections hkl are present without restriction, the lattice is P. But if the general reflections hkl present are only those for which $-h + k + l = 3n$, the lattice is R in obverse setting; and if the general reflections are only those for which $h - k + l = 3n$, the lattice is R in reverse setting.

Alternative derivations. The criteria which have just been derived for identifying the lattice type from the x-ray reflections that the crystal gives can also be established in several other ways. An alternative way is to recognize that, when referred to a primitive cell, all general planes exist, and each one, in general, produces a reflection. With the aid of a transformation of axes, the translations of the primitive cell, and also the reflections, can be referred to any multiple cell such as a centered cell. When this is done, it is seen that the reflection indices transform to new ones which conform to the criteria derived above.

Another approach is to determine the lattice type which is reciprocal to each type P, I, A, B, C, and F. When this is done, it turns out that all lattice types but two are their own reciprocals. The exception is that the reciprocal of I is F (and, of course, the reciprocal of F is I). The reciprocal-lattice points have h, k, l coordinates which are the same as the indices of the reflections. The centered reciprocal lattices have systematically missing indices which conform to the rules for the systematically missing reflections that have been derived for the lattice to which they are reciprocals.

Criteria for symmetry elements. Glide planes and screw axes leave records of themselves by causing certain classes of specialized reflections (but not the general reflections) to vanish. This occurs in the following ways.

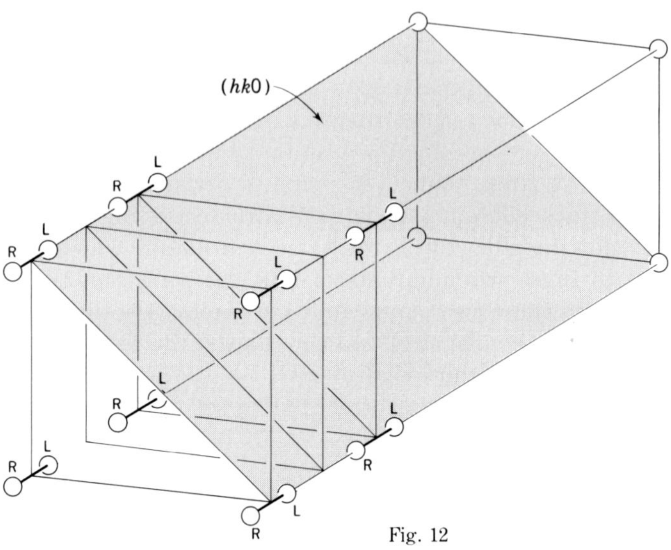

Fig. 12

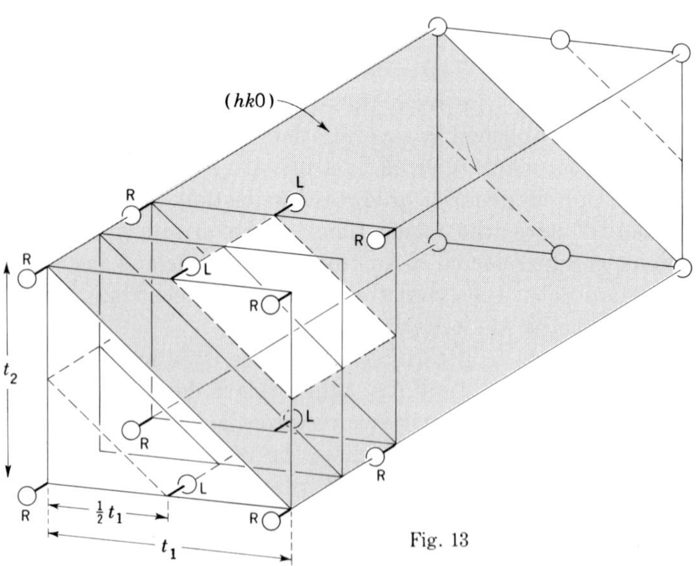

Fig. 13

Glide planes. Any stack of planes whose members are perpendicular either to a reflection plane or to a glide plane have some interesting characteristics. In the case of any stack whose members are perpendicular to a pure reflection plane (Fig. 12), each plane of the stack contains an equal number of left and right motifs. In the direction of the reflection operation (that is, in a direction perpendicular to the reflection plane), the projected left and right motifs superpose, and the projection has the same period as the cell. But in the case of the glide plane (Fig. 13), the submultiple-translation component separates the left and right motifs so that they constitute two different stacks, the planes of one containing only left motifs and the planes of the other, only right motifs. The planes of these two stacks are interleaved exactly halfway between each other, resulting in a composite stack having a half-normal spacing. The motifs of both stacks project in the direction of the reflection component of the glide (that is, in a direction perpendicular to the glide plane) so that the projected motifs cannot be distinguished. As a consequence, each of these two stacks contributes equal amplitudes to an x-ray reflection. The composite stack, therefore, behaves as an ordinary stack having half-normal spacing. The results for typical kinds of glide planes are indicated in Table 2. The reflections affected are specialized in that they are those from planes which are normal to the glide plane and are therefore functions of two index parameters only (for example, k and l in Table 2). This effect does not occur for ordinary mirrors because the left and right remain together in the same planes, as shown in Fig. 12.

Screw axes. The particular stack of planes whose orientation is perpendicular to a screw axis provides a criterion for the translation component of the screw. When this translation component is not zero, as in Fig. 14ii, it repeats each plane of the stack at an interval τ. This requires one or more duplicate planes to occur at regular intervals between the planes of the stack whose lattice interval is t. Only the planes perpendicular to the screw axis have this subnormal interval (Fig. 14ii); planes in other orientations (Fig. 14i) do not have it. Planes of all

Table 2
Reflection conditions for typical glide planes

Glide plane	Translation component	Space-group symbol	Affected planes normal to glide plane	Affected reflections	Conditions for reflections
$\perp a$	$\frac{1}{2}b$	b	$(0kl)$	$0kl$	$k = 2n$
$\perp a$	$\frac{1}{2}c$	c	$(0kl)$	$0kl$	$l = 2n$
$\perp a$	$\frac{1}{2}b + \frac{1}{2}c$	n	$(0kl)$	$0kl$	$k + l = 2n$
$\perp a$	$\frac{1}{4}b + \frac{1}{4}c$	d	$(0kl)$	$0kl$	$k + l = 4n$

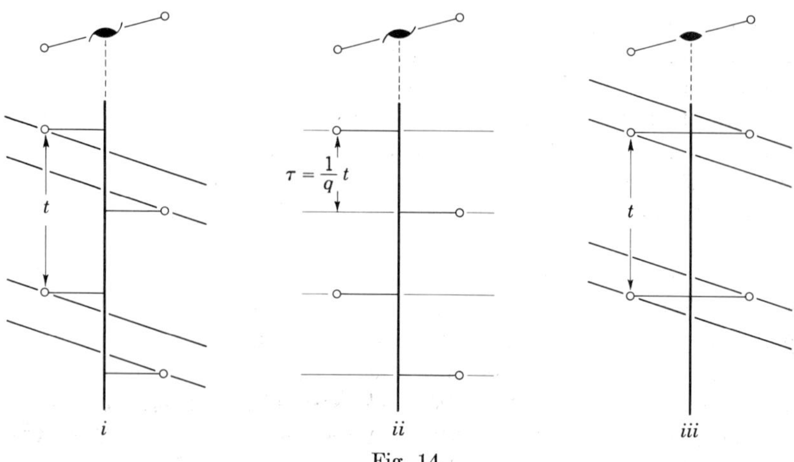

Fig. 14

orientations have their normal period for screws with $\tau = 0$, that is, for pure rotation axes, as seen in Fig. 14iii.

Extinctions. The central theme of the foregoing discussions was that, because certain symmetry operations contain translations which are submultiples of the cell edges, these submultiple translations permit only certain reflections to occur and so reveal the presence of these symmetry operations. From another point of view, some of the reflections which might be expected from the cell dimensions are absent or missing. These missing reflections are usually called *extinctions*, because the reflections are utterly extinguished.† Since they are due to features which characterize a space group, they are often referred to as *space-group extinctions*. Such reflections are systematically missing for certain predictable classes of indices. The class in which the missing reflections occur reveals the kind and location of the coincidence operation which controls them, and the specific features of the absence give further information about the translation component of the operation. These characteristics are brought together in Table 3. This brings out the facts that the general class hkl gives complete information about the lattice type; that the special classes with two independent indices, like $hk0$, $0kl$, $h0l$, and hhl, give information about glide planes; and that the special classes with one independent index, like $00l$, $h00$, and $hh0$, give information about screw axes.

It is important to note that the presence of an extinction in a class of reflections reveals the symmetry element which produces it. But the

† Unfortunately, the word "extinction" is also used to refer to the reduction of the expected intensity of an x-ray reflection when the perfection of the crystal causes attenuation of the primary beam, as noted in Chapter 11.

Table 3
Extinction criteria for lattices and symmetry elements

Class of reflections	Condition for presence	Interpretation	Symbol	Occurrence in lattice type
		Lattice type:		
	none	primitive	P	
	$h + k + l = 2n$	body-centered	I	
	$h + k = 2n$	centered on the C face	C	
	$k + l = 2n$	centered on the A face	A	
hkl	$h + l = 2n$	centered on the B face	B	
	$\left.\begin{array}{l} h + k = 2n \\ k + l = 2n \\ h + l = 2n \end{array}\right\}$	centered on all faces	F	
	$-h + k + l = 3n$	rhombohedral, obverse	R	
	$h - k + l = 3n$	rhombohedral, reverse	R	
	$h = 2n$		a	P, B, I
$hk0$	$k = 2n$	glide plane $\parallel$ (001)	b	P, A, B
	$h + k = 2n$		n	P
	$h + k = 4n$		d	F
	$k = 2n$		b	P, B, C
$0kl$	$l = 2n$	glide plane $\parallel$ (100)	c	P, C, I
	$k + l = 2n$		n	P
	$k + l = 4n$		d	F
	$h = 2n$		a	P, A, I
$h0l$	$l = 2n$	glide plane $\parallel$ (010)	c	P, A, C
	$h + l = 2n$		n	P
	$h + l = 4n$		d	F, B
	$l = 2n$		c	P, C, F
hhl	$h = 2n$	glide plane $\parallel$ $(1\bar{1}0)$	b	C
	$h + k = 2n$		n	C
	$2h + l = 4n$		d	I
	$l = 2n$		$2_1, 4_2, 6_3$	
$00l$	$l = 3n$	screw axis $\parallel$ c	$3_1, 3_2, 6_2, 6_4$	
	$l = 4n$		$4_1, 4_3$	
	$l = 6n$		$6_1, 6_5$	
$h00$	$l = 2n$	screw axis $\parallel$ a	$2_1, 4_2$	
	$l = 4n$		$4_1, 4_3$	
$0k0$	$k = 2n$	screw axis $\parallel$ b	$2_1, 4_2$	
	$k = 4n$		$4_1, 4_3$	
$hh0$	$h = 2n$	screw axis $\parallel$ [110]	2_1	

Table 4
Diffraction symbols and the implied space groups

Triclinic

Friedel symmetry	Lattice type	Diffraction symbol	Space groups isogonal with point group	
			1	$\bar{1}$
$\bar{1}$	P	$\bar{1}\,P\,-$	$P\,1$	$P\,\bar{1}$

Monoclinic†

Friedel symmetry	Lattice type	Diffraction symbol	Space groups isogonal with point group		
			2	m	$2/m$
$2/m$	P	$2/m\ P\ -/-$	$P\ 2$	$P\ m$	$P\ 2/m$
		$2/m\ P\ -/a$		$P\ a$	$P\ 2/a$
		$2/m\ P\ 2_1/-$	$P\ 2_1$		$P\ 2_1/m$
		$2/m\ P\ 2_1/a$			$P\ 2_1/a$
	A	$2/m\ A\ -/-$	$A\ 2$	$A\ m$	$A\ 2/m$
		$2/m\ A\ -/a$		$A\ a$	$A\ 2/a$

† In the monoclinic system there are several ways of describing the same space group by symbols; these alternatives depend on which axial symbol is chosen to represent the unique crystallographic axis and also on the way the remaining two axes of the cell are chosen from the permissible translations of the lattice. If the unique crystallographic axis is designated c (as it is in the tetragonal and hexagonal systems), the crystal is said to be described in the *first setting;* if it is designated b (as was the practice in older crystallographic literature), the crystal is said to be described in the *second setting.* When the lattice is centered, these other two cell axes can be chosen in three different ways, as a result of which the lattice type is described by one of three different symbols. These are all represented by A in the monoclinic table, but the alternatives are noted below.

Furthermore, if the symmetry includes a glide plane, its glide direction may be parallel to either of the two nonunique crystallographic axes or to the cell diagonal. When the cell is selected, therefore, the glide plane may turn out to be described by any of three symbols in either of the two possible settings. These are all represented by a in the monoclinic table above, but the alternatives are as follows:

	Alternative permissible lattice symbols for centered cells	Alternative permissible glide-plane symbols
First setting	$A,\ B,\ I$	$a,\ b,\ n$
Second setting	$A,\ C,\ I$	$a,\ c,\ n$

absence of extinction in a class does not necessarily imply the presence of some alternative symmetry element. It merely shows that a submultiple translation is not present; this may be caused by the absence of *any* symmetry element in the position in question.

Diffraction symbols. All the symmetry information concerning a crystal which can be derived qualitatively† from x-ray diffraction experiments can be set down neatly in a *diffraction symbol.* The total information consists of the Friedel class, the lattice, the glide planes, and the screw axes. To form the diffraction symbol, the symbol for the Friedel class (which is the maximum point-group information directly observable) is written down first. Next comes the symbol for the lattice type (which is uniquely determinable from the extinctions in the general class hkl). Last, a sequence of symbols is set down in the same order as the symbols of the Friedel class, each one of which adds possible space-group information to the corresponding symbol in the Friedel-class sequence. If an extinction is observed, the proper symbol from Table 3 is entered; but if no extinction is observed, a dash is entered. The dash means that a translation component in the space-group symbol was sought but not found. The total number of diffraction symbols is 122. As examples of these symbols, those for the triclinic and monoclinic systems are given in Table 4.

Since there are 230 space groups but only 122 diffraction symbols, it is apparent that some diffraction symbols correspond to several space groups. This occurs because there are many instances in which several space groups within the same Friedel-symmetry class are without extinctions. For example, Friedel symmetry $2/m$ includes the point groups 2, m, and $2/m$. The space groups isogonal with these, and based upon a primitive lattice, include $P\,2$, $P\,m$, and $P\,2/m$. None of these three has any extinctions, and so all belong to the diffraction symbol $2/m\,P\,-/-$. These three cannot be distinguished by qualitative diffraction effects. On the other hand, of the 122 diffraction symbols, 58 characterize only one space group each (or an enantiomorphic pair of

---

A cell which provides any one of these alternatives can be chosen readily for any monoclinic crystal. But a desirable standard practice is to select from the translations of the lattice those three translations which define the reduced cell; for monoclinic crystals, this is the cell whose axes a and b (in the first setting) are the shortest in the ab plane. After these two axes are assigned labels so that $a < b$, then a glide plane, if present, will turn out to have a particular one of the three possible symbols a or b or n, and the cell, if centered, will turn out to have a particular one of the three possible symbols A or B or I.

† The Patterson function, which can be computed from the reflection intensities, contains additional symmetry information which is accordingly available from the reflections, as shown in Chapter 13, but to derive this information a measurement of the intensities of the reflections is required.

space groups), which is therefore uniquely determined by the diffraction symbol.

Other, less useful, kinds of symbols for extinction effects have been suggested. The importance of the diffraction symbol just discussed lies in the fact that it is a complete record of the results of (*a*) an examination of the Friedel symmetry, (*b*) an examination of extinctions for lattice type, and (*c*) an examination of extinctions for glide planes and screw axes. It carries a record of what has been looked for and what has been found.

Summary

It was seen in Chapter 4 that when a crystal is studied by x-ray diffraction, each diffraction maximum with index *hkl* corresponds to a point in the reciprocal lattice whose coordinates are also *h*, *k*, and *l*. The reciprocal lattice, therefore, is a convenient device for filing information about diffraction by the crystal. In addition to an index, each diffraction maximum has a complex amplitude F_{hkl}. Now the reciprocal of a lattice, namely, the reciprocal lattice, is a simple case of a *Fourier transform* applied to a set of points. If each point is assigned a weight F_{hkl}, the resulting array of points is the Fourier transform of the crystal structure. Such a set of weighted points could also be called a *reciprocal structure*.

The Fourier transform of a crystal conforms to the point-group symmetry of the crystal. It also has further symmetry (provided none of its atoms has appreciable anomalous dispersion), because the amplitudes F_{hkl} and $F_{\bar{h}\bar{k}\bar{l}}$ have equal but opposite phases. This relation is called *conjugate symmetry*. Unfortunately, the phases cannot be experimentally observed, so that this symmetry, as observed in any x-ray diffraction experiment, merely appears as if F_{hkl} and $F_{\bar{h}\bar{k}\bar{l}}$ were equal. This circumstance gives rise to *Friedel's law*, which states that the absence of a center of symmetry cannot be observed by x-ray diffraction. Instead, every crystal (unless it contains an atom which has strong anomalous dispersion) appears to have its actual symmetry, to which is added an inversion center if not already present. As a result, only 11 centrosymmetrical symmetries, called the *Friedel symmetries*, can be distinguished by x-ray diffraction.

The lattice types, namely, *P*, *I*, *A*, *B*, *C*, *F*, and *R*, can be readily distinguished by x-ray diffraction, because each type is accompanied by systematic absences which are characteristic of it in the general class of reflections *hkl*. Such systematic absences are called *extinctions*. Glide planes reveal their presence by extinctions in the special classes *hk*0, *h*0*l*, 0*kl*, and *hhl*. Screw axes are characterized by extinctions in the classes 00*l*, 0*k*0, *h*00, and *hh*0. Neither mirrors nor simple rotation axes leave evidence of their existence by extinctions. The absence of extinctions corresponding to a glide plane cannot, however, be interpreted as the

the presence of a corresponding mirror, nor can the absence of extinctions corresponding to a screw be interpreted as the presence of a corresponding rotation axis. The absence of an extinction merely means that no extinction characterizing a symmetry element has been observed.

What actually is observed about the symmetry of a crystal from its qualitative diffraction effects can be neatly set down in a sequence of symbols which together make up the *diffraction symbol* of the crystal. This consists of the standard symbol for the symmetry of the Friedel class; followed by the symbol of the lattice type (as revealed by extinctions); followed by a sequence of symbols, one for each item in the Friedel class, which indicates (by the appropriate letter or numeral) each glide plane or screw axis that has revealed itself by its characteristic extinction or, alternatively, which indicates by a dash that the corresponding item in the Friedel symbol showed no extinction. This sequence displays the complete result of the search for symmetry in the crystal by the qualitative examination of the diffraction by the crystal.

There are 122 diffraction symbols. Of these, 58 each correspond to one space group only, or to an enantiomorphic pair. Each of the remaining 64 symbols corresponds to two or three space groups. The space groups of crystals having these diffraction symbols cannot be determined by a qualitative examination of their diffraction.

Notes on history

The earliest crystal-structure investigations proceeded without the formal use of symmetry information. For example, in Bragg's determination of the structure of NaCl, it was known that the crystal was isometric, and this was assumed in adopting Barlow's model of the packing. But no attempt was made to deduce the Friedel symmetry or space group from the diffraction effects. Had these been understood and used at the time, the diffraction symbol and the knowledge of the density would have immediately reduced the possible structures to two, namely, what are now known as the NaCl type and the sphalerite type. But an understanding of the determination of space groups and how to make use of them in crystal-structure investigation was to await a book by Niggli.

There was little understanding of the symmetry of diffraction in the first year of x-ray diffraction. Laue was confused by the fact that sphalerite, which had only 2-fold axes, should give a Laue pattern showing a 4-fold axis, and that both right and left quartz should give the same pattern. It was against this background that Georges Friedel, only a year and a half after the discovery of x-ray diffraction, announced Friedel's law and established the 11 symmetries that diffraction effects could show.

Space-group theory had been developed well before the opening of the twentieth century, and Harold Hilton's book *Mathematical crystallography and the theory of groups of movements* made this available in compact form as early as 1903. But this information was not used in the study of crystal structures until after 1919, when Niggli's book *Geometrische Kristallographie des Diskontinuums* appeared. This not only contained graphical and analytical descriptions of the space groups but pointed out how they could be determined by x-ray diffraction effects (essentially what we now describe as absent reflections or extinctions). In 1924 Astbury and Yardley published a long paper in the Philosophical Transactions entitled *Tabulated data for the examination of the* 230 *space-groups by homogeneous x-rays,* in which the space groups were again pictured and their abnormal spacings listed. This classic work was widely read and led to the custom of determining the space group before attempting a crystal-structure investigation.

In the early days of x-ray crystallography Friedel symmetry was always determined with the aid of Laue photographs, since powder photographs give no symmetry information whatever, and the oscillating-crystal method very little. Although the Weissenberg method was invented in 1924, very little use was made of it, and it was not until about 1934 that the technique of using this apparatus was simplified and commercial apparatus became available. In 1935 Buerger called attention to the fact that all the symmetry information about the crystal could be determined from a mere inspection of a set of Weissenberg photographs. He proposed that the sum of such information be listed in a diffraction symbol consisting of the Friedel symmetry followed by symmetry information from extinction data. This represented the maximum amount of symmetry information which could be extracted from diffraction experiments by qualitative means. It developed later, as shown in Chapter 13, that the set of intensities of all the *hkl* reflections contained further symmetry information which could be extracted by appropriate treatment.

Additional reading

M. J. Buerger. *X-ray crystallography.* (Wiley, New York, 1942) 55–91.

Significant literature

Friedel symmetry

G. Friedel. *Sur les symétries cristallines que peut révéler la diffraction des rayons Röntgen.* Compt. rend. **157** (1913) 1533–1536.

Use of abnormal spacings

Paul Niggli. *Geometrische Kristallographie des Diskontinuums.* (Gebrüder Borntraeger, Leipzig, 1919) 463–503.

W. T. Astbury and Kathleen Yardley. *Tabulated data for the examination of the* 230 *space-groups by homogeneous x-rays.* Phil. Trans. London (A) **224** (1924) 221–257 and plates 5–24.

Diffraction symbols

M. J. Buerger. *The application of plane groups to the interpretation of Weissenberg photographs.* Z. Kristallogr. (A) **91** (1935) 255–589, especially 287–288.

M. J. Buerger. *X-ray crystallography.* (Wiley, New York, 1942) 510–516.

M. J. Buerger. *Diffraction symbols.* Chapter 3 of *Physics of the solid state,* edited by S. Balakrishna, M. Krishnamurthy, and B. Ramachandra Rao. (Academic, London, 1969) 27–42.

6

Introduction to the use of x-rays in the study of crystals

Available radiation

The fact that diffraction provides a way of studying the reciprocal of a pattern, that is, its Fourier transform, and that the pattern can be derived from this reciprocal, has made diffraction the chief physical tool for the study of the patterns of atoms in crystals. Diffraction requires some kind of radiation having wave properties and with a wavelength of about the order of magnitude of the periods of the pattern to be investigated. In the case of most crystals of ordinary inorganic and organic interest this is about 3 to 30 Ångström units† (commonly about 10 Å), although many crystals of biological interest such as the proteins, and a few polytypes of inorganic substances, have larger repeat periods.

It turns out that x-rays emitted by some of the metals have wavelengths which are suitable for investigating patterns with these periods, and it was this fact that led to the discovery of x-ray diffraction by crystals. X-rays are still the kind of radiation most commonly used to explore crystal patterns, but beams of neutrons, available in certain favored localities since the last war, are also used. Neutron beams have certain advantages (as well as many disadvantages) because the scattering by neutrons occurs from the nucleus of the atom rather than from its electrons. Beams of electrons are also occasionally used, but these do not offer decided advantages over x-rays. Whether x-rays, neutron beams, or electron beams are used, the general features of their use in the study of crystals are the same. Accordingly, the methods of studying crystal patterns outlined in this book are discussed in terms of x-ray methods since they are used in almost every center where crystals are studied.

In this chapter a brief introduction is given to some of the peculiarities of x-rays. The objective of this book is not to present diffraction methods in such detail that the book is suitable as a guide to the actual practice and techniques of the methods, but rather to outline the general theory and features of the methods so that the student scientist who does not necessarily plan to become a practicing crystallographer may nevertheless obtain an intelligent grasp of what methods are in common use, to what crystallographic problems they can be applied, and how each of them contributes to the data of present-day crystallography.

Production of x-rays

The continuous spectrum. Only enough need be known about x-rays to understand some of the problems encountered in their use in x-ray diffraction. X-rays are produced by an evacuated tubular vessel made of glass and metal and called an x-ray tube (shown diagrammatically in Fig. 1). The x-rays are generated by causing electrons, each having an electrical charge e, to fall through a large voltage drop V. The electrons are supplied by a glowing filament at the cathode, and are

† One Ångström unit (abbreviated Å) is 10^{-10} meter or 10^{-8} centimeter. This unit of linear measure is named after Anders Jonas Ångström (1814–1874), a pioneer in spectroscopy, who used this unit in his work.

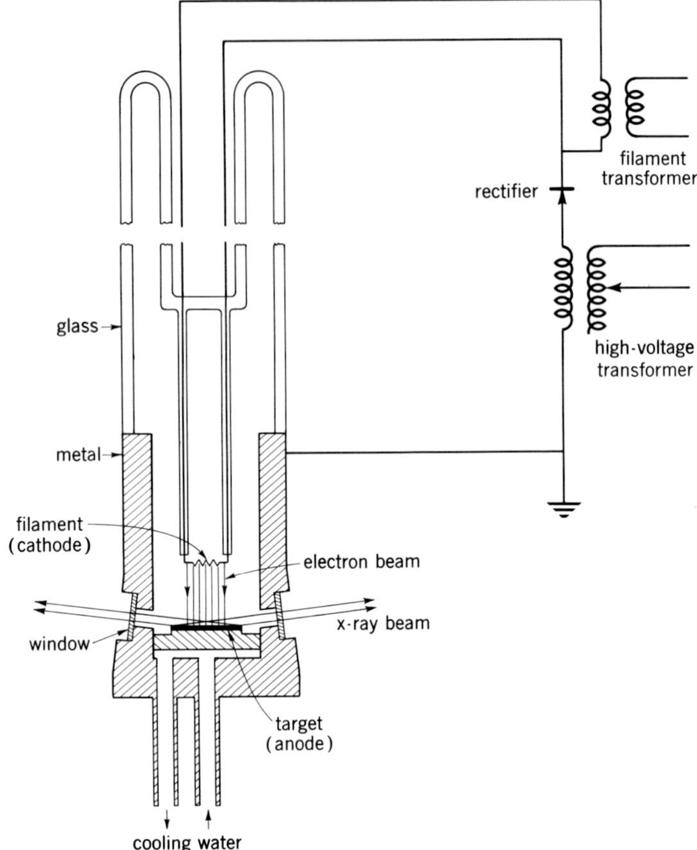

Fig. 1. Diagram of a modern x-ray tube. (*Much simplified but based on the Machlett tube, with permission of the Machlett Laboratories.*)

caused to be accelerated by applying the potential V between cathode and anode. In this process the electron finally achieves an energy eV. The electrons are eventually stopped by impinging on the atoms of the metal of the anode, where they impart their energies to one or more atoms of the metal. In being stopped by the metal atoms the electrons suffer deceleration in one or more stages, and in each such deceleration the energy loss ΔE is converted into a photon of radiation whose frequency and wavelength are given by

$$h\nu = h\frac{c}{\lambda} = \Delta E, \tag{1}$$

where h is Planck's constant, c is the velocity of light, and ν is the frequency and λ is the wavelength of the resulting radiation. The maximum energy transfer occurs when the entire energy of the electron is given up in one step; under these circumstances

$$h\frac{c}{\lambda} = Ve. \tag{2}$$

The term on the left is a maximum, and, since h and c are constants, λ is a minimum. A convenient form of (2) is one in which the universal constants are on one side of the equation, namely,

$$\lambda_{\min} V = \frac{hc}{e} \tag{3}$$

$$= 12.398 \qquad \text{when } \lambda \text{ is expressed in Å, and } V \text{ in kV.} \tag{4}$$

This shows that the minimum wavelength of the x-rays produced is inversely proportional to the potential across the x-ray tube. Provided that a critical voltage (characteristic of the anode metal) has not been exceeded, the distribution of intensity of the x-rays with wavelength is that illustrated in Fig. 2.

Characteristic x-radiation. If the voltage across the x-ray tube exceeds the critical voltage just mentioned, the electrons arriving at the anode have large enough energies to dislodge the electrons in the atoms of the anode from their normal states. When this occurs, the state becomes vacant, and electrons in other, higher-energy, states in the atom can drop into the vacant state, and so lose an amount of energy ΔE. Each such energy loss is transformed into radiation according to relation (1). Since the energy of each of the two states is sharply defined, so is their difference ΔE, as well as λ of (1). Such wavelengths are characteristic of the element which emits them, and they appear in the wavelength spectrum emitted by the metal as sharp peaks superposed on the continuous intensity distribution, as shown in Fig. 3*A*.

The anode of an x-ray tube is usually known as the "target" because the electron beam is stopped by it. From what has been said, it is seen that the x-ray target emits noncharacteristic x-rays whose wavelength

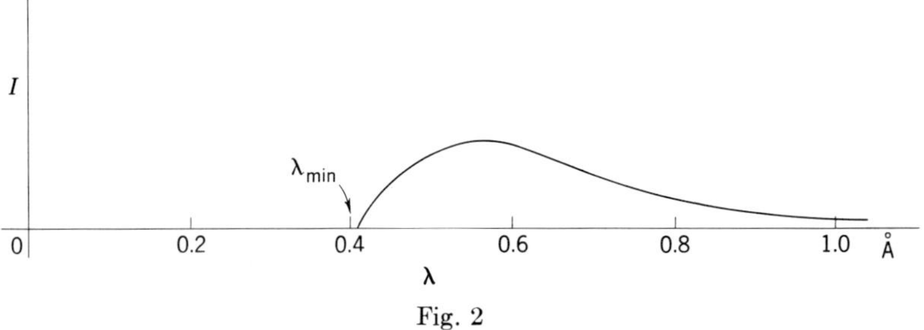

Fig. 2

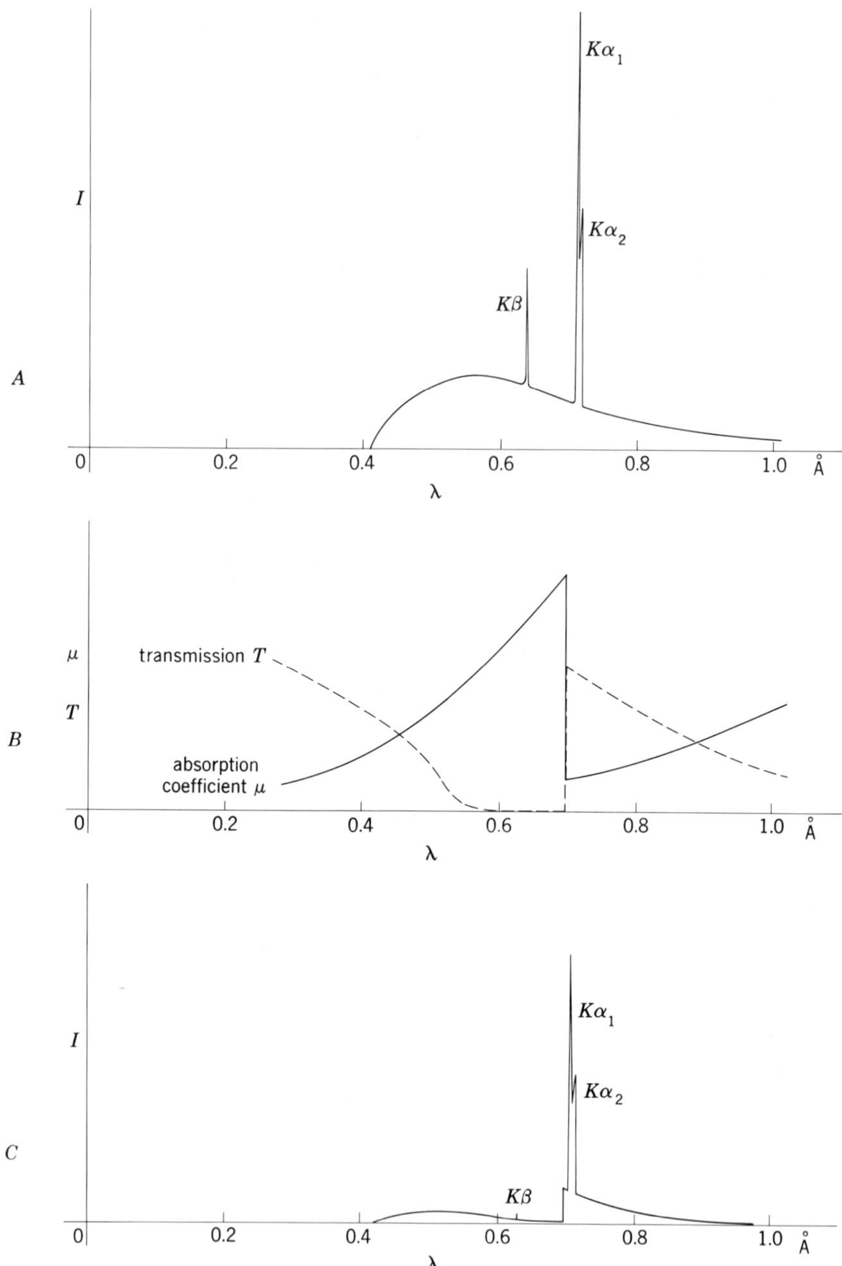

Fig. 3. The action of a filter for x-rays. Figure 3*A* shows the distribution of intensity in the x-rays produced by the x-ray tube. Figure 3*B* shows the absorption and transmission of a thin foil of appropriate filtering material. If the height of the intensity curve in Fig. 3*A* is multiplied by the height of the transmission curve in Fig. 3*B*, the result, shown in Fig. 3*C*, is the distribution of intensity in the x-rays passing through the filter. This is dominated by the characteristic $K\alpha$ radiation, but still contains a little continuous background, especially at wavelengths larger than $K\alpha$.

varies continuously from the minimum wavelength permitted by the potential across the tube, and in accordance with relation (3). If the potential is sufficient to permit the transitions noted above, there is superposed upon this continuum a series of sharp, nearly monochromatic peaks characteristic of the element producing them, as illustrated in Fig. 3*A*. These characteristic wavelengths decrease with increasing atomic number Z. For diffraction purposes, the most important characteristic series is the K series, consisting, in order of decreasing wavelength, of the $K\alpha_2$, $K\alpha_1$, and $K\beta$ radiations. The wavelengths of these lines for elements of certain atomic numbers Z are listed in Table 1.

Monochromatic x-radiation

Rigorously monochromatic x-radiation probably does not exist, but an approximation to it can be obtained by isolating a part of the characteristic radiation of an element. There are two simple means of achieving this isolation: filtering, and reflection from a crystal. Filtering is the method most commonly used because of its simplicity.

Filtered x-radiation. Each element is characterized by a sharp discontinuity in its absorption of x-rays, illustrated in Fig. 3*B*. On the short-wavelength side of such an absorption "edge" the absorption suddenly increases. The relation of x-ray emission to x-ray absorption is such that all radiation shorter than $K\alpha$ from an element of atomic number Z can be highly absorbed by an element of atomic number $Z - 2$ or $Z - 1$. This can be seen by comparison of emissions and absorptions in Table 1. By using a foil of a properly chosen absorber, the K radiation of an element, such as shown in Fig. 3*A*, can be filtered so that the $K\beta$ and the short-wavelength general radiation are substantially removed, as shown in Fig. 3*C*. The radiation which comes through the filter is essentially $K\alpha_1 + K\alpha_2$ (which is a close doublet, as seen in Table 1) plus a trail of long-wavelength general radiation.

Crystal-monochromated x-radiation. A method of isolating the $K\alpha$ radiation more perfectly than by filtration is to reflect it from a crystal set so that a stack of planes of spacing d makes the proper angle θ to fulfill the Bragg equation for the λ of the $K\alpha$ radiation. Of course a crystal to be used for this purpose must be chosen so that the intensity of its reflection is very strong, but this characteristic alone is not sufficient, because the angle θ chosen for reflection hkl for the desired wavelength λ is the same as that for reflection $2h\ 2k\ 2l$ for wavelength $\frac{1}{2}\lambda$, so that λ is likely to be accompanied by $\frac{1}{2}\lambda$. To avoid this harmonic wavelength, it is desirable either to find a crystal with a strong hkl reflection and a weak $2h\ 2k\ 2l$ reflection or to operate the x-ray tube at a voltage so low that the $\frac{1}{2}\lambda$ radiation cannot be excited.

Crystal-monochromated x-radiation is partially plane-polarized because

Table 1

Useful characteristic x-ray wavelengths and appropriate filters

Radiation						Filter		
Z	Element	Average $K\alpha$	$K\alpha_2$ (strong)	$K\alpha_1$ (very strong)	$K\beta_1$ (moderate)	$Z - 2$	Element	K absorption edge
47	Ag	0.5608Å	0.56377Å	0.55936Å	0.49701Å	45	Rh	0.5338Å
46	Pd	0.5869	0.589801	0.585415	0.52052	44	Ru	0.5605
45	Rh	0.6147	0.617610	0.613245	0.54559	43	Tc	0.5891
44	Ru	0.6445	0.64736	0.64304	0.57246	42	Mo	0.6198
43	Tc	......	0.67927	0.67493	0.60141	41	Nb	0.6529
42	Mo	0.7107	0.71354	0.70926	0.63225	40	Zr	0.6888
41	Nb	0.7476	0.75040	0.74615	0.66572	39	Y	0.7276
40	Zr	0.7873	0.79010	0.78588	0.70171	38	Sr	0.7697
						$Z - 1$		
30	Zr	1.4364	1.43894	1.43511	1.29522	29	Cu	1.3804
29	Cu	1.5418	1.54433	1.54051	1.39217	28	Ni	1.4880
28	Ni	1.6591	1.66169	1.65784	1.50010	27	Co	1.6081
27	Co	1.7902	1.79278	1.78892	1.62075	26	Fe	1.7433
26	Fe	1.9373	1.93991	1.93597	1.75653	25	Mn	1.8964
25	Mn	2.1031	2.10568	2.10175	1.91015	24	Cr	2.0701
24	Cr	2.2909	2.29351	2.28962	2.08480	23	V	2.2690

of the reflection; therefore the diffraction results must be corrected for this effect if intensities are important.

Selection of appropriate radiation. As pointed out at the beginning of Chapter 5, the objective of an x-ray diffraction experiment is to provide data with which to determine the space group, the cell, and eventually the arrangement of atoms in a crystal structure. These data consist of the locations and intensities of the x-ray reflections by the crystal. In the language of reciprocal space, a sufficient number of reciprocal-lattice points must be explored to assure that the correct reciprocal-lattice cell can be selected, that the Friedel symmetry can be fixed, and that a sufficient number of the weights of these points (the reflection intensities) can be measured to solve the number of variable parameters of the structure. Table 1 shows that the practical available wavelengths vary from 0.56 to 2.29 Å, with a gap (due to inappropriate physical properties of elements of atomic numbers 31 to 39) in the range 0.79 to 1.44 Å. If a mean value, say 1.5 Å, is substituted in Bragg's law [(7), Chapter 3], it is seen that the number of orders of a reflection which can be observed is something more than the number of Ångström units in the cell edge. Even for crystals with the very smallest cells, then, several orders of reflections can be observed, and many more can be observed for crystals with large cells. All this simply means that the available practical x-ray wavelengths are adequate to explore the structures of crystals, which is a fortunate circumstance of nature.

The most commonly used radiation is Cu$K\alpha$, for which $\lambda = 1.5418$ Å, which is about the center of the available wavelength range. This radiation has a number of all-around advantages. Copper is an excellent heat conductor with a comparatively high melting point, so that it dissipates heat well and so makes an excellent x-ray target material. Its wavelength gives an adequate number of reflections for most purposes, except possibly for the refinement of certain crystals with small cells and many refinable parameters.

Radiations with wavelengths larger than Cu$K\alpha$ have the advantage of being "cleaner" because the accompanying general-radiation background decreases with wavelength; but unfortunately, longer wavelengths are highly absorbed, both by the air in the beam path and, usually, by the crystal itself. On the other hand, shorter-wavelength radiations, while having the advantages of being little absorbed by the air and usually by the crystal, are "dirtier," that is, have high general-radiation backgrounds.

In selecting a wavelength, any radiation which is highly absorbed by an element in the crystal is to be avoided, because this radiation is incoherently scattered by these atoms of the crystal and this scattered radiation contributes a high unwanted background to the desired diffracted radiation. Ordinarily an intense background is caused by using radiation from an element of atomic number $Z + 3$, where Z is the atomic number of an element of the crystal. For example, Cu ($Z = 29$) is intensely scattered by a crystal containing Fe ($Z = 26$) atoms.

Controlling x-rays

When traveling in solid matter, x-rays display a very small refractive index; in fact, this differs from unity by only a few parts per million. Accordingly, it is impracticable to control x-rays the way visible light is controlled—by means of lenses. About the only way to control x-rays is to block them with an absorber. This method is used in making an x-ray *collimator*, as noted below.

All x-ray methods depend on what amounts to measuring the deviation angle 2θ or certain of its components. This normally requires defining a narrow beam of x-rays whose direction is the origin direction for measuring 2θ. The piece of accessory apparatus which does this is given the

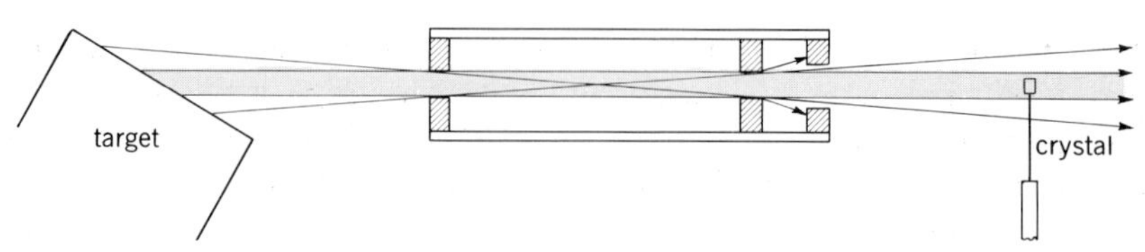

Fig. 4

generic name collimator, one of its special varieties being known as a *pinhole system.* A diagrammatic cross section of a pinhole system is shown in Fig. 4. X-rays derived from the target of an x-ray tube are limited by two circular apertures (called *pinholes,* of the order of $\frac{1}{2}$ mm in diameter) drilled in lead plugs. These are mounted in a brass tube whose centerline defines the direction of the x-ray beam which the collimator permits to pass. Actually, the x-ray beam consists of all rays which can pass through the two apertures, and these usually make up a central, more or less uniform, strong section surrounded by a cone of weaker radiation. Of course the lead of the second pinhole diffracts the x-rays impinging on it, and so it is arranged that this radiation is caught, as shown in Fig. 4, by what amounts to the inside of a third aperture. The last aperture has a larger diameter than the other two, specifically, a size which is designed to be large enough to be missed by all the rays of the beam passed by the collimator, yet small enough to catch the radiation diffracted from all sides of the second aperture.

There are, of course, many details of the design of collimator systems which need not be considered in this book.

Detection of diffracted radiation

It was known before the original diffraction experiment by Friedrich and Knipping that x-rays could be recorded on a photographic film. It is natural, therefore, that these experimenters should have used photographic detection of diffracted x-rays in their experiments. The same method of detection has been used ever since, although it has recently been supplemented by photon-counter methods for certain purposes.

Photographic recording methods are simple to use, and they require no supervision while the recording is going on. They are routinely used in practically all circumstances in the preliminary stages of examining the pattern of a crystal, specifically, in determining its symmetry and cell. They have the advantage that, since they are able to record the diffraction corresponding to all accessible reciprocal space, they reveal any imperfections in a crystal, so that their photographs often serve as a test as to whether a crystal is sufficiently perfect to be used with counter methods.

During the Second World War, x-ray photon detectors and counters were developed to such an extent that, at the end of the war, a powder diffractometer employing a Geiger counter as detector became commercially available. It quickly became evident that photon counters offered a very precise way of determining x-ray diffraction intensities, which previously had been estimated by the blackening on a photographic film. Shortly after the development of the powder diffractometer, therefore, the single-crystal diffractometer came into use for accurate measurement of intensities. Although intensities are still often judged by film blackening using photographic methods, it is usual now to use the single-crystal diffractometer for crystal-structure investigations where great accuracy is desired. Recently such single-crystal diffractometers have been automated so that they, too, collect intensity data without supervision.

Absorption by the specimen

The sample which is set in the x-ray beam for the experimental study of its diffraction not only diffracts x-rays but also, to a greater or lesser extent, absorbs them. Absorption of a ray during transmission through a thickness t of material reduces the original intensity I_0 to a new intensity I according to the usual law of absorption,

$$I = I_0 e^{-\mu t}, \tag{5}$$

in which μ is a constant for the absorbing material and is known as its linear-absorption coefficient. This quantity can be computed for any crystalline compound of known chemical composition and density. When it is found, relations like (5) must be summed for all paths contributing to each diffracted ray if an allowance is needed for the effects of absorption.

Absorption affects the x-ray record by producing errors of two sorts: It reduces the total diffracted intensity, and it shifts the center of the record so that 2θ appears a little larger than it actually is. Allowance must be made for both these kinds of error in an accurate interpretation of x-ray diffraction data.

Summary

X-radiation suitable for the examination of crystals by x-ray diffraction is currently derived from a sealed evacuated tubular vessel of glass and metal called an x-ray tube. Within this tube, electrons from an electrically heated filament in the cathode assembly are caused to fall through a large potential difference of some 30 to 60 kV. On reaching the anode (called the target) they are decelerated as they hit the metal target, and their energy is converted into radiation. This contains a continuous spectrum of wavelengths whose minimum value is inversely proportional to the potential across the x-ray tube. Provided that the voltage also exceeds a critical value, a series of intense monochromatic components characteristic of the target metal also occur in the radiation. Certain metals with atomic numbers in the range 24 to 47 produce characteristic radiations in the wavelength range of 2.3 to 0.56 Å; these are useful in studying crystal structures, but unfortunately this desirable radiation is accompanied by the continuous spectrum. The unwanted radiation can be reduced by filtering through a metal foil which has a high absorption at wavelengths below a critical value characteristic of the metal foil. Alternatively, it can be substantially eliminated by making use of Bragg's law, reflecting only the radiation having the desired wavelength from a special crystal.

The only practical way of controlling x-rays is blocking off the radiation except where it is wanted. In routine x-ray diffraction work a beam of almost parallel x-rays about a millimeter or less in diameter is normally used. This is arranged by a device called a *collimator*, or *pinhole system*, which blocks off all radiation except that which can pass through the circular apertures (pinholes) of two pieces of sheet lead.

Notes on history

X-rays were discovered by Röntgen in 1895; he received the first Nobel prize in physics for this in 1901. In 1912, the diffraction experiments of Friedrich and Knipping under Laue's supervision proved that these rays were wavelike in nature. Shortly thereafter (1914), Duane and Hunt showed that the minimum wavelength was given by the Planck-Einstein quantum condition in the form $\lambda_{min} = hc/Ve$.

That the x-rays contained monochromatic components characteristic of the target metal was discovered by Barkla in 1909, well in advance of the discovery of x-ray diffraction, and the K and L series of characteristic lines were named by him. Immediately after the discovery of x-ray diffraction, Moseley made use of this new tool to study the variation of wavelength of these characteristic lines with atomic number. He found that with increasing atomic number the wavelengths decreased in a regular fashion. He also discovered and named the $K\alpha$ and $K\beta$ components, as well as the L series of components.

The general laws of absorption of radiation by matter had been established in the eighteenth century. The existence of absorption edges in the absorption of x-rays was discovered by de Broglie in 1916.

X-rays were first detected with a fluorescent screen by Röntgen, who also noted that they could be detected by the blackening they produced on developed photographic plates. It was this photographic method of detection that was used by Friedrich and Knipping in their discovery of x-ray diffraction. The early investigations of the Braggs were carried out by using an ionization chamber as a quantitative detector of x-rays. Because of experimental difficulties, this method declined in popularity and gave way almost entirely to photographic detection until recently, when the development of the Geiger counter made quantitative measurement of scattered radiation easy. More recently, proportional counters and scintillation counters have come into common use for the measurement of diffracted x-radiation.

Additional reading

M. J. Buerger. *X-ray crystallography.* (Wiley, New York, 1942) 166–167, 172–182.

Wayne T. Sproull. *X-rays in practice.* (McGraw-Hill, New York, 1946) 1–156, 168–197.

N. F. M. Henry, H. Lipson, and W. A. Wooster. *The interpretation of x-ray diffraction photographs.* (Macmillan, London, 1951) 25–33.

B. D. Cullity. *Elements of x-ray diffraction.* (Addison-Wesley, Reading, Mass., 1956) 1–26.

Significant literature

Charles G. Barkla. *Phenomena of x-ray transmission.* Proc. Cambridge Phil. Soc. **15** (1909) 257–268.

W. Friedrich, P. Knipping, and M. Laue. *Interferenz-Erscheinungen bei Röntgenstrahlen.* Sitzungsber. math.-phys. Klasse der Königlich Bayer., Akad. Wiss. München (1912) 303–322; reprinted in Naturwiss. **39** (1952) 361–368.

H. G. J. Moseley. *The high-frequency spectra of the elements.* Phil. Mag. **26** (1913) 1024–1034; **27** (1914) 703–713.

William Duane and Franklin L. Hunt. *On x-ray wave-lengths.* Phys. Rev. **6** (1915) 166–171.

M. de Broglie. *La spectrographie des phénomènes d'absorption des rayons x.* J. Phys. **6** (1916) 161–168.

W. C. Röntgen. *Über die Elektrizitätsleitung in einigen Kristallen und über den Einfluss einer Bestrahlung darauf.* Ann. Phys. **64** (1921) 1–195.

7

The less powerful diffraction methods

X-ray diffraction methods in general

There are more than a dozen ways of using the diffraction of x-rays in studying various features of crystals. Of interest here are only those which lead to a knowledge of the symmetry, cell, and arrangement of atoms in the crystal. Only a few methods have been used for this purpose, and of these, only two or three are extensively used.

The most fundamental differences between the various methods depend on the kind of radiation used and the kind of sample employed. Only the Laue method uses the continuous x-ray spectrum, and only the powder method makes use of a polycrystalline sample. The methods using characteristic radiation with a single-crystal sample include the rotating-crystal method, the oscillating-crystal method, the Weissenberg method, and the precession method. All these methods record the diffraction on photographic film. (In the final stages of the study of a crystal structure it is important to have precise diffraction intensities. These are currently obtained by supplementary methods which record the intensities with the aid of photon counters instead of photographically. Such methods are sophisticated modifications of the rotating-crystal and Weissenberg methods; they are outlined in Chapter 11.)

The most powerful methods for determining the symmetry and cell are the Weissenberg method and the precession method. One or both of these two methods are almost universally used in crystal-structure investigations, at least in their beginning stages; indeed, an entire investigation may be carried out by using data supplied by either of these methods. Because of their importance in present-day routine these methods are treated separately in Chapters 8 and 9.

On the other hand, the data supplied by the Laue, powder, rotating-crystal, and oscillating-crystal methods have severe shortcomings for crystal-structure investigations; therefore they are not currently used except when equipment designed for the more powerful methods is not available. Some brief comments on the less desirable methods follow.

The powder method

The powder method was published by Debye and Scherrer in Germany in 1916, and independently by Hull in the United States in 1917. The method was little used until after the First World War.

The sample for this method is a polycrystalline powder. When a crystal in this powder happens to be so oriented that one of its planes makes an angle θ (which satisfies Bragg's law) with the x-ray beam, this crystal reflects the small pencil of x-rays which reaches it. All other crystals in the sample for which this plane makes the same angle θ with the x-ray beam also reflect the x-rays which reach them. Each such crystal sends out a diffracted pencil of x-rays which is inclined at angle 2θ with the x-ray beam. The loci of all such rays are the generators of a cone which is coaxial with the x-ray beam.

Because of the random orientation of the many crystal fragments in the powder sample, every plane (hkl) of the crystal is represented by a cone of diffraction which has a characteristic half-opening angle $2\theta_{hkl}$. Part of each cone is ordinarily caught on a photographic film which is held in the form of a cylinder whose axis is normal to the x-ray beam, as seen in Fig. 1. From the known diameter of the film cylinder, a measure-

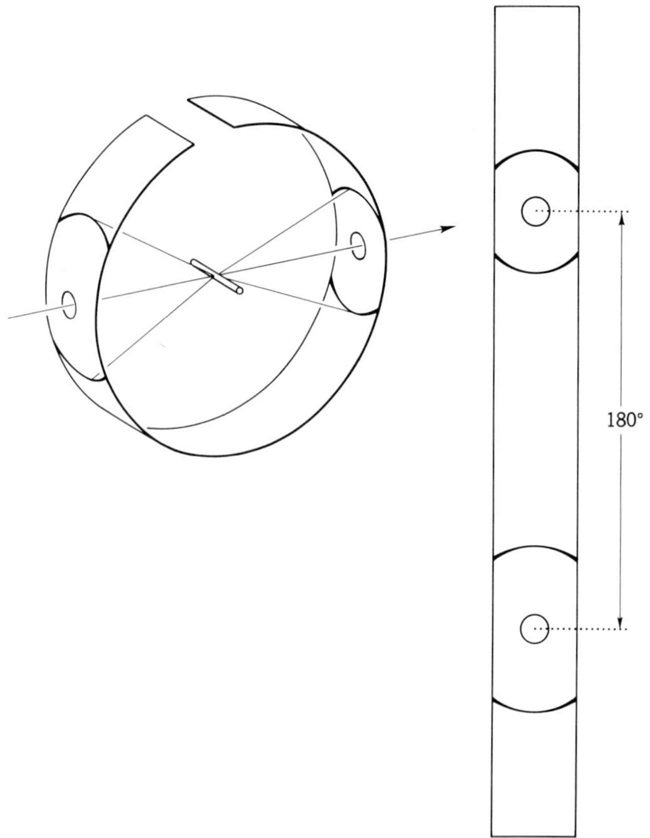

Fig. 1. [*From Leonid V. Azároff and Martin J. Buerger: The powder method in x-ray crystallography.* (*McGraw-Hill, New York*, 1958) 25.]

ment of the length of the film intercepted by the diffraction cone allows θ_{hkl} to be computed for each powder "line" on the film. From each θ_{hkl}, the d_{hkl} of the reflecting plane can be computed with the aid of Bragg's equation (7), Chapter 3. The result of such an investigation is a set of d's for the various planes (hkl) of the crystal, plus the relative intensities of the corresponding reflections hkl.

Unfortunately, it is not easy, nor even always possible, to derive the unit-cell dimensions of a crystal from the set of its d_{hkl}'s. Thus the powder method commonly fails to give adequate data even to start a crystal-structure investigation. Secondly, in many cases reflections having different indices coincide approximately or identically in the same powder line. In fact, identical coincidences are inescapable features for all crystal classes less symmetrical than the class with maximum symmetry (called the holohedral class) of each crystal system. Furthermore, a powder photograph inherently gives no symmetry information.

Because of (*a*) the difficulty or impossibility of determining the unit cell, (*b*) the coincidences of nonequivalent reflections, and (*c*) the lack of symmetry information, the powder method is most unsuitable for any of the stages of crystal-structure determination. Its chief advantage is that it supplies a fingerprint-like empirical characterization of a crystalline phase. This feature is useful in identification and is responsible for the employment of powder photographs in much routine testing.

The rotating-crystal method

Relation to the Laue cones. The rotating-crystal method was devised in the early years of x-ray diffraction. It is fundamentally a simple way of recording the diffraction of monochromatic radiation by a single crystal, with special attention to one of its sets of Laue cones.

In Chapter 3 it was seen that if the direct x-ray beam makes an angle $\bar{\mu}$ with some rational direction in the crystal along which the translation is t, then the direction $\bar{\nu}$ which the diffracted beams make with that rational direction is given by the Laue equation

$$\cos \bar{\nu} = -\cos \bar{\mu} + \frac{m\lambda}{t}, \tag{1}$$

where m is an integer. Thus the many diffracted beams corresponding into a particular value of m all make the same angle $\bar{\nu}$ with the rational direction, and are accordingly the generators of a cone coaxial with that direction and known as a *Laue cone*.

It was also seen in Chapter 3 that, in general, no diffracted beams arise for an arbitrary orientation of the crystal if λ is held constant, provided monochromatic x-radiation is used. To give rise to diffracted beams the crystal orientation must be adjusted so that the Laue condition (1) is satisfied simultaneously for three different rational directions of the crystal. The same conclusions may be reached by considering the solutions of the Bragg equation: In general, no reflection occurs when the angle θ between the x-ray beam and a crystal plane is fixed at an arbitrary value; to cause a reflection, the crystal orientation must be varied so that θ passes through one of the values which satisfy the Bragg equation.

Determination of the translation period t. Now the Laue equation (1) involves the translation period t. This important crystallographic parameter could be readily determined if the distinctive feature of a Laue cone, namely, its angle $\bar{\nu}$, could be determined. This can be

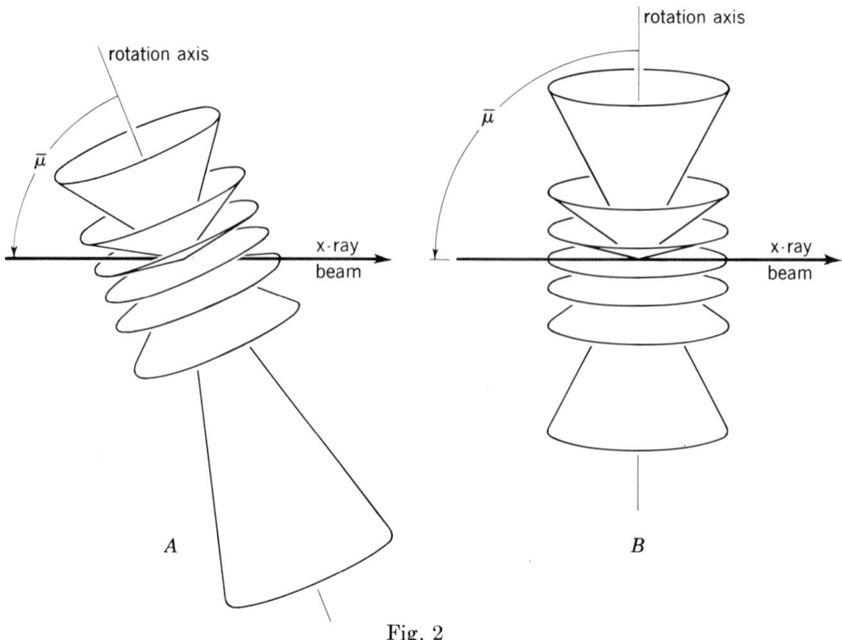

Fig. 2

readily accomplished if the orientation of the crystal with respect to the x-ray beam can be changed while the angle $\bar{\nu}$ is kept constant. Relation (1) shows that, for a particular experiment in which m, λ, and t are constant, $\bar{\nu}$ remains constant provided $\bar{\mu}$ is fixed. There are two mechanical ways of keeping $\bar{\mu}$ fixed while the orientation of the crystal is varied. An obvious way† is to rotate the crystal about the axis of the Laue cone. This maintains $\bar{\mu}$ constant for that rational axis but permits the angle between all other rational axes and the x-ray beam to vary. In general, then, the whole nest of Laue cones shown in Fig. 2*A* remains fixed as the crystal is rotated about the common axis of the cones. A particularly simple arrangement of x-ray beam and rotation axis is the symmetrical one, shown in Fig. 2*B*, where $\bar{\mu} = 90°$, in which case (1) degenerates to

$$\cos \bar{\nu} = \frac{m\lambda}{t}. \tag{2}$$

In order to determine the angle of the Laue cone, the diffraction beams arising from the rotating crystal may be recorded on a photographic film. The most useful film placement is a cylindrical film coaxial with the axis of the Laue cones, as suggested in Fig. 3*A*. With this film arrangement

† The second way, discussed in Chapter 9, leads to the cone-axis method.

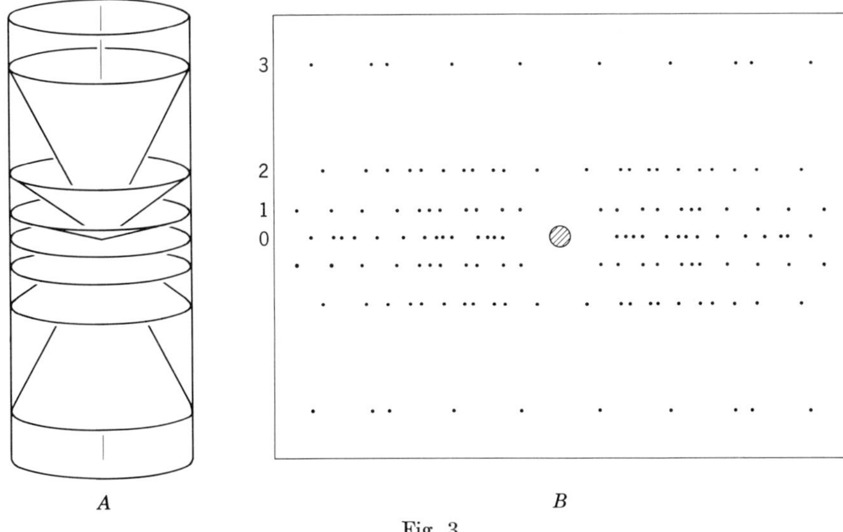

Fig. 3

the cones intersect the cylindrical film in circles. When the film is flattened out, as in Fig. 3*B*, the circles become straight lines which are called *layer lines*. Actually, the lines are merely marked out by spots where reflections directed along the generators of the Laue cones reached the film.

A diagram showing a simple design of the apparatus used to make such photographs is shown in Fig. 4. The crystal is usually stuck to a glass fiber mounted on a device known as a *goniometer head*, which permits the orientation of the crystal to be adjusted. When the orientation has been adjusted, a selected rational direction, ordinarily a crystallographic axis, is parallel to a shaft about which the crystal can be rotated by means of a small motor. A collimator, whose axis intersects the rotation axis at right angles, directs a fine beam of monochromatic x-rays at the crystal. The photographic film, forced to take a cylindrical form by being pressed against the inside of a metal cylinder, is placed so that its cylinder axis coincides with the axis of crystal rotation.

An example of an actual rotating-crystal photograph made with such a camera is shown in Fig. 6. The photograph displays a feature which is characteristic of a photograph made with a well-adjusted crystal: All reflections are neatly concentrated in a small number of parallel straight lines—the layer lines.† The lines are symmetrically disposed above and

† If the adjustment of the rational axis of the crystal to the mechanical axis of the instrument is not precise, the layer lines become broadened into bands whose widths depend on the error in adjustment.

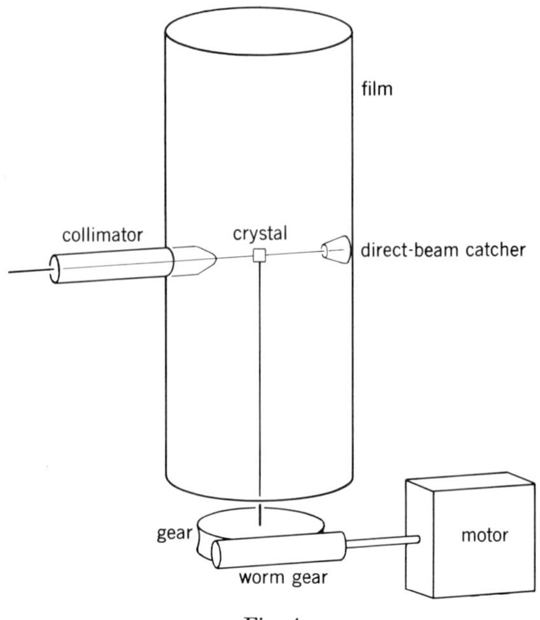

Fig. 4

below a central layer line, which is sometimes called the equatorial layer line, or the equator. The equatorial line is a record of the reflections of the Laue cone of order $m = 0$, while the first layer line above the equator is a record of the reflections of the Laue cone of order $m = 1$, etc.

The geometrical characteristic of a Laue cone can be computed from a measurement of the height y of its record above the equator. Figure 5 shows that y_m, the height of the mth layer, is given by

$$\tan \nu_m = \frac{y_m}{r}, \tag{3}$$

where ν is the complement of $\bar{\nu}$:

$$\nu = \frac{\pi}{2} - \bar{\nu}.$$

In terms of ν, (2) becomes

$$\sin \nu = \frac{m\lambda}{t}. \tag{4}$$

If (3) and (4) are combined, the period is found to be given by

$$t = \frac{m\lambda}{\sin \tan^{-1}(y_m/r)}. \tag{5}$$

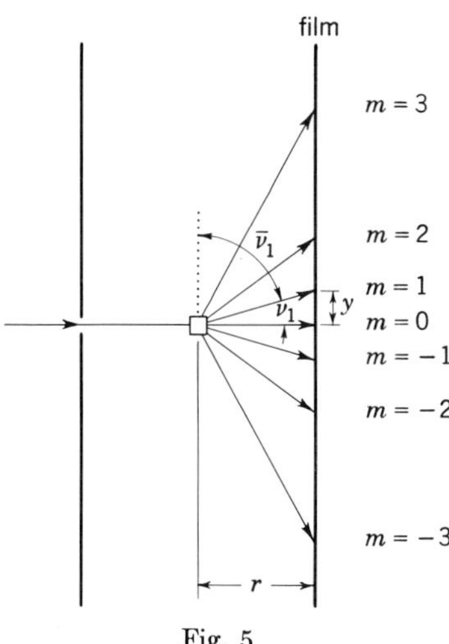

Fig. 5

Symmetry information. Examination of any rotating-crystal photograph, such as Fig. 6, shows that each is symmetrical with respect to the equator and with respect to a centerline perpendicular to the equator. The reason for this is that, as the crystal is rotated through one cycle, the upper sides of a stack of crystallographic planes of arbitrary orientation reflect to the upper-right and the upper-left parts of the film, whereas the lower sides of the same stack reflect to the lower-left and lower-right parts of the film. Since all four reflections are based upon the same spacing d, and since the reflections occur in symmetrically related orientations, their records occur in symmetrical locations on the film.

This symmetry is the essential symmetry of the apparatus and its motion. The rotating-crystal photograph of any crystal must show this symmetry regardless of the symmetry of the crystal. It follows that a lower symmetry of the crystal can produce no inferior symmetry of the record, and a higher symmetry of the crystal can produce no higher symmetry of the record. As a consequence, a *rotating-crystal photograph displays no symmetry information about the crystal which produced it.*

Present-day use of the rotating-crystal method. There are methods of indexing the reflections on each layer line of a rotating-crystal photograph, but there are many disadvantages to such indexing procedures. Fundamentally, the problem of indexing is indeterminate for this method. For example, if a crystal is rotated about its c axis, the reflections on the third (say) layer line all have indices $hk3$. To assign the unknown indices h and k to spots along a line calls for mapping the two variables h and k on the one dimension of the layer line. This

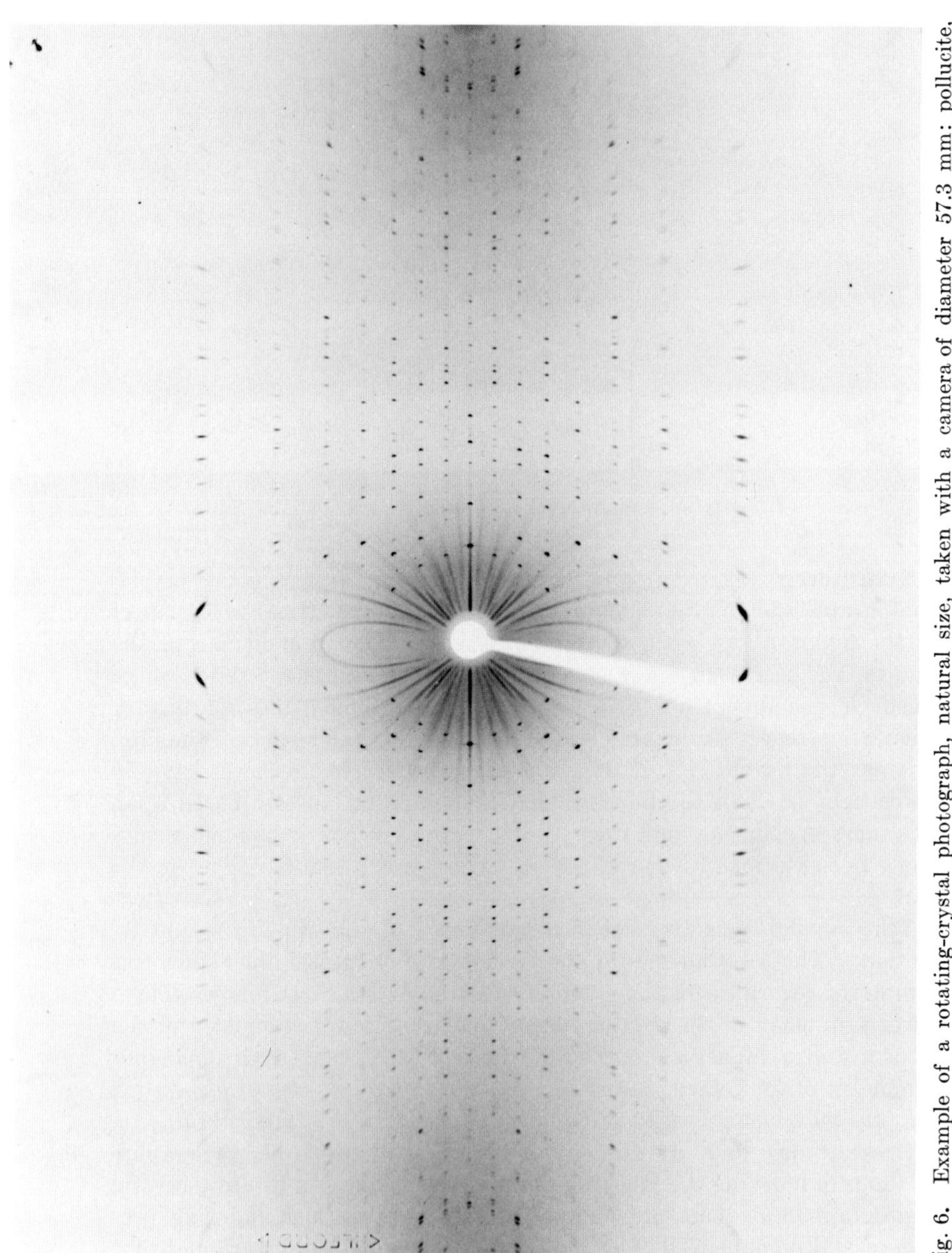

Fig. 6. Example of a rotating-crystal photograph, natural size, taken with a camera of diameter 57.3 mm: pollucite, $(Ca,Na)AlSi_3O_6 \cdot xH_2O$, rotated about an a axis ($CuK\alpha$ radiation, 35 kV, 15 mA, 3 hr).

cannot be done, in general, although a partial assignment of indices can be made when the two dimensions are related by symmetry, as they are in tetragonal and hexagonal crystals. Even in these cases, however, reflections such as $hk0$ and $kh0$, which are not equivalent unless they are related by a mirror between the a_1 and a_2 axes, reflect to the same point and therefore cannot be resolved.

Because other more powerful x-ray diffraction methods are now known, the primitive rotating-crystal method is little used except for one purpose—to determine the period along a crystallographic (or other rational) axis. It shares with the cone-axis method, described in Chapter 9, the property of providing this kind of information superbly, and no other methods can give this information so unequivocally. Once a crystal is oriented, it is routine procedure to determine its identity period along at least one axis with the aid of the rotating-crystal (or cone-axis) method in order to know where all the Laue cones are, so that the reflections they contain can be explored.

What has just been said applies specifically to the x-ray examination of single crystals. There is a crystallographic intergrowth known as a *fiber*, however, which can be examined only by the rotating-crystal (or cone-axis) method. A few crystal species, such as asbestos and certain amphiboles, occur naturally as bundles of parallel, very thin filaments from which a single crystal cannot be separated. The axis of the filament is a rational crystallographic direction, but the orientations of the filaments in the bundle, called a fiber, are random. Such a sample, placed in a fixed position in a monochromatic x-ray beam, gives an x-ray photograph called a *fiber diagram*, which looks exactly like a rotating-crystal photograph. From measurements of the layer-line spacings, the period along the fiber axis can be computed, as in the standard rotating-crystal photograph. Since the sample is not a single crystal, but a bundle, it cannot yield a better x-ray photograph by another method, so that, if the indices of the reflections are required, an attempt must be made to index the reflections in each layer line.

The oscillating-crystal method

The rotating-crystal method can be improved somewhat if the rotation is limited to a small known angle, usually 5 to 25°. In order to obtain sufficient exposure of the film, the crystal is oscillated back and forth through this range. So modified, the rotating-crystal method becomes the *oscillating-crystal* method. A classic paper published in 1926 by Bernal, showing how the rotating-crystal method and oscillating-crystal method can be interpreted easily by the use of the reciprocal lattice, led to the popularization of the latter method. Meanwhile, however, the oscillating-crystal method had already been rendered obsolete by the Weissenberg method, invented in 1924. This is discussed in Chapter 8.

Additional reading

The Laue method

Ralph W. G. Wyckoff. *The structure of crystals.* (Chemical Catalog, New York, 1931), 124–150.

E. Schiebold. *Methoden der Kristallstrukturbestimmung mit Röntgenstrahlen,* vol. I, *Die Lauemethode.* (Akademische Verlagsgesellschaft, Leipzig, 1932).

Wheeler P. Davey. *A study of crystal structure and its applications.* (McGraw-Hill, New York, 1934) 54–85.

Franz Halla and Hermann Mark. *Leitfaden für die Röntgenographische Untersuchung von Kristallen.* (Barth, Leipzig, 1937) 193–207.

N. F. M. Henry, H. Lipson, and W. A. Wooster. *The interpretation of x-ray diffraction photographs.* (Macmillan, London, 1951) 71–86.

B. D. Cullity. *Elements of x-ray diffraction.* (Addison-Wesley, Reading, Mass., 1956) 89–92, 138–148, 215–237, 502–505.

The powder method

Wheeler P. Davey. *A study of crystal structure and its applications.* (McGraw-Hill, New York, 1934) 111–138, 596–613.

N. F. M. Henry, H. Lipson, and W. A. Wooster. *The interpretation of x-ray diffraction photographs.* (Macmillan, London, 1951) 168–193, 212–224.

Harold P. Klug and Leroy E. Alexander. *X-ray diffraction procedures.* (Wiley, New York, 1954) 162–585.

H. S. Peiser, H. P. Rooksby, and A. J. C. Wilson. *X-ray diffraction by polycrystalline materials.* (Institute of Physics, London, 1955).

B. D. Cullity. *Elements of x-ray diffraction.* (Addison-Wesley, Reading, Mass., 1956) 149–214, 299–401, 431–453.

Leonid V. Azároff and Martin J. Buerger. *The powder method in x-ray crystallography.* (McGraw-Hill, New York, 1958).

R. W. M. D'Eye and E. Wait. *X-ray powder photography in inorganic chemistry.* (Academic, New York, 1960).

A. Taylor. *X-ray metallography.* (Wiley, New York, 1961) 152–209.

Lev Iosifovich Mirkin (translated by J. E. S. Bradley). *Handbook of x-ray analysis of polycrystalline materials.* (Consultants Bureau, New York, 1964).

The rotating-crystal method

M. J. Buerger. *X-ray crystallography.* (Wiley, New York, 1942) 92–106, 133–213.

8

The Weissenberg method

Background

It was seen in Chapter 7 that the indexing of a rotating-crystal photograph is indeterminate because the one dimension of each layer-line record cannot resolve two variables such as the indices h and k. Weissenberg recognized this problem and neatly avoided it by devising a new method of recording. Although presented in 1924, his method, which came to be known as the Weissenberg method, was not generally used until a dozen years later, when it began to replace all earlier ways of recording x-ray diffraction from single crystals. Since each x-ray reflection could be examined separately, not only could the Friedel symmetry of the crystal be determined, but it could be treated in a more elegant manner than by the Laue method.

General theory

The Weissenberg method, like the rotating- and oscillating-crystal methods, is based upon rotating a single crystal in a beam of monochromatic x-rays. In Weissenberg instruments it is customary to have the rotation axis of the crystal horizontal, so that the arrangement of crystal, rotation axis, film, and x-ray beam are as shown in Fig. 2*A*. (The horizontal rotation axis facilitates varying μ, the angle which the rotation axis makes with the x-ray beam, a feature which later discussion will show to be very desirable.) As in the rotating-crystal method, therefore, the reflections of the Weissenberg method are confined to Laue cones coaxial with the rotation axis. But instead of allowing all Laue cones to record on a fixed film, one cone at a time is permitted to pass through a slot in a cylindrical *layer-line screen*, as illustrated in Fig. 1. The location of the slot is variable, so that any desired Laue cone can be isolated. If an ordinary rotating-crystal photograph were taken with the layer-line screen interposed between crystal and film, the photograph would show a single layer line only.

The new feature of the Weissenberg method is that the angle ω, which is the rotation required to bring a plane into reflecting position, is recorded. The way this is accomplished can be appreciated from Fig. 2. In the older rotating-crystal method, the reflection would be recorded as

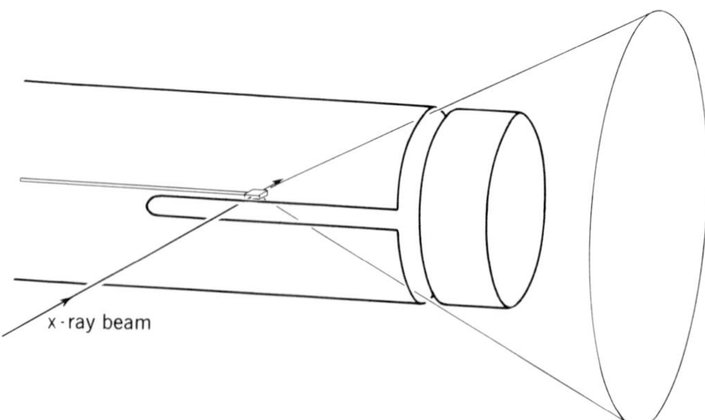

Fig. 1. The isolation of one cone from a nest of Laue cones by the cylindrical layer-line screen of a Weissenberg instrument.

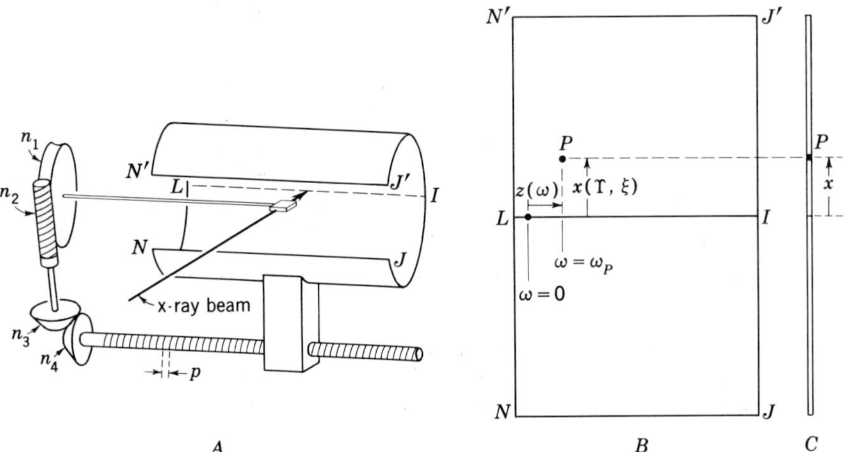

Fig. 2. Diagram of a Weissenberg instrument showing (*A*) the coupling of the camera translation to the crystal rotation; (*B*) the resulting position of a reflection on the unrolled film; (*C*) the position of the corresponding spot along a layer line of a rotating-crystal photograph. [*From M. J. Buerger: X-ray crystallography.* (*Wiley, New York,* 1942) 222.]

a spot on a layer line and located by one measurable coordinate x, as in Fig. 2*C*. In the Weissenberg method the layer line is caused to migrate along the surface of the film by translating the film parallel to the rotation axis. The film translation is coupled to the crystal rotation, as shown in Fig. 2*A*, so that the position of any reflection in a direction parallel to the motion of the film is directly proportional to the amount of the rotation ω required to bring the plane into Bragg reflecting position. In this neat way the coordinate ω, which is lost in the rotating-crystal method and merely bracketed between inequalities in the oscillating-crystal method, is completely preserved in the Weissenberg method. Thus the two coordinates of the unknown indices within a layer line are represented on the two dimensions of the flat surface of the photographic film. Of course, this simply implies that indexing is determinate with the Weissenberg method. It remains to find the transformation from film coordinates to reciprocal-lattice indices.

Apparatus

As noted already, it is desirable to be able to vary the angle μ between the direction of the rotation axis and the x-ray beam. This is accomplished by a double-base construction for the Weissenberg instrument, as suggested in Fig. 3. All present-day apparatus is based upon this design, and an actual instrument is shown in Fig. 4.

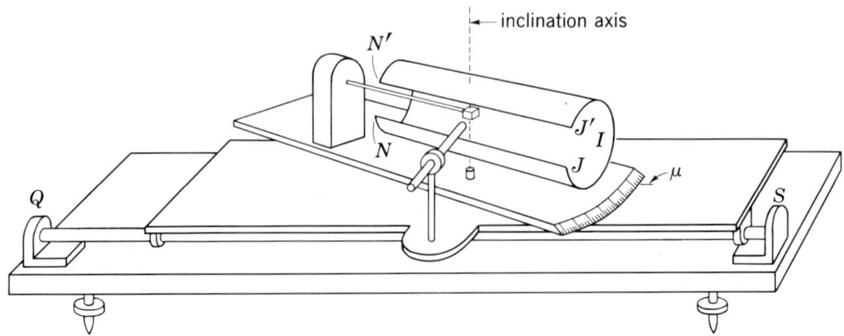

Fig. 3. The multiple-base construction of the Weissenberg instrument which permits the angle μ (or $\bar{\mu}$) to be varied. [*From M. J. Buerger: The Weissenberg reciprocal-lattice projection. . . . Z. Kristallogr.* **88** (1934) 358.]

The ratio of film translation to crystal rotation and its relation to camera radius are important to the practical interpretation of Weissenberg photographs. The coordinate x in Fig. 2*B*, measured normal to the direction of film translation, is proportional to the projection of the Bragg glancing angle 2θ on a plane normal to the rotation axis, as shown in Figs. 5 and 6. The projection is designated Υ (Greek upsilon), and its value is provided by the proportion

$$\frac{\Upsilon}{x} = \frac{360°}{2\pi r_F}, \tag{1}$$

$$\Upsilon = \left[\frac{360°}{2\pi r_F}\right] x$$

$$= C_1 x, \tag{2}$$

where the term in brackets is an instrumental constant C_1. It is common practice to give this constant a value of 2°/mm, in which case the camera radius r_F turns out to be 28.648 mm. (For the precision back-reflection instrument mentioned later, C_1 is preferably 1°/mm, so that $r_F = 57.296$ mm.)

The mechanism of the Weissenberg instrument causes the linear motion of camera travel z to be directly proportional to the crystal rotation ω (Fig. 2*A*). The relation is given by

$$\omega = C_2 z, \tag{3}$$

where C_2 is another instrumental constant expressed in degrees per millimeter. It is a convenience to be able to have the same transformation constant for (2) and (3), that is, $C_1 = C_2$, and instruments with this relation are said to have an undistorted scale.

(A)

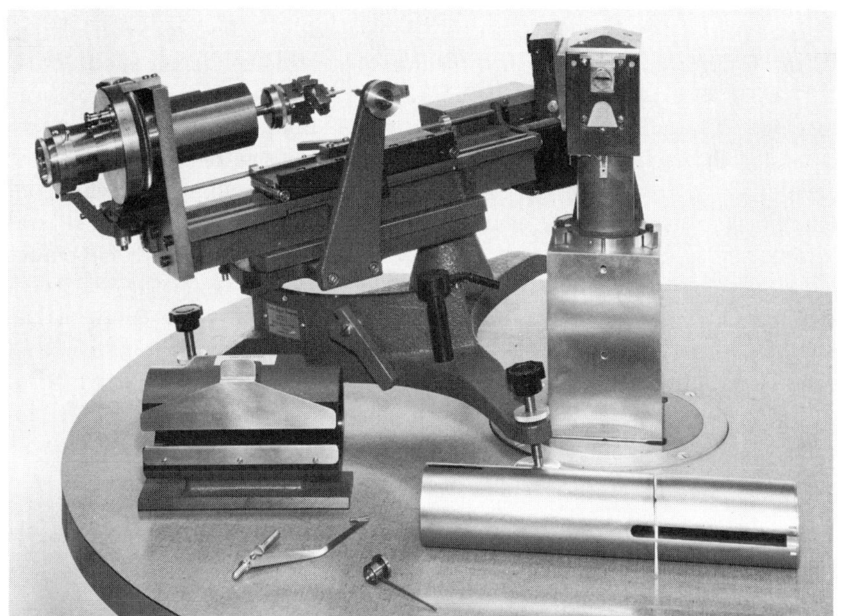

(B)

Fig. 4. Modern version of the Weissenberg instrument. 4*A*. Assembled instrument. 4*B*. Upper part of instrument rotated away from x-ray tube to permit a direct visual check that the crystal is in the center of the beam path. [*The Charles Supper Company.*]

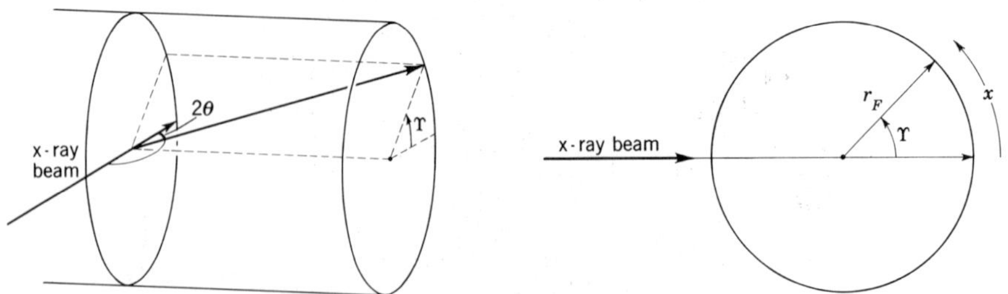

Figs. 5 and 6. The relation between 2θ and Υ.

Interpretation using the reciprocal lattice

There are a number of ways of interpreting Weissenberg photographs, and substantially all of these involve a transformation between coordinates in the reciprocal lattice and film coordinates. It turns out that the most natural and therefore the most simple methods involve mapping reciprocal space on the film, and not the reverse. It also turns out that great similarity in mapping the various levels is achieved if the x-ray beam is inclined to the rotation axis in such a way that the x-ray beam approaches the crystal along the generator of the Laue cone to be recorded. This requires setting $\mu = \nu$, a condition which is called *equi-inclination*. In the following discussion it is assumed that this condition is observed.

Use of cylindrical coordinates. Figure 7 gives views along the rotation axis and normal to it, showing the sphere of reflection and two layers of the reciprocal lattice. The origin O of the reciprocal lattice is always the point where the x-ray beam leaves the sphere of reflection. Since the rotation axis is parallel to a rational direction of the crystal (ordinarily a crystallographic axis), there are reciprocal-lattice planes perpendicular to this direction. These are called levels of the reciprocal lattice. The equi-inclination condition requires the Laue cones of the upper level and the zero level to have equal cone angles; the upper-level cone opens upward in Fig. 7, while the zero-level cone opens downward. This causes the intersections of both cones with the sphere to be equal small circles with radius R. When the radius of the sphere is taken as unity, the geometry of Fig. 7 shows that

$$\cos \nu = R \tag{4}$$

and

$$\sin \nu = \frac{\zeta}{2}, \tag{5}$$

while

$$R = \sqrt{1 - \left(\frac{\zeta}{2}\right)^2}. \tag{6}$$

The upper part of Fig. 7 shows a reciprocal-lattice point P', whose radial coordinate is ξ, touching the circle where the upper level intersects the sphere. This circle will be called the reflecting circle; its radius R was given by (4) and (6). It is seen in Fig. 7 that

$$\frac{\Upsilon}{2} = \omega. \tag{7}$$

This leads to the following useful feature of equi-inclination Weissenberg photographs: Suppose that the line $O'P'$ contains a sequence of reciprocal-

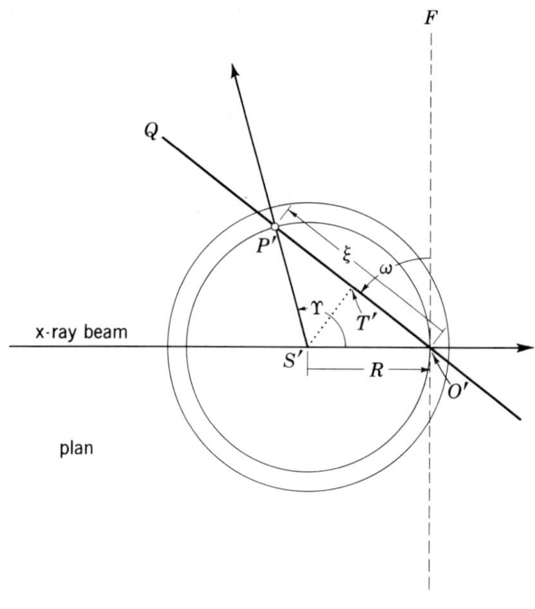

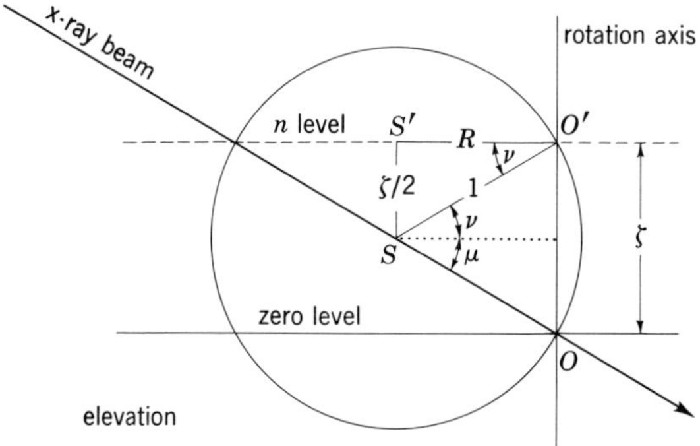

Fig. 7. [*From M. J. Buerger: X-ray crystallography.* (*Wiley, New York,* 1942) 255.]

lattice points. (Since such a line contains the origin O' of the level, it is called a *central reciprocal-lattice line.*) Suppose further that, as the camera translation motion starts, this central line $O'P'$ has the initial position $O'F$. When the crystal starts to rotate, ω starts at zero and begins to increase, bringing successive lattice points of the line $O'F$ into contact with the reflecting circle, where they generate reflections which all follow the law noted in (7), namely,

$$\Upsilon = 2\omega. \tag{8}$$

Now, it was shown that the two dimensions of the film, x and z, are proportional (through C_1 and C_2) to Υ and ω, respectively, so that the trail of these reflections on the film is a straight line whose slope is 2, expressing (8). This would appear as the left sloping line of Fig. 8.

If the initial position of central line $O'P$ is not $O'F$ but is some other central-line location like $O'F_1$ (which is located along a direction which makes a clockwise angle $\bar{\phi}_1$ with $O'F$), then lattice points along the line segment $O'F_1$ cannot generate reflections until the line has been rotated through a counterclockwise angle $-\bar{\phi}_1$; at this point the line has the position $O'F$, and a reflection becomes just possible. Continued rotation causes a relation which is expressed by (8), in which the relation between

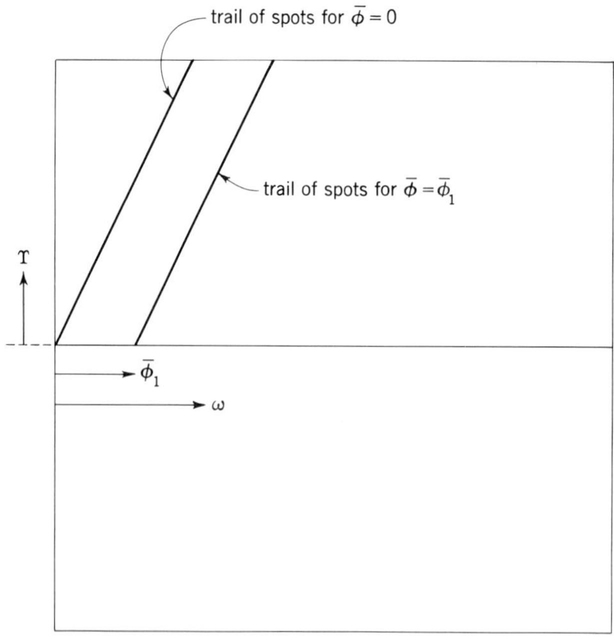

Fig. 8

angular coordinates is

$$\omega = \omega_1 - \bar{\phi}_1. \tag{9}$$

On the photographic film (Fig. 8), the central line, initially in position $O'F_1$, makes a trail like that of $O'F$, except that it is displaced along the ω axis by amount $\bar{\phi}_1$.

Now $\bar{\phi}$ is simply the complement of the more usual angular coordinate ϕ. In any level of the reciprocal lattice, a reciprocal-lattice point is completely fixed by its two coordinates ξ and ϕ, while the level is fixed by the coordinate ζ. Thus every spot on a Weissenberg film can be placed in the reciprocal lattice by its coordinates ξ and ϕ (or $\bar{\phi}$). The relation of ξ to Υ is easily derived from the upper part of Fig. 7, which shows that

$$\sin\frac{\Upsilon}{2} = \frac{\xi/2}{R}. \tag{10}$$

If this is combined with (4) and solved for ξ, the result is

$$\xi = 2\cos\nu\sin\frac{\Upsilon}{2}. \tag{11a}$$

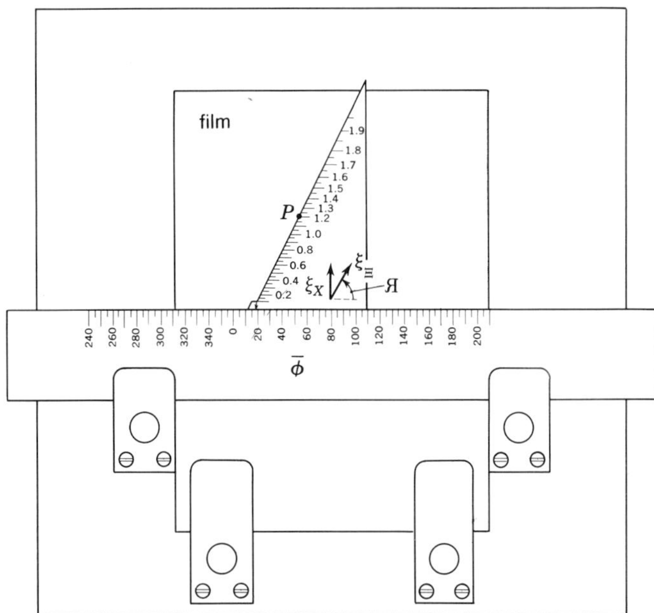

Fig. 9. [*From M. J. Buerger: X-ray crystallography.* (*Wiley, New York,* 1942) 267.]

Alternatively, it can be solved for Υ to give

$$\Upsilon = 2 \sin^{-1} \frac{\xi}{2 \cos \nu}. \tag{11b}$$

By assuming specific values of ξ, lines can be drawn on a film for these equal values of ξ; by making use of Fig. 8, lines can also be drawn on the film where $\bar{\phi}$ has specific constant values. In this way, a film can be mapped in the cylindrical coordinates ξ and $\bar{\phi}$. If a Weissenberg photograph is placed on a properly reduced chart of this kind, the coordinates ξ and $\bar{\phi}$ can be read for each of the spots on the photograph.

An easier practical procedure is to place the film on a device like that

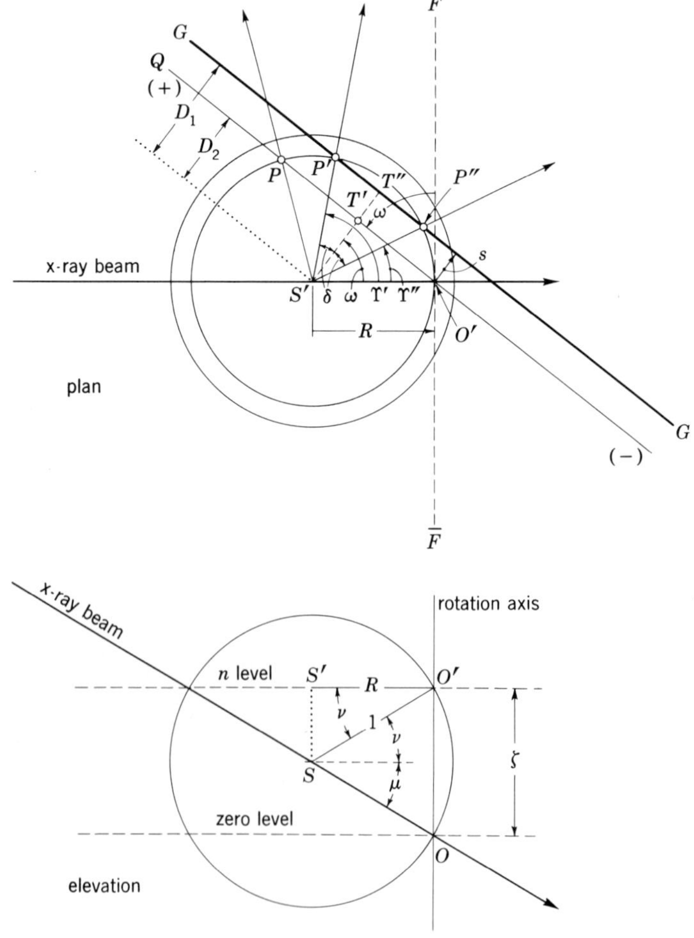

Fig. 10. [*From M. J. Buerger: X-ray crystallography.* (*Wiley, New York,* 1942) 268.]

shown in Fig. 9. In this coordinate-reading device a transparent triangle, one of whose edges is a sloping line as in Fig. 8, moves along a horizontal strip. The edge of the triangle is divided into ξ units, and the edge of the horizontal strip into degrees $\bar{\phi}$ or ω. If the triangle is moved so that any spot on the Weissenberg film comes to the edge of the triangle, the ξ and $\bar{\phi}$ coordinates of the spot can be read directly. From these two values the corresponding reciprocal-lattice point can be plotted in reciprocal space by using polar coordinates ξ and ϕ for r and θ.

Reciprocal-lattice coordinates. Since the most natural coordinate system for the reciprocal lattice is the network of its axial translations, this would offer the neatest device for indexing a Weissenberg photograph. It was seen in the last section that when a line of the reciprocal lattice is a central line, it appears on the Weissenberg photograph as a straight line of slope 2. How does a parallel but noncentral lattice line appear? This question can be readily answered by the following analysis:

Figure 10 shows a view along the rotation axis corresponding to the upper part of Fig. 7, but with a line parallel to the central line and at a distance s from it. The illustration shows that this parallel line produces two reflections in directions SP' and SP''. These are symmetrical about direction $S'T''$, so that the angles Υ of these two reflections are

$$\begin{aligned} \Upsilon' &= \omega + \delta, \\ \Upsilon'' &= \omega - \delta. \end{aligned} \tag{12}$$

The angle δ is easily derived as follows:

$$S'T'' = S'T' + s, \tag{13}$$

$$\cos \omega = \frac{S'T'}{R}, \tag{14}$$

$$\cos \delta = \frac{S'T''}{R} \tag{15}$$

$$= \frac{S'T' + s}{R}$$

$$= \frac{R \cos \omega + s}{R}$$

$$= \cos \omega + \frac{s}{R}. \tag{16}$$

This provides a value for δ which gives (12) the detailed values

$$\begin{aligned} \Upsilon' &= \omega + \cos^{-1}\left(\cos \omega + \frac{s}{R}\right), \\ \Upsilon'' &= \omega - \cos^{-1}\left(\cos \omega + \frac{s}{R}\right). \end{aligned} \tag{17}$$

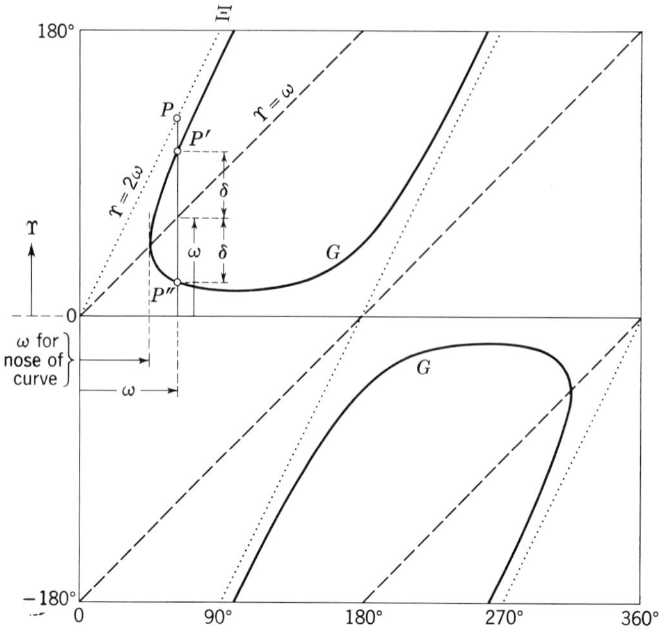

Fig. 11. [*From M. J. Buerger: X-ray crystallography.* (*Wiley, New York,* 1942) 269.]

This function $\Upsilon(\omega)$ is plotted in Fig. 11, which therefore shows the shape of a noncentral reciprocal-lattice line on the Weissenberg photograph. A family of such curves is shown in Fig. 12 for a set of parallel reciprocal-lattice lines whose separation is $s/R = 0.05$ wavelength unit. An interesting feature of (17) is that the function is the same for equal values of s/R. This means that the shapes of the curves in Fig. 12 are the same for Weissenberg films of all reciprocal-lattice levels. This invariance is one of the important results of the equi-inclination condition, and is one of the reasons this condition is used whenever possible.

Some Weissenberg photographs are shown in Figs. 13 and 14. One sees that the spots line up along "festoons," one above the other, which are the curves of Fig. 12. If a transparent template on which the curves of Fig. 12 are drawn is laid under a Weissenberg photograph and then translated left and right, several locations will be found where the spots of the photographs obviously line up on the curves. The most obvious matches occur for lines of greatest density and separation; these are probably reciprocal-lattice axes (symmetry may dictate that they *must* be axes). If festoons are sketched on the photograph for two such acceptable locations, every spot will be found on two curves, one of which

Fig. 12. Template for use with equi-inclination Weissenberg photographs made with an instrument having a coupling ratio between camera translation and crystal rotation of 2°/mm and a camera diameter of 57.3 mm. The curves show the shapes of reciprocal-lattice lines on the Weissenberg photograph. [*From M. J. Buerger: Z. Kristallogr.* **88** (1934) 375.]

gives one index (say h), the other another index (say k); and in this manner the photograph may be easily indexed.

Examination of the diffraction record

The Weissenberg method was the first technique devised which permitted a complete study of important features of the diffraction record of a crystal. It is always used in conjunction with the rotating-crystal method in a manner which will be appreciated shortly.

Like the rotating-crystal method, the Weissenberg method requires the crystal to be rotated about a rational direction of the crystal; this is nearly always a crystallographic axis. If the crystal shows the normal plane faces of a well-developed crystal, the adjustment of a rational axis to the rotation axis of the instrument is ordinarily carried out on an optical goniometer. If it does not have faces well enough developed, it

A

Fig. 13. Examples of zero- and first-level Weissenberg photographs, natural size, taken with a camera of diameter 57.3 mm: pollucite, $(Ca,Na)AlSi_3O_9{\cdot}xH_2O$, rotated about an *a* axis (Cu$K\alpha$ radiation, 35 kV, 15 mA, 24 hr). *A*: zero level, symmetry $4\,m\,m$. *B*: first level, symmetry $4\,m\,m$.

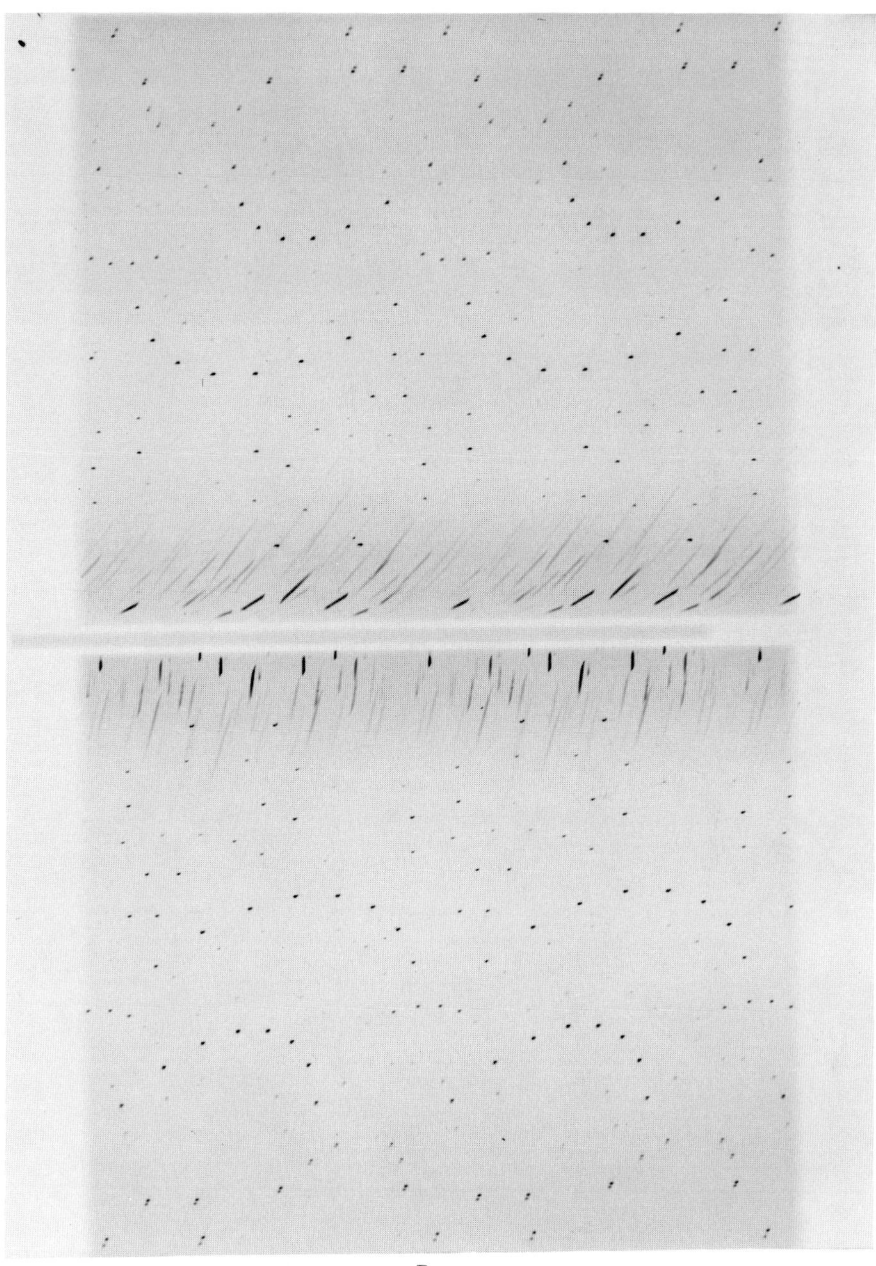

B

Fig. 13. (Continued)

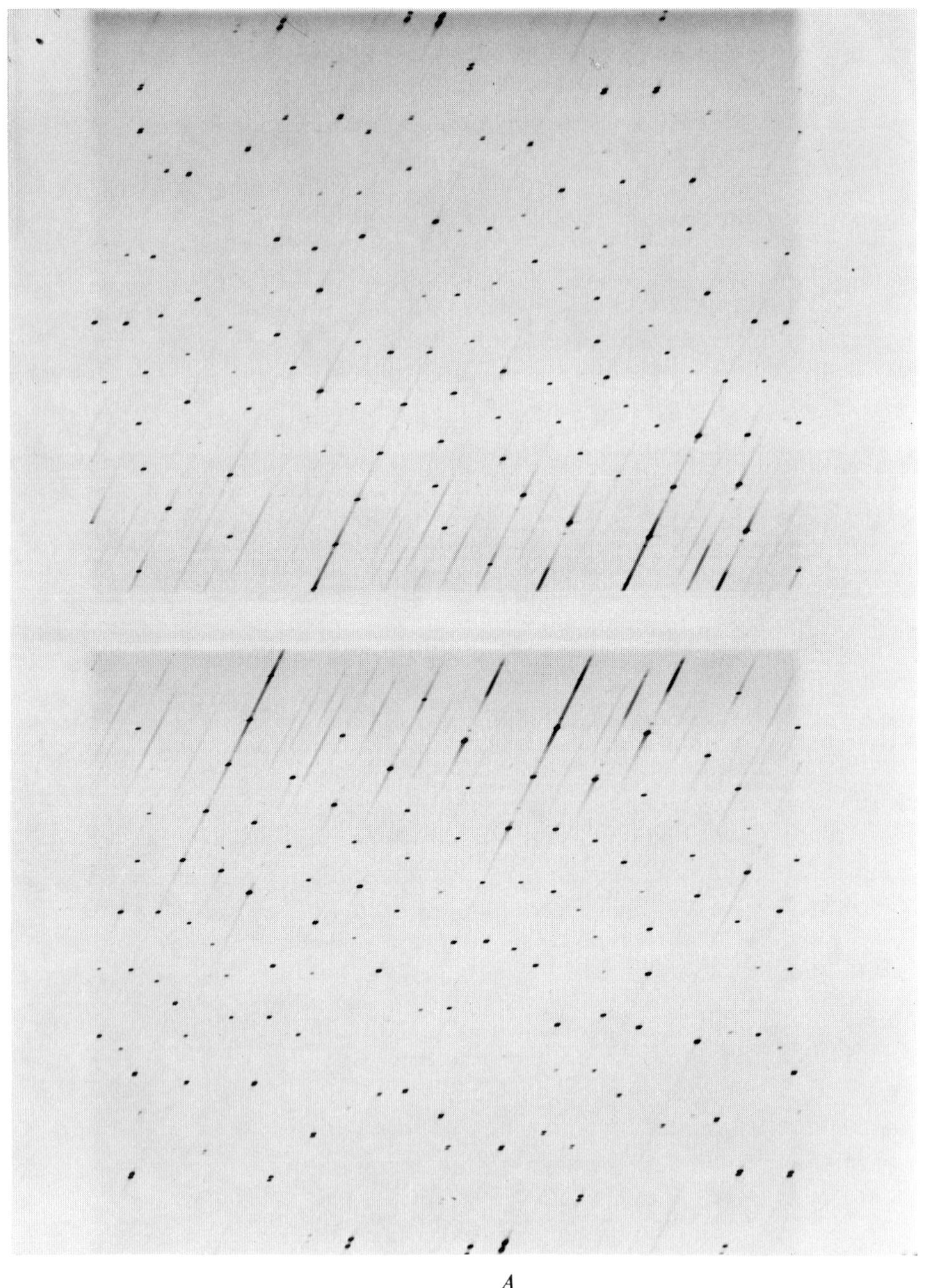

A

Fig. 14. Examples of zero- and first-level Weissenberg photographs, natural size, taken with a camera of diameter 57.3 mm: parawollastonite, $CaSiO_3$, rotated about the *b* axis (Cu$K\alpha$ radiation, 35 kV, 15 mA, 24 hr). *A*: zero level, symmetry 2. *B*: first level, symmetry 2.

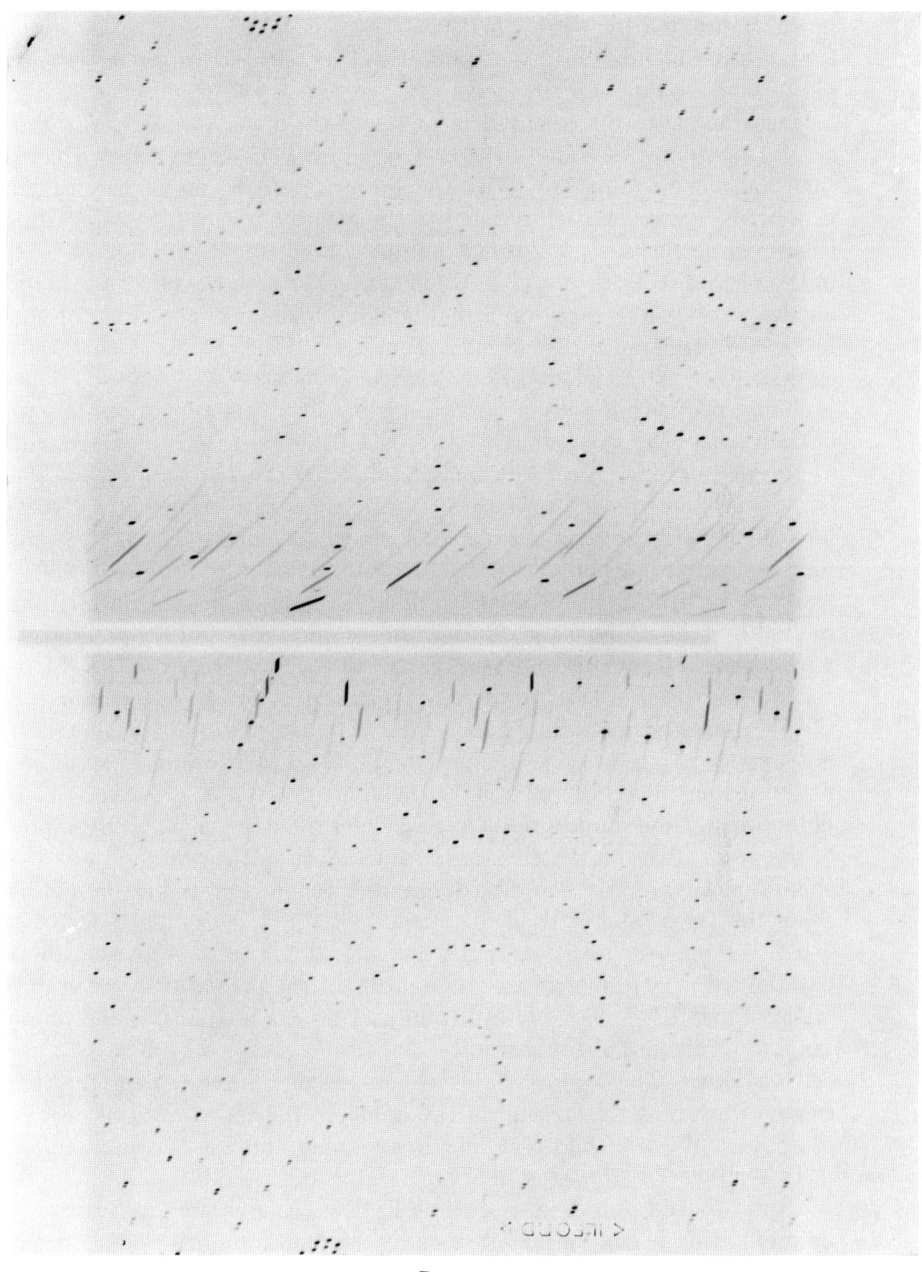

B

Fig. 14. *(Continued)*

must be oriented by some other means; for example, if it is transparent, it may often be oriented by making use of its optical properties with the aid of a petrographic microscope.

Once the crystal is oriented, it is necessary first to make a rotating-crystal photograph. This fixes the sizes of all the Laue cones, that is, their half-opening angles. With the aid of (5) of Chapter 7, the translation of the crystal lattice parallel to the rotation axis can thus be established, while the use of (20) of Chapter 4 predicts the spacing of levels in the reciprocal lattice. Then (5) of this chapter shows how to compute all the settings $\mu = \nu$ required for the equi-inclination recording of each level. The setting s of the layer-line screen (Fig. 1) for each level is simple: $s = r \tan \mu$, where r is the radius of the layer-line screen. These preliminaries permit taking an equi-inclination Weissenberg photograph of each level of the reciprocal lattice within the range of the instrument.

The photographs must be examined first for Friedel symmetry. This is a little inconvenient because of the curious distortion of reciprocal-lattice geometry as it appears on a Weissenberg photograph. In particular, a symmetry plane parallel to the rotation axis intersects a level along a central reciprocal-lattice line; symmetrical spots which are related by such a plane accordingly appear on a Weissenberg photograph on a horizontal line and at equal distances left and right of the sloping central line (such as in Fig. 8) marking the position of the mirror. In the Weissenberg photograph, therefore, spots which are mirror images do not appear to obey the ordinary mirror law, $i = r$. But in spite of a little inconvenience the complete Friedel symmetry of a crystal can be determined from appropriate Weissenberg photographs. Indeed, one of the great advances which came with the Weissenberg method was that this Friedel symmetry could be determined for the first time by a method other than the Laue method.

The systematic absences must be looked for next. These can be determined very easily by an examination of the points of the reciprocal lattice which have been plotted from coordinates of spots on the photographs. This plot of the reciprocal lattice also makes the choice of a unit cell obvious. The approximate cell dimensions measured on this plot can be improved by measuring the angle 2θ for certain reflections, for example, and can be highly refined in a manner noted in the next section.

This much of the examination establishes the diffraction symbol (Chapter 5) and dimensions of the cell. If a complete crystal-structure determination is contemplated, there is needed, in addition, an estimate or measurement of the intensity of each reflection hkl. Since these reflections occur as discrete spots on the set of Weissenberg photographs, the only real problem is how to relate intensity to blackness of a spot. This is discussed briefly in Chapter 11.

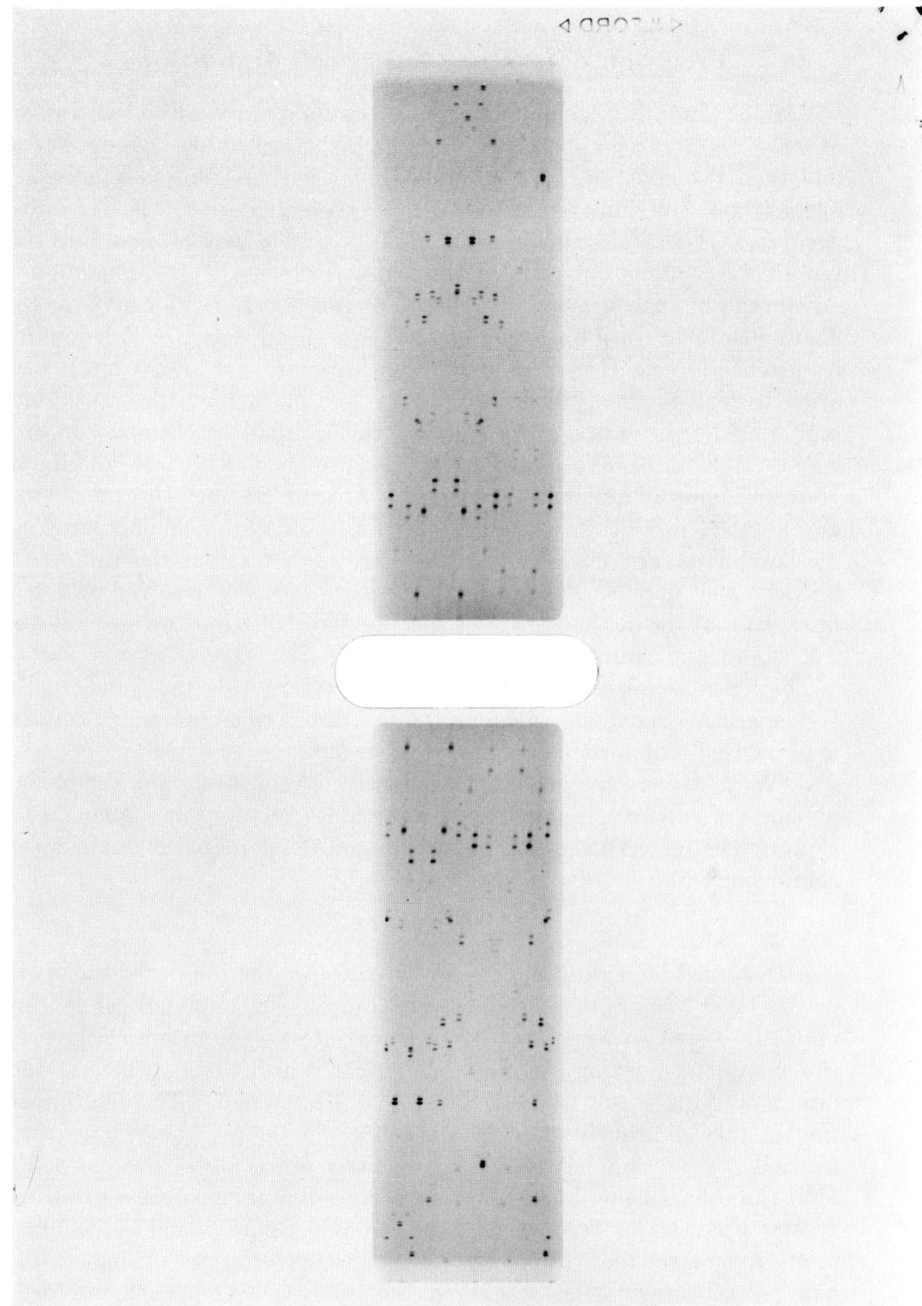

Fig. 15. Example of a back-reflection Weissenberg photograph, natural size, taken with a camera diameter of 114.6 mm: parawollastonite, $CaSiO_3$, rotated about the b axis (Cu$K\alpha$ radiation, 35 kV, 15 mA, 14 hr).

Precision determination of cell dimensions

Various kinds of systematic errors in recording x-ray diffraction vanish at $2\theta = 180°$. If the locations of a number of reflections are measured and then the apparent d's computed from them are plotted against an appropriate function, the value of d extrapolated to $\theta = 90°$ is quite accurate. The Weissenberg method offers a good way of recording the required reflections since all reflections are resolved. To take maximum advantage of such a method a special Weissenberg instrument is used. An unusual feature of the camera is that the reflections are recorded in the back-reflection field on a double-radius camera ($r = 57.296$ mm). In order to lead the x-ray beam through a hole in the film, the translation motion is shortened so that the coupling constant is $C_2 = 8°/\text{mm}$. In this way the collimator may enter the camera through a fairly short slot (about $1\frac{3}{8}$ inches long) in a film 5 inches wide. A complete Weissenberg photograph made in this way is shown in Fig. 15. On such photographs, measurements are made across the slot to determine the difference between 360° and 4θ. The $\alpha_1\alpha_2$ doublets are very well resolved on these films, so that the method provides spacing determinations good to at least five significant figures.

The spacings are, of course, reciprocals to reciprocal-lattice translations. By measuring axial and diagonal translations, trigonometric formulae may be used to determine the interaxial reciprocal-cell angles α^*, β^*, and γ^*. With the reciprocal cell accurately established, the direct-cell dimensions can be computed with the aid of relations in Table 1B of Chapter 4 and certain other relations somewhat more convenient for computing.

Summary

As recorded on a rotating-crystal photograph, the indices of the spots in any layer line are indeterminate because the single cylindrical coordinate ξ of a spot corresponds to two coordinates in the reciprocal lattice. For complete indexing, another cylindrical coordinate ϕ, which is the direction of the vector whose length is ξ, is also needed. The determination of this additional coordinate is made possible by the Weissenberg method. To accomplish this, only one layer line is allowed to reach the film, the others being blocked off by a layer-line screen. If a rotating-crystal photograph were then taken, the only record would be confined to one line on the film, any point on which is characterized by one coordinate ξ. In the Weissenberg method, the film is translated back and forth in synchronism with a back-and-forth oscillation of the crystal. The result of this is to spread the diffraction record of the layer line over the two dimensions of the film. The displacement of a spot in the direction of the film translation is a measure of the required additional coordinate

ϕ. Since the key to unambiguous indexing by the Weissenberg method is to spread the reflection which would occur on one layer of a rotating-crystal photograph over the useful area of the film, the Weissenberg method requires one film for each level of the reciprocal lattice to be explored.

While rotating-crystal photographs are customarily made by directing the x-ray beam at right angles to the rotation axis (that is, so that $\bar{\mu} = 90°$ and its complement $\mu = 0$), it is easier to interpret Weissenberg photographs if the x-ray beam is directed at an angle μ characteristic of each level, namely, such that $\mu = \nu$. This setting, known as *equi-inclination*, is a function of the cylindrical coordinate ζ of the level to be recorded and is given by $\sin \mu = \sin \nu = \zeta/2$.

Two methods of indexing Weissenberg photographs are in common use. In one method the use of appropriately engraved scales makes it possible to determine the cylindrical coordinates ξ and ϕ of each spot on a film. From these polar coordinates, the corresponding point can be mapped on a sheet of paper representing a level of the reciprocal lattice.

In a second method, advantage is taken of the fact that, in recording by equi-inclination, a set of parallel reciprocal-lattice lines appears on the Weissenberg photograph as a set of curves having the same shapes regardless of the level. This set of shapes in the form of a chart, when placed under a Weissenberg film, permits connecting the spots on the film by curves which correspond to lines parallel to the axes of the reciprocal lattice, thus revealing the indices of each spot.

A set of Weissenberg photographs can be regarded as a set of distorted maps of the reciprocal lattice, level by level. When one becomes accustomed to the nature of the distortion, the Friedel symmetry of the rational axis about which the crystal has been rotated can be judged. When these symmetries have been determined for the several appropriate rotation axes, this information and the extinctions observed lead to the diffraction symbol of the crystal.

Accurate cell dimensions can be derived from measurements of films made with a special back-reflection Weissenberg camera.

Notes on history

The Weissenberg method was invented by K. Weissenberg, who had had considerable experience with the examination of fibers by the fiber method. When Schiebold pointed out that the fiber method had the same geometry as the rotating-crystal method, this presumably drew Weissenberg's attention to the single-crystal case, in which (unlike the fiber method) the angular location of a reflecting plane could be controlled. He clearly recognized that indexing by the rotating-crystal method was indeterminate because the layer line in which a reflection occurs has only one dimension. Accordingly, he arranged to spread the layer line over the surface of the film by translating the film. In this way Weissenberg was able to devise a method whereby the angular parameter ϕ could be determined. Other methods, in particular one by Schiebold and another by Sauter, were developed when this philosophy was recognized. The Weissenberg method was so relatively simple and straightforward, however, that it maintained its position as the chief method for studying x-ray diffraction by single crystals for many years.

Initially, Weissenberg photographs were interpreted with the aid of the parameters 2θ (the Bragg glancing angle) and ϕ (the direction of the normal to the reflecting plane). But as early as 1928, Schneider, influenced by Bernal's 1926 classic on the rotating-crystal method, showed how Weissenberg photographs could be interpreted graphically with the aid of the reciprocal lattice. In 1933 W. A. Wooster and Nora Wooster showed how cylindrical coordinates could be mapped on the Weissenberg film.

Up to this time it had been standard practice to make Weissenberg photographs of upper levels with the same geometric relations of rotation axis to x-ray beam as with the rotating-crystal method, namely, the beam normal to the rotation axis (called the *normal-beam technique*). In 1934 Buerger demonstrated the many advantages of the equi-inclination method, which has been used ever since.

The precision back-reflection Weissenberg camera was presented in 1937.

Additional reading

M. J. Buerger. *X-ray crystallography.* (Wiley, New York, 1942) 214–311, 435–464.

Significant literature

K. Weissenberg. *Ein neues Röntgengoniometer.* Z. Phys. **23** (1924) 229–238.

W. Schneider. *Über die graphische Auswertung von Aufnahmen mit dem Weissenbergschen Röntgengoniometer.* Z. Kristallogr. **69** (1928) 41–48.

W. A. Wooster and Nora Wooster. *A graphical method of interpreting Weissenberg photographs.* Z. Kristallogr. **84** (1933) 327–331.

M. J. Buerger. *The Weissenberg reciprocal lattice projection and the technique of interpreting Weissenberg photographs.* Z. Kristallogr. **88** (1934) 356–380.

M. J. Buerger. *An apparatus for conveniently taking equi-inclination Weissenberg photographs.* Z. Kristallogr. **94** (1936) 87–99.

M. J. Buerger. *The precision determination of the linear and angular lattice constants of single crystals.* Z. Kristallogr. **97** (1937) 433–468.

M. J. Buerger. *The development of methods and instrumentation for crystal-structure analysis.* Z. Kristallogr. **120** (1964) 3–18.

9

The precession method

Background

It was pointed out in Chapter 7 that only a limited amount of information about a crystal can be determined from oscillating-crystal photographs because the low symmetry of the oscillation motion degrades the symmetry of the diffraction effects from the crystal. This degradation of symmetry is avoided by a method somewhat related to the oscillating-crystal method, known as the *precession method*, which is discussed in this chapter.

In the older oscillating-crystal method, the crystal is oscillated about an axis at right angles to the x-ray beam, as shown in Fig. 1. The statistical symmetry of this motion, as projected on the photographic film, is $2\,m\,m$, as suggested in Fig. 1. The symmetry of the diffraction record is a combination of this motion symmetry and the Friedel symmetry at the midpoint of the oscillation range of the crystal. If the Friedel symmetry contains any symmetry parallel to the x-ray beam other than 2, m_1, and m_2, it is degraded by the motion symmetry, which means it is suppressed. But in the precession method, the manner of varying the orientation of the crystal with respect to the x-ray beam has radial symmetry about the x-ray beam. This is accomplished by causing a rational direction of the crystal to precess uniformly about the x-ray beam while the crystal is prevented from rotating about the beam direction. The mechancial means for arranging this motion in precession instruments, as currently manufactured, is to support the crystal on a gimbal, as suggested diagrammatically in Fig. 2, and to cause a point on the rational axis of the crystal to rotate around the x-ray beam.

The Laue cones

In the early form of the precession apparatus a rational direction of the crystal was caused to precess about the x-ray beam as just described, and the photographic film was placed normal to the x-ray beam, as suggested in Fig. 2. But it was discovered that there was a great advantage in maintaining the plane of the film normal to the direction of the precessing crystal axis. The reason is as follows:

The directions of the diffracted beams coming from a crystal are limited to the generators of a nest of Laue cones coaxial with a rational axis of the crystal, as outlined in Chapter 3. The rational axis may be any of the infinite number of such axes of the crystal, but an inspection of equation (3) of Chapter 3 shows that if the period along the rational axis is small, the number of possible solutions for $\bar{\nu}$ is small, which means that the number of cones in the nest is small. Rational axes with small periods are also the kind chosen for crystallographic axes. Suppose then,

m_1
film
m_2
x-ray beam
axis of
oscillation

Fig. 1

film
rational direction
$\bar{\mu}$
x-ray beam
gimbal

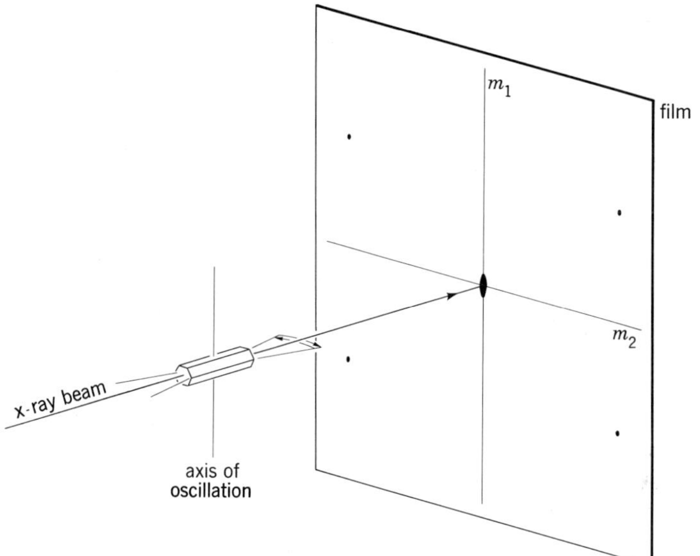

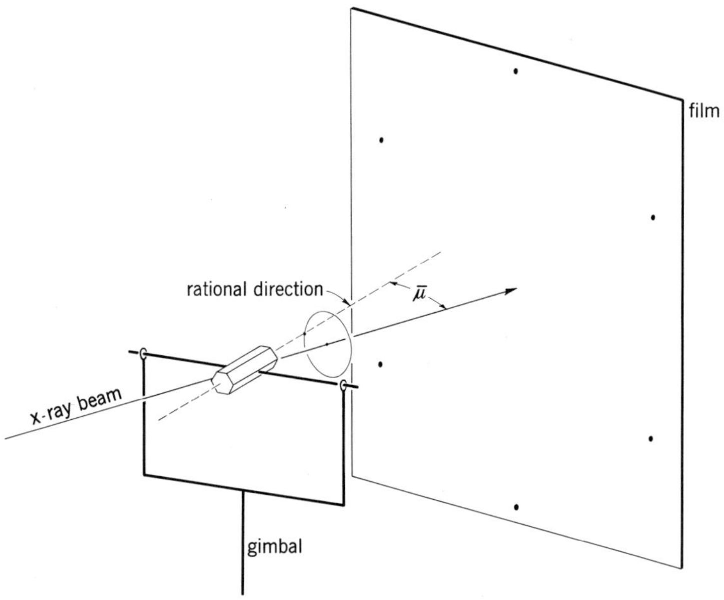

Fig. 2

for definiteness, that the rational axis is c. Then the generators of the zero-order Laue cone are the directions of the reflections $hk0$, the generators of the first-order cone are the directions of the reflections $hk1$, and, in general, the generators of the pth-order cone are the directions of the reflections hkp.

The zero-order Laue cone is of particular interest in this preliminary discussion. For this cone, p is zero in the third Laue equation of (4) in Chapter 3; from this it follows that $\bar{\nu} = \pi - \bar{\mu}$. Accordingly, the direct x-ray beam continues beyond the crystal as one of the generators of the cone, as shown in Fig. 3. When the crystal axis is oriented so that it makes an angle $\bar{\mu}$ with the direct x-ray beam, then the reflections having index zero for this crystal axis occur along the generators of a Laue cone whose half-opening angle is also $\bar{\mu}$.

Now, when the photographic film is oriented normal to the axis ST of the Laue cone, as shown in Fig. 3, the records of the reflections fall along a circle. This simplifies the geometry. Consider any specific generator SP'. This reflection direction makes a deviation angle 2θ with the x-ray beam. The symmetry of the figure is such that the distance R of the record P' from the direct beam O' is given by

$$\sin \theta = \frac{R/2}{M}. \tag{1}$$

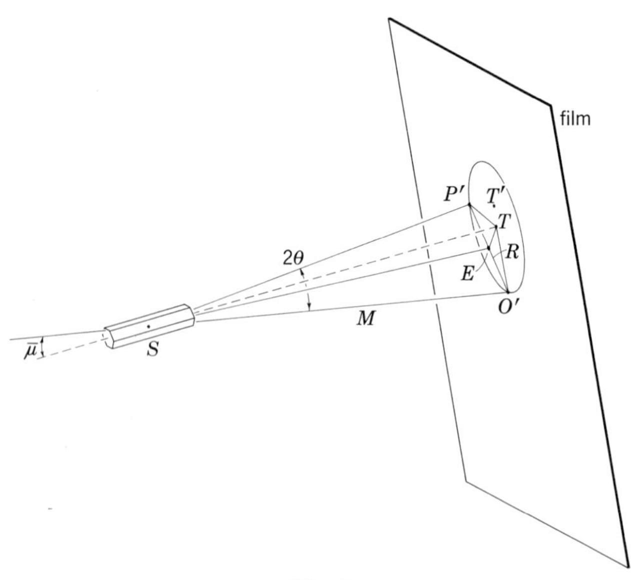

Fig. 3

But sin θ is also given by Bragg's relation for reflection $hk0$ as

$$\sin \theta = \frac{\lambda}{2d_{hk0}}. \tag{2}$$

Comparison of these shows that

$$\frac{R/2}{M} = \frac{\lambda}{2d_{hk0}}, \tag{3}$$

so that

$$R = M\lambda \frac{1}{d_{hk0}} \tag{4}$$

$$= M\xi_{hk0}. \tag{5}$$

Since M is an instrumental constant (the distance of the film from the crystal, measured along the x-ray beam) and λ is another experimental constant, it is apparent that, when the film is oriented normal to the Laue cone, the position of a reflection is proportional to the distance of the corresponding reciprocal-lattice point $hk0$ from the origin point 000. The symmetry of the figure shows that plane EST is perpendicular to plane $O'SP'$, so that EST is the plane responsible for the reflection, and R is normal to it. Thus R has both the direction and the (scaled) magnitude of the vector from the origin to reciprocal-lattice point $hk0$.

The advantage of maintaining the photographic film normal to the Laue cone is obvious: When this is done, the photographic record of the reflections on the zero level of the reciprocal lattice is an undistorted map of the zero-level reciprocal-lattice points. The precession motion assures that the intensity distribution in these lattice points is the same as the projected Friedel symmetry of that level of the reciprocal lattice.

Reciprocal-lattice interpretation

In Chapter 3 it was seen that in order to produce reflections of monochromatic x-radiation by the planes of a crystal, the orientation of the crystal with respect to the x-ray beam must be varied so that the planes may pass through the Bragg condition for such reflections. In the last section it was shown that if a flat photographic film is used to record the reflections from the zero-order Laue cone, it does so in an especially elegant way when the plane of the film is maintained normal to the axis of the Laue cone. Under these circumstances (with further obvious restrictions which will be apparent shortly) the reflections are recorded at the points of the reciprocal lattice. The general nature of the motion is immaterial, but a precessing motion of the axis of the Laue cone allows the symmetry of the record to reveal the Friedel symmetry of the crystal.

Since the record of the precession method is in the form of an arrangement of spots at the points of one level of the reciprocal lattice, a powerful way of explaining and understanding the precession method is to start with a level of the reciprocal lattice and examine the consequences of causing it to undergo a precessing motion. Figure 4*A* shows a horizontal x-ray beam with its enveloping sphere of reflection. The origin of the reciprocal lattice is at O, where the x-ray beam leaves the sphere. Assume that, initially, a zero-level reciprocal-lattice plane is placed normal to the x-ray beam, as in Fig. 4*A*; the plane is then tangent to the sphere at O. Next, the orientation of this plane is changed so that its normal makes an angle $\bar{\mu}$ with the x-ray beam, as shown in Fig. 4*B*; the plane then intersects the surface of the sphere in a circle, which is seen in Fig. 4*C*. Any reciprocal-lattice point in the plane which is on this circle of intersection, such as point P, is in a position to give rise to an x-ray reflection. If the normal to the plane in Fig. 4*B* is caused to precess about the x-ray beam, then the circle in Fig. 4*C* rotates about point O, and as each reciprocal-lattice point in the plane is touched by the circle, a reflection occurs.

Another view of this geometry is shown in Fig. 5. It is seen there that if the photographic film is parallel to the reciprocal-lattice plane, then the vector OP, from the origin to the reciprocal-lattice point responsible for the reflection, is projected to a parallel enlargement of itself

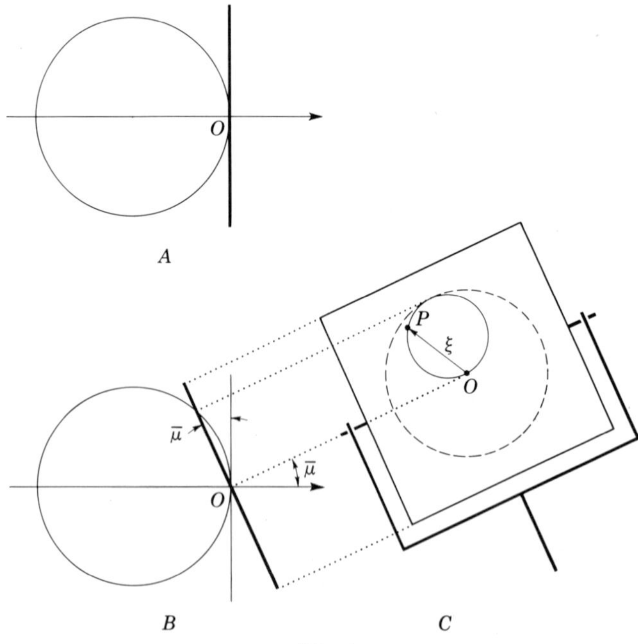

Fig. 4

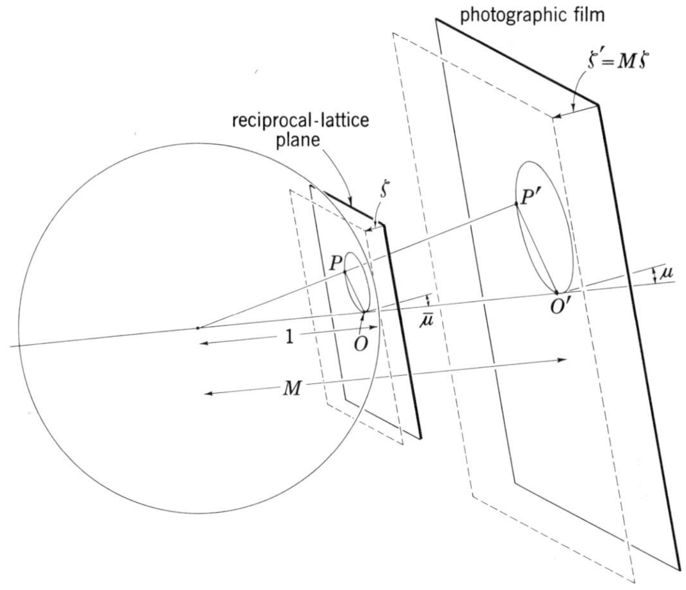

Fig. 5

$O'P'$ on the photographic film. This relation holds for all vectors from the origin to any reciprocal-lattice point. The pattern of points photographed on the film, therefore, is an enlarged map of the points on a level of the reciprocal lattice, provided that, during the precession motion, the photographic film undergoes a motion which duplicates that of the reciprocal-lattice plane. In this way not only are the two planes kept parallel but any line on one plane is also kept parallel to a corresponding line on the other, while a point O in one plane and O' in the other are at fixed locations in the x-ray beam. This condition can be attained by mounting the reciprocal-lattice plane and the film on duplicate suspensions and coupling the suspensions to maintain parallelism. Under these conditions it may be said that the reciprocal-lattice plane is photographed on the film.

Figure 5 shows that the photographic film records the level of the reciprocal lattice on an enlarged scale. The dimensions of the reciprocal-lattice plane were discussed in Chapter 4: If the radius of the sphere is taken as unity, then the origin distances of the reciprocal-lattice points are $\lambda(1/d)$; in this case the reciprocal lattice is scaled by the wavelength. Thus the origin of the wavelength-scaled reciprocal lattice is located at a unit distance from the center of the sphere, and the photographic film is placed at the crystal-to-film distance M in the precession instrument. The magnification of the photographed reciprocal lattice, accordingly,

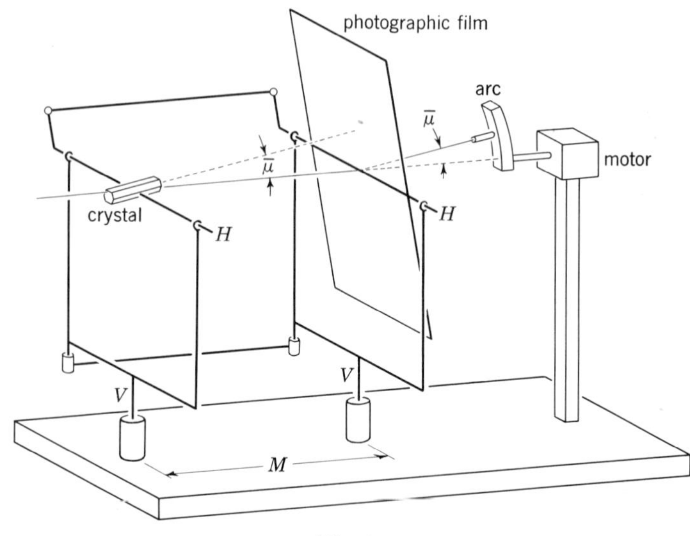

Fig. 6

is given by the proportion

$$\frac{O'P'}{OP} = \frac{M}{1}. \tag{6}$$

This same projective geometry holds for a nonzero level of the reciprocal lattice, as suggested by the dashed lines in Fig. 5. The upper-level plane of the reciprocal lattice is seen to be displaced toward the center of the sphere by a spacing ζ. This displaced plane intersects the sphere in another circle (not shown) corresponding to another Laue cone belonging to the same coaxial set as the zero-level cone. To photograph this upper level, the photographic film must also be displaced in the same direction from its zero-level position as the upper level of the reciprocal lattice, and by an amount which is proportional (through the magnification factor) to ζ. The film displacement is therefore $M\zeta$.

The precession instrument

A diagram of the mechanism of the precession apparatus, as currently manufactured, is given in Fig. 6. A photograph of the actual apparatus is shown in Fig. 7. In Fig. 6 the crystal and photographic film are seen supported on duplicate gimbals. Each has a vertical axis V and horizontal axis H about which the crystal or film can be rotated, but neither crystal nor film can be rotated about an axis normal to the plane of this pair of axes. The two gimbals are coupled by linkages, so that any

Fig. 7A. A modern version of the precession instrument. [*The Charles Supper Company.*]

motion of one is duplicated by that of the other. The normal to the film is tipped away from the line of the x-ray beam along an arc on which $\bar{\mu}$ can be measured. The linkage causes the crystal axis to be tipped from the x-ray beam direction by a duplicate amount and in the same direction. A motor rotates the film normal around the x-ray beam (that is, makes the normal precess about the beam), and the linkage requires the crystal axis to follow this motion exactly.

In the actual apparatus (Fig. 7), there are several modifications and additions to this diagrammatically shown scheme. In the first place, the bearing for the horizontal axis on the near side of the crystal gimbal is omitted, the entire alignment of this axis being taken care of by a large bearing on the far side. This permits mounting the crystal on a standard goniometer head. Secondly, the rotational position of this H axis is indicated on a large dial, which aids in adjusting the orientation of the crystal. On the horizontal axis is also mounted a device for blocking all Laue cones of the nest parallel to the crystal axis except the one chosen to record. This serves the same function as the layer-line screen of the Weissenberg camera; in the precession apparatus this screen has the form of a flat metal plate with an annular hole through which the desired Laue

Fig. 7*B*. Instrument in Fig. 7*A* viewed from another direction.

cone is permitted to reach the film. The opening angle of the cone is controlled by selecting an annulus with approximately the desired radius and then adjusting its distance from the crystal.

Uses of the precession motion

The precession motion has two major applications in the study of crystals, and some additional minor applications. It has been shown in the foregoing sections that the motion provides a simple means of photographing the reciprocal lattice, level by level. Such photographs are called *precession photographs.*

Given a crystal one of whose axial directions is known, so that an axis can be oriented to coincide with the precessing axis of the instrument, it is always possible to photograph the zero level of the reciprocal lattice of the crystal without any further knowledge about that crystal. This follows from the foregoing discussion because all zero-order Laue cones have the same half-opening angle as the precession angle $\bar{\mu}$. On the other hand, in order to photograph any of the upper levels of the reciprocal lattice, the half-opening angle of the Laue cone to which each level corresponds must be known. These can be determined by using the precession motion with the same mounting of the crystal, but by using a different arrangement of photographic film which results in a *cone-axis* photograph. This is described in a subsequent section.

A subsidiary but very useful application of precession motion is to adjust the orientation of a crystal so that a rational axis comes into coincidence with the precessing axis of the instrument. Photographs taken for this purpose are called *orientation photographs.* The technique is briefly outlined in a subsequent section.

In studying a crystal by the precession method the general plan follows this sequence: First the crystal must be mounted on a goniometer head so that, when the latter is placed on the precession instrument, some crystallographic axis (preferably the axis of highest symmetry, but at least a rational axis) is oriented exactly along the precessing axis of the instrument. A rough orientation can usually be arranged by eye, making use of the morphology or cleavage displayed by the specimen. This can often be made exact by adjusting the goniometer-head arcs while examining the crystal on an optical goniometer. If the crystal shows no external features useful for orientation, it can often be oriented approximately with the aid of a polarizing microscope, provided the crystal is transparent. Any of these means leads to a more or less exactly oriented crystal. The orientation can often be improved with the aid of an autocollimator attached to the precession instrument, but in any case the exactness of the orientation should be finally tested, and possibly improved, by one or more orientation photographs. (If no preliminary orientations prove feasible, the entire orientation can be carried out with the aid of such orientation photographs.)

With the orientation assured, the second step is to take a cone-axis photograph. Measurements made on this photograph provide data for computing the translation period along the rational axis (in crystal space) or the spacing between layers of the reciprocal lattice (in reciprocal space). This information, in turn, permits computing the layer-line screen settings and the film-cassette settings for each level of the reciprocal lattice to be photographed.

With these settings known, the third step can be carried out. This consists in making precession photographs of the zero level of the reciprocal lattice, and as many of the upper levels as lie within the range of the instrument.

Ordinarily it is desirable to repeat these three steps with another rational axis. The first step can be very quickly accomplished because the reciprocal lattice of the crystal is already known from the first sequence of photographs. If the investigation has been properly planned in advance, a simple rotation of the crystal about the dial axis of the precession instrument will bring another desired rational axis into the precessing axis. After a simple check of the resulting orientation with an orientation photograph, steps 2 and 3 are then repeated for the new rational axis.

Some highlights of these three steps are given in the following sections.

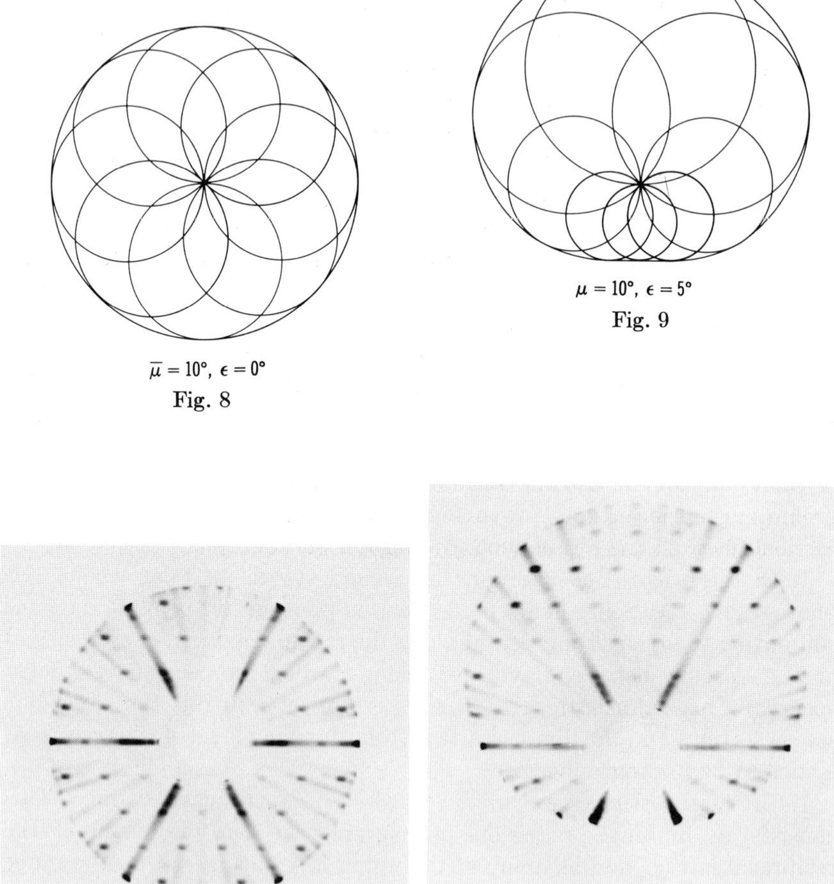

Fig. 8

Fig. 9

Fig. 10

Fig. 11

Figs. 8 and 9. Positions of the circle where the Laue cone of Fig. 3 intersects the film at various stages of the precession cycle. In Fig. 8 there is no orientation error, while in Fig. 9 the axis of the Laue cone points 5° too high.

Figs. 10 and 11. Examples of orientation photographs corresponding to Figs. 8 and 9: beryl, $Be_3Al_2(SiO_3)_6$, c axis. (Unfiltered MoK radiation, 45 kV, 15 mA, 15 min.)

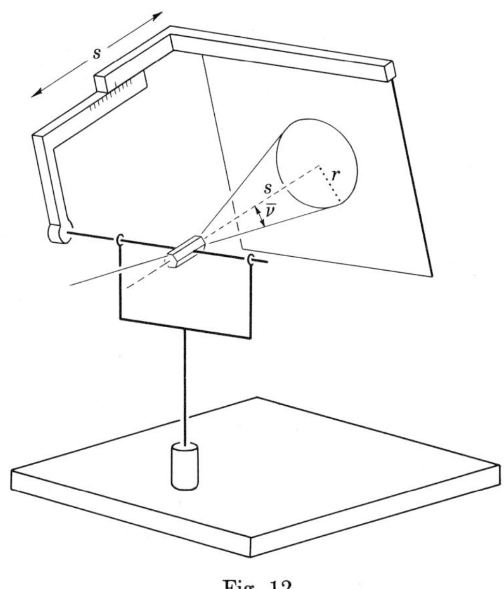

Fig. 12

Orientation photographs

The manner in which a precession photograph records reflections can be understood with the aid of Fig. 3. For present purposes, the illustration is interpreted as follows: A decision has been made as to the magnitude of the precession angle $\bar{\mu}$; therefore the instrument has been set for this value. Thus the precessing axis is ST, and $\angle O'ST$ is set equal to $\bar{\mu}$; the plane of the photographic film is always maintained perpendicular to ST by the coupling mechanism of the instrument.

First consider the recording when the crystal axis is accurately coincident with the precessing axis ST; this is described by saying that the orientation error is zero, that is, $\epsilon = 0$. Figure 3 shows the precession cycle when ST is directly above the direct beam SO'. As the precession motion proceeds, ST precesses around SO' and the circle on the film sweeps around the center of the film, occupying successive positions on the film suggested by the circles in Fig. 8. Each reflection records on the film as the circle sweeps by its location on the film. After a full cycle of precession, the circle has swept out the area within the envelope outlined by the larger circle. All recorded reflections lie within this envelope.

For the zero layer the half-opening angle of the Laue cone is $\bar{\mu}$, and this is independent of all other variables, including the wavelength. Thus the envelope of the circles on the film is dependent only on $\bar{\mu}$. But if the x-radiation is not strictly monochromatic, the radiations of other wavelengths diffracted by the crystal also fall within this envelope area. As a consequence, the area within the envelope usually becomes some-

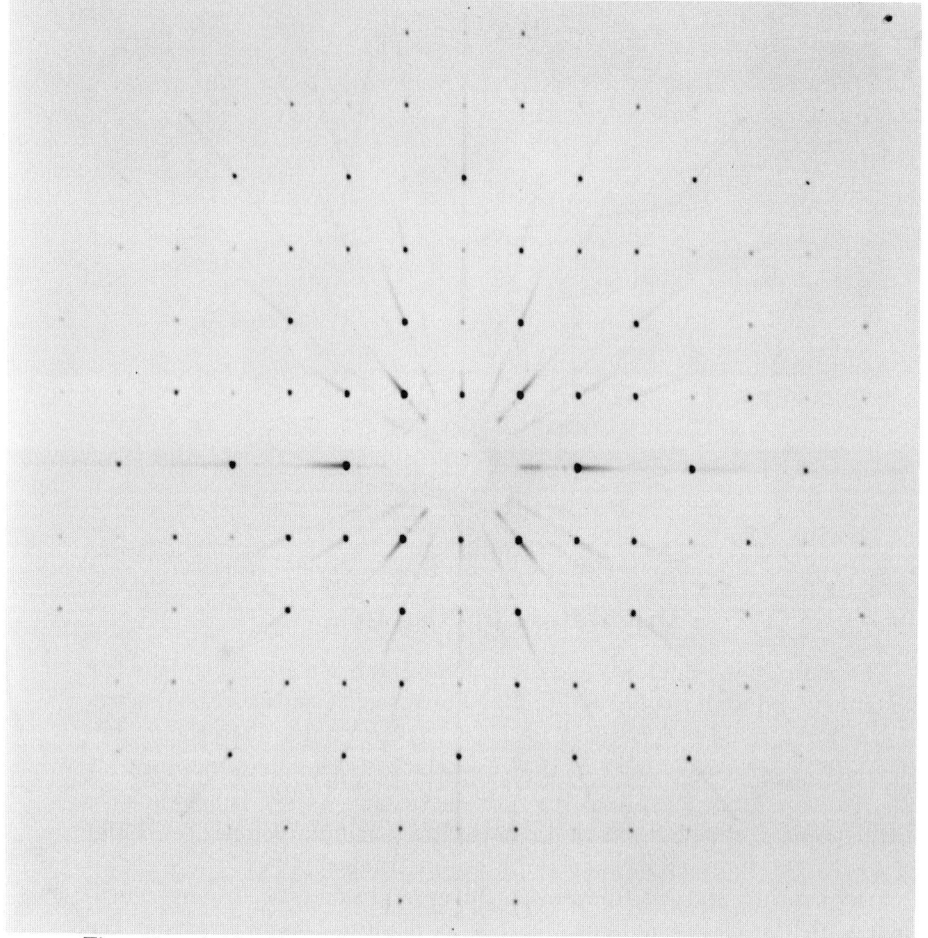

Figs. 13 and 14. Examples of zero-level and second-level precession photographs: barite, $BaSO_4$, c axis (Mo$K\alpha$ radiation, 45 kV). Fig. 13. Zero level (3mA, 12 hr.)

what darker than the rest of the photograph. This shading effect, an example of which is shown in Fig. 10, can be enhanced by using unfiltered radiation, especially radiation from elements of reasonably high atomic number. Unfiltered MoK radiation is commonly used to bring out this effect.

Now, if there is an error in the orientation of the crystal to the precessing axis, this shaded area is eccentric, and the eccentricity is a delicate measure of the magnitude and direction of the angular error ϵ. The reason for the eccentricity can be understood by reference to Fig. 3. Suppose that the rational crystal axis, instead of being accurately aligned along the precessing axis ST, is actually along ST', so that the angular error ϵ is $O'ST' - O'ST$. Now, the half-opening angle of the Laue cone

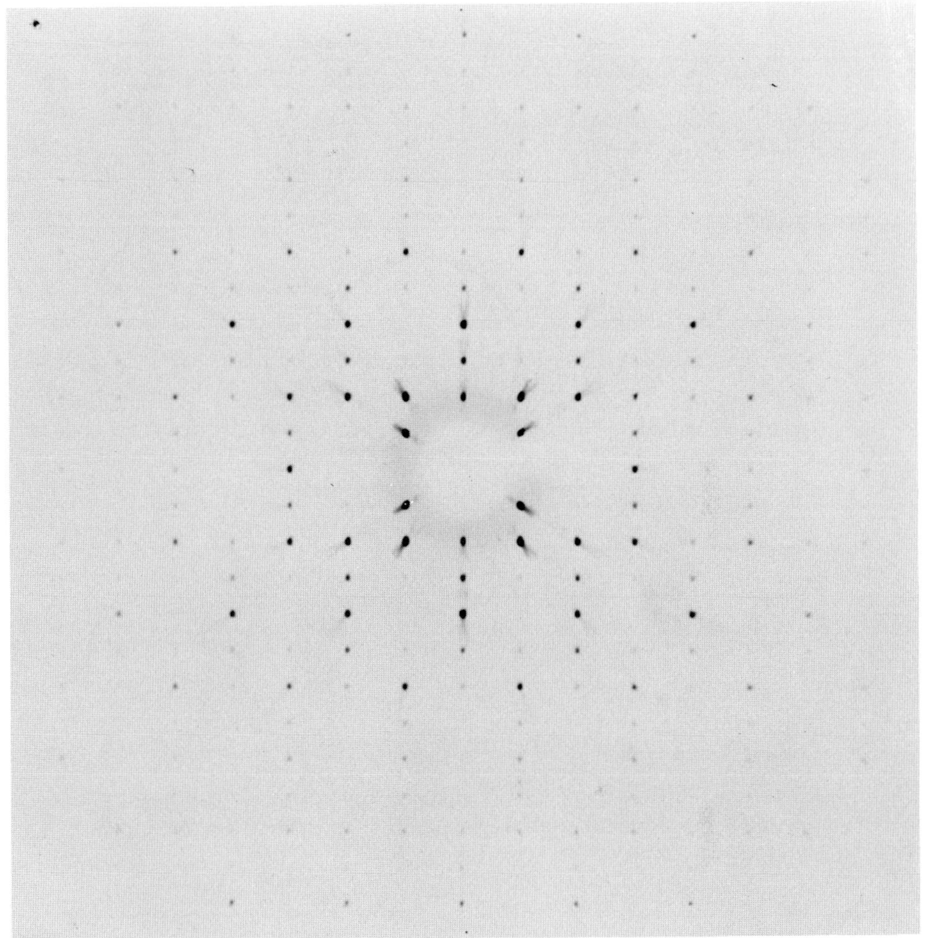

Fig. 14. First level (10 mA, 11 hr.) The additional horizontal rows of spots imply that these rows are missing on the zero level. This shows that there is a glide plane $\perp c$ with translation component parallel to the a axis.

is the angle between the x-ray beam and the actual rational crystal axis. If the Laue-cone angle is called $\bar{\mu}'$, it is seen to be equal to $\bar{\mu} + \epsilon$ in the phase of the precession motion shown in Fig. 3. In the opposite phase, when ST slopes down to the right, ST' lies above ST, so that the Laue-cone angle is $\bar{\mu}' = \bar{\mu} - \epsilon$. The Laue-cone angle is larger than $\bar{\mu}$ in the upper phase shown in Fig. 3, which is the phase in the direction of the error, and smaller than $\bar{\mu}$ in the opposite phase. These cones, and those of intermediate phases, determine circles on the film which are mapped for various phases in Fig. 9.

This variation of the size of the circle causes the shaded area to be eccentric with respect to the origin of the photograph, an example of which is shown in Fig. 11. The maximum radius of the area is in the

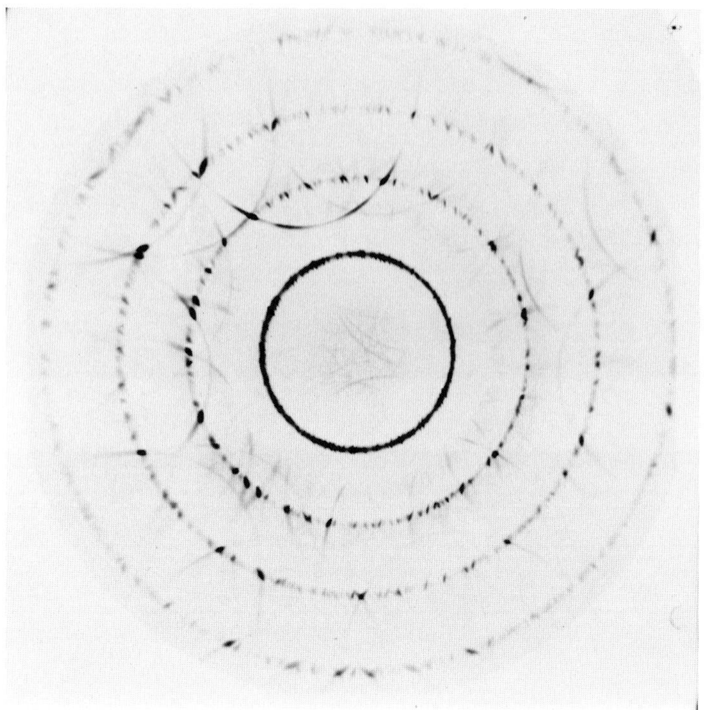

Fig. 15. Symmetry 1: axinite.

Figs. 15 to 34. Examples of cone-axis photographs (*left-hand pages*) and precession photographs (*right-hand pages*) displaying the 10 symmetries in a plane. The minerals used to display these symmetries are

1: axinite	m : adularia
2: kaliborite	$2\,m\,m$: beryl
3: dolomite	$3\,m$: calcite
4: scapolite	$4\,m\,m$: garnet
6: apatite	$6\,m\,m$: beryl

direction of the error. There are several ways of assessing the error from measurements made of geometrical features of the envelope. Some are useful when the error is large; others are sensitive to small errors. These relations make it possible to adjust a misoriented crystal to the correct orientation by rotation of the crystal along the two arcs of the goniometer head and about the dial axis.

Cone-axis photographs

Apparatus. The precession instrument shown in Fig. 7 has a frame for holding a flat layer-line screen so that the plane of the screen is

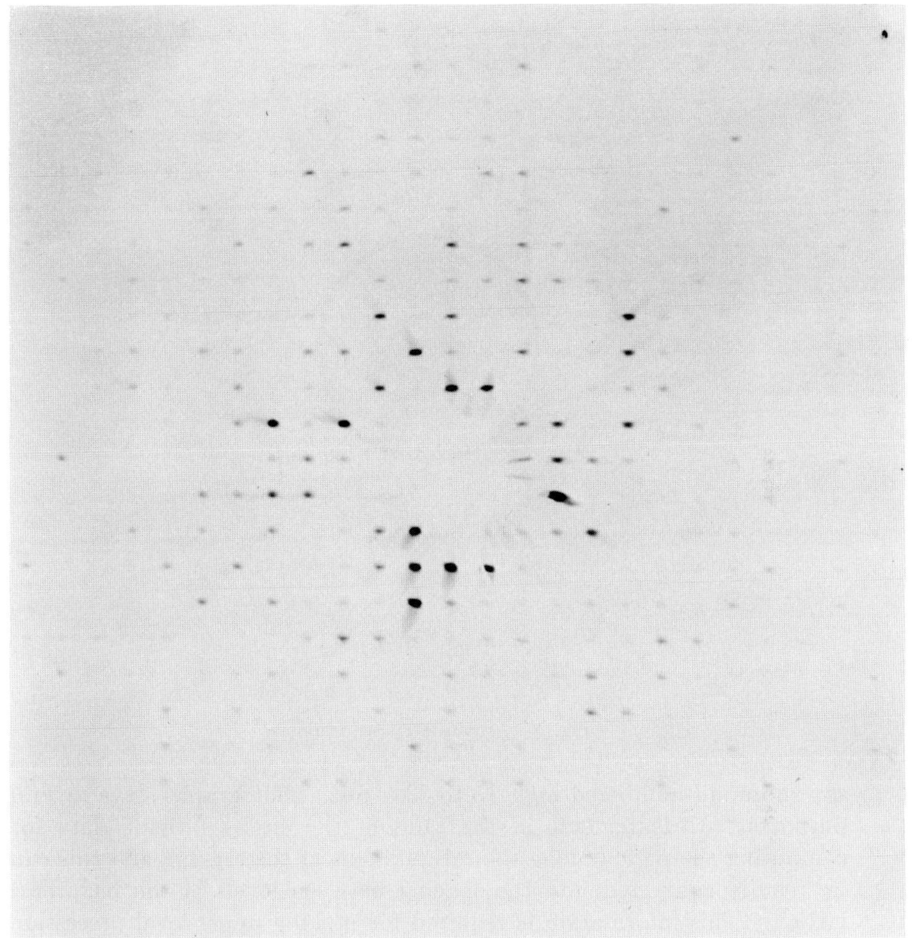

Fig. 16. Symmetry 1: axinite.

always normal to the precessing axis. The relation of this frame to the precessing crystal is shown diagrammatically in Fig. 12. The frame is rigidly locked to the dial axis. When the crystal is properly oriented, the rational direction of the crystal is always perpendicular to the plane of the frame at its center. Whatever motion the crystal undergoes, the frame does also.

Into this frame, which travels with the crystal, may be inserted a photographic film in a thin cassette called the *cone-axis cassette*. When the crystal undergoes a precessing motion, each Laue cone, which is always coaxial with the crystal axis and maintains the same size during the precessing motion if the crystal is properly oriented, records as a circle on the photographic film. Examples of cone-axis photographs are

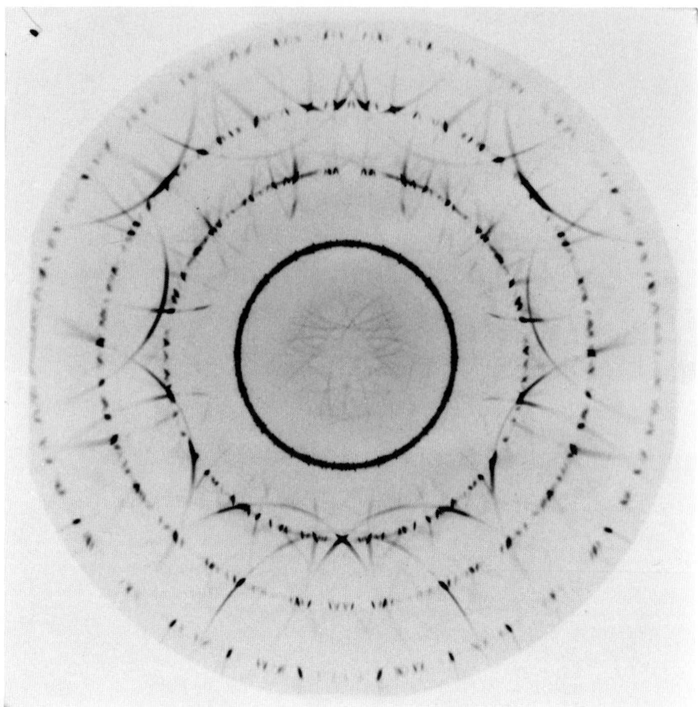

Fig. 17. Symmetry m: adularia.

shown in odd-numbered Figs. 15 to 33. Such photographs serve several purposes. Measurements of the radii of the circles provide data for computing the period along the rational axis of the crystal, and this can be readily converted into the spacing between levels of the reciprocal lattice. This information is required for making upper-level precession photographs. The cone-axis photographs also reveal the Friedel symmetry of the axis, as noted later.

Determination of level spacing. The half-opening angle of each Laue cone can be computed from the geometry shown in Fig. 12, which provides that

$$\tan \bar{\nu}_n = \frac{r_n}{s}. \tag{7}$$

The Laue equation relates $\bar{\nu}$ to the translation period along the crystal axis, as seen in (2) of Chapter 3. For present purposes this relation can be rewritten as

$$\cos \bar{\mu} + \cos \bar{\nu}_n = \frac{n\lambda}{t}, \tag{8}$$

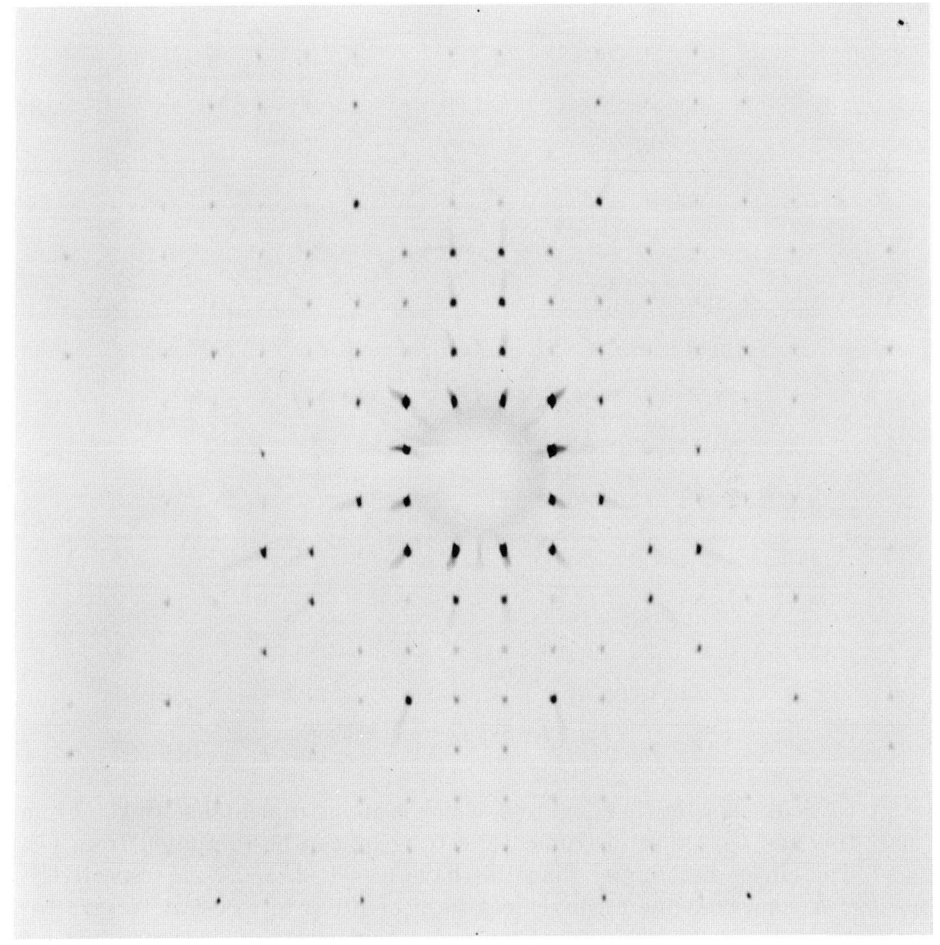

Fig. 18. Symmetry m: adularia.

where t is the period of the axis, and $\bar{\nu}_n$ is the half-opening angle of the Laue cone of order n.

Ordinarily the value of t corresponding to a circle of a given radius r_n is desired. This can be obtained by combining (7) and (8) and solving for t, which gives

$$t = \frac{n\lambda}{\cos \bar{\mu} + \cos \tan^{-1}(r_n/s)}. \tag{9}$$

To make use of this relation requires the ability to identify the order n of the Laue cone which produces the circular record whose radius is measured. This identification can be made because the zero-order cone

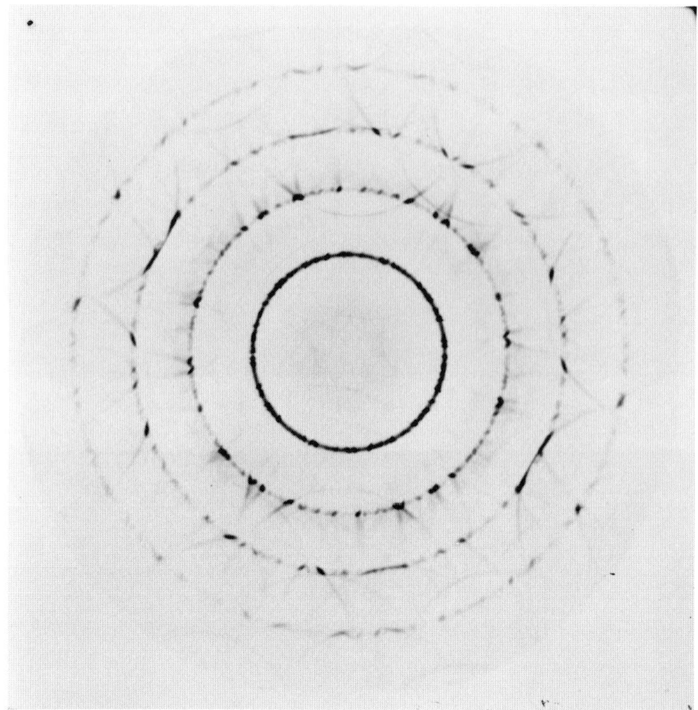

Fig. 19. Symmetry 2: kaliborite.

($n = 0$) has $\bar{\nu}_n$ of (7) equal to $\bar{\mu}$, which is an instrumental setting. From this and the other instrumental setting, r_0 can be computed from (7). The circles next larger than this have $n = 1, 2, 3, \ldots$, respectively.

A somewhat more convenient form of (9) is to recast it to give the spacing d^* between levels of the reciprocal lattice. This transformation can be readily made by using the fundamental relation given by (20) in Chapter 4, but expressed in λ-scaled units of the reciprocal lattice:

$$t = \lambda \frac{1}{d^*}. \tag{10}$$

By combining this with (9), it is found that

$$d^* = \frac{\cos \bar{\mu} + \cos \tan^{-1}(r_n/s)}{n}. \tag{11}$$

Determination of Friedel symmetry. An ideal precession motion gives all members of a set of equivalent planes equal opportunities to

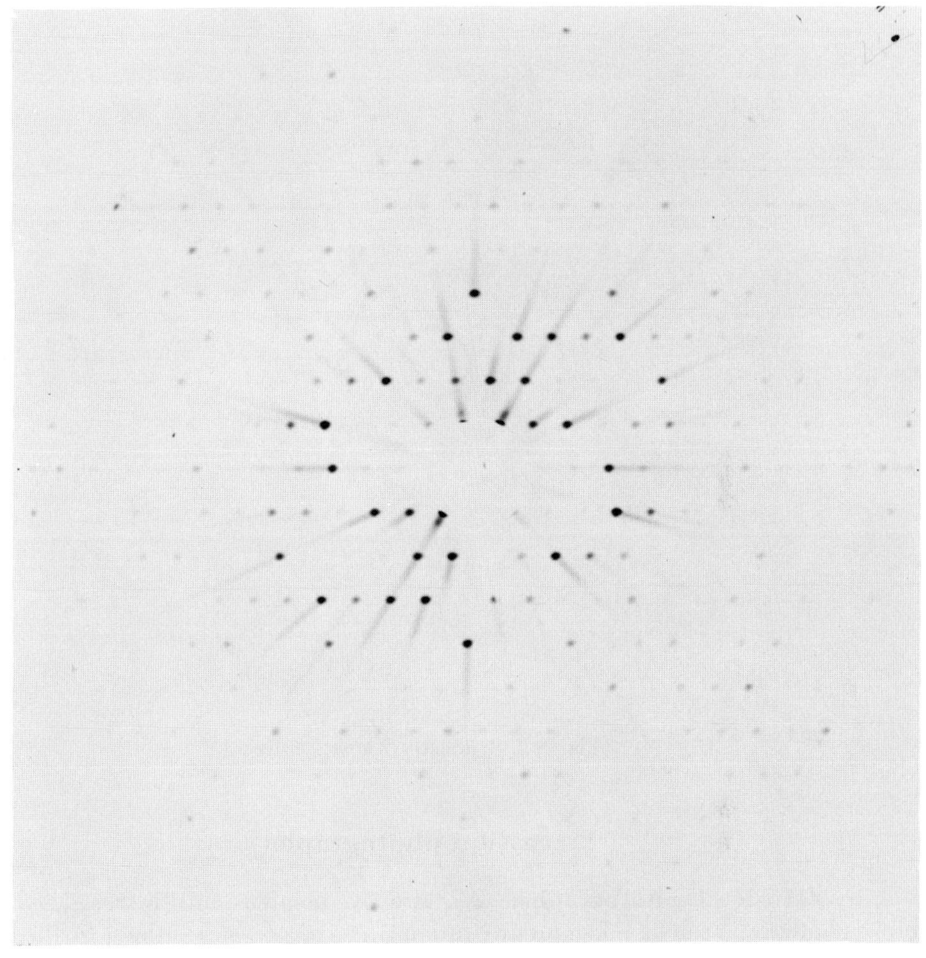

Fig. 20. Symmetry 2: kaliborite.

reflect, provided they make equal angles with the rational axis of the crystal. As a consequence, a set of reflections by planes which are equivalent by a symmetry element that contains the rational axis (an n-fold axis or mirror) displays the Friedel symmetry of that axis in the cone-axis ring.

This symmetry of the reflections in a cone-axis ring is the same as the Friedel symmetry of the level of the reciprocal lattice. It can also be determined from precession photographs, as noted later. The cone-axis photograph provides a compact tool for determining the symmetry of a rational axis, level by level. Examples are shown in odd-numbered Figs. 15 to 33.

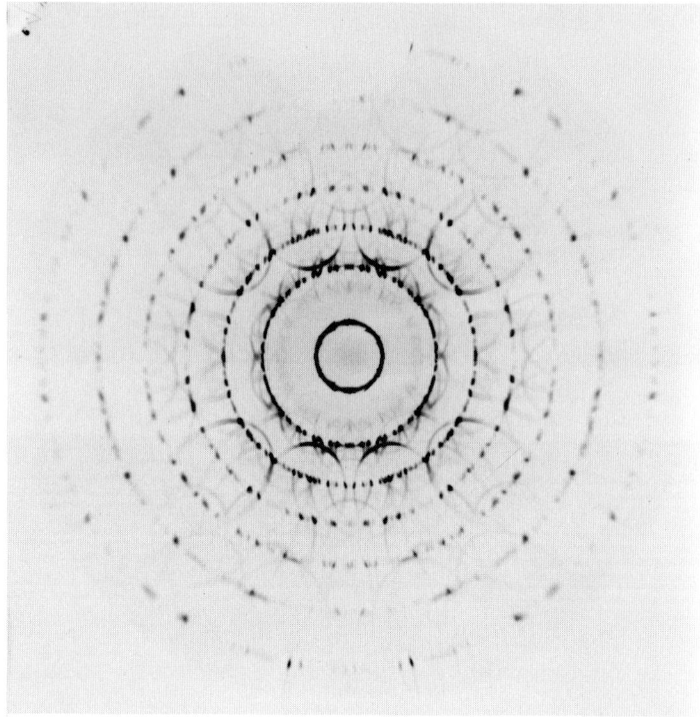

Fig. 21. Symmetry 2 *m m*: beryl.

Precession photographs

Zero-level photographs. As already pointed out, a zero-level photograph can be taken for any rational axis of a crystal without further data. The only exception to this occurs when the rational axis has such a large period that the Laue cones are very close together; then one cone cannot be isolated from the others with the layer-line screen. The upper permissible value of the period is about 50 Å for Mo$K\alpha$ radiation, and about twice this for Cu$K\alpha$ radiation. Isolation for such large periods requires a slit width of about 2 mm.

It is common practice to take zero-level precession photographs with a precession angle of $\bar{\mu} = 30°$. This is the maximum angle available in instruments currently manufactured. This provides a radial range of $\xi = 1.000$ wavelength unit. If Mo$K\alpha$ radiation is used, the range is about the same as that recorded by the Weissenberg method using Cu$K\alpha$ radiation.

An example of a zero-level precession photograph is shown in Fig. 13.

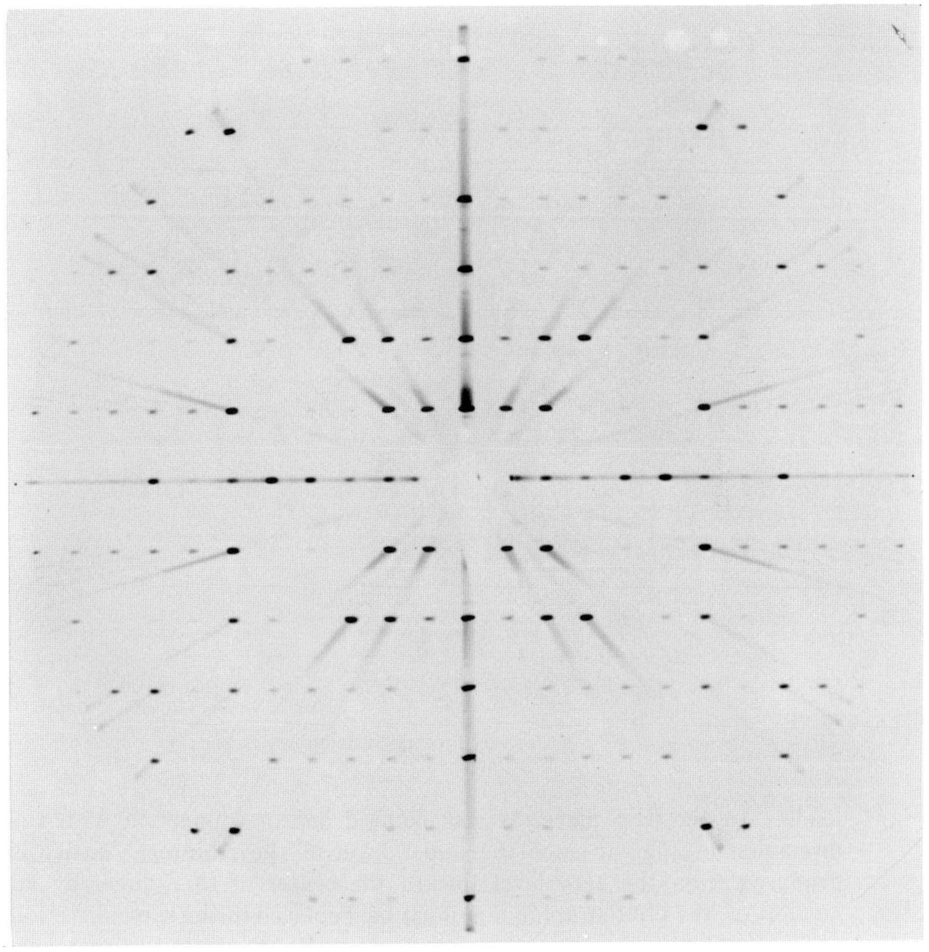

Fig. 22. Symmetry $2\,m\,m$: beryl.

Upper-level photographs. An example of an upper-level precession photograph is shown in Fig. 14. To make such a photograph requires a knowledge of the height ζ of the upper level above the zero level. If d^* is the spacing between levels, then this distance is simply

$$\zeta_n = nd^*. \tag{12}$$

The value of d^* is provided by relation (11) from measurements made on a cone-axis photograph.

In order to take an upper-level photograph, two settings must be

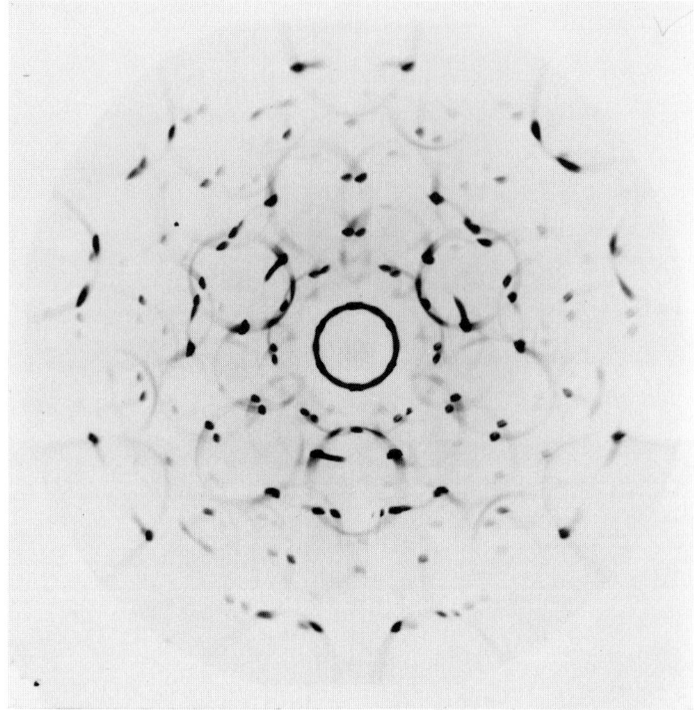

Fig. 23. Symmetry 3: dolomite.

made. In the first place, as was pointed out in connection with the discussion of Fig. 5, since the upper level of the reciprocal lattice is displaced from the zero level toward the center of the sphere by an amount ζ_n, the photographic film must be displaced in the same direction by the proportional amount $M\zeta_n$, where M is the crystal-to-film distance. This setting is independent of everything but nd^* of (12).

Secondly, the layer-line screen must be set to block out all but the nth-order Laue cone. This requires solving (7) for r_n. By combining (7), (8), and (12), the following relation results:

$$r_n = s \tan \cos^{-1}(\zeta - \cos \bar{\mu}). \tag{13}$$

This shows that, for a given value of ζ, the radius r_n of the annular opening of the layer-line screen depends on the choice of the precession angle $\bar{\mu}$ and on the choice of the distance s of the layer-line screen from the crystal. The solutions of r as a function of ζ for various convenient values of $\bar{\mu}$ and s are available in tabular and chart form, and so (13) can be quickly solved by inspection.

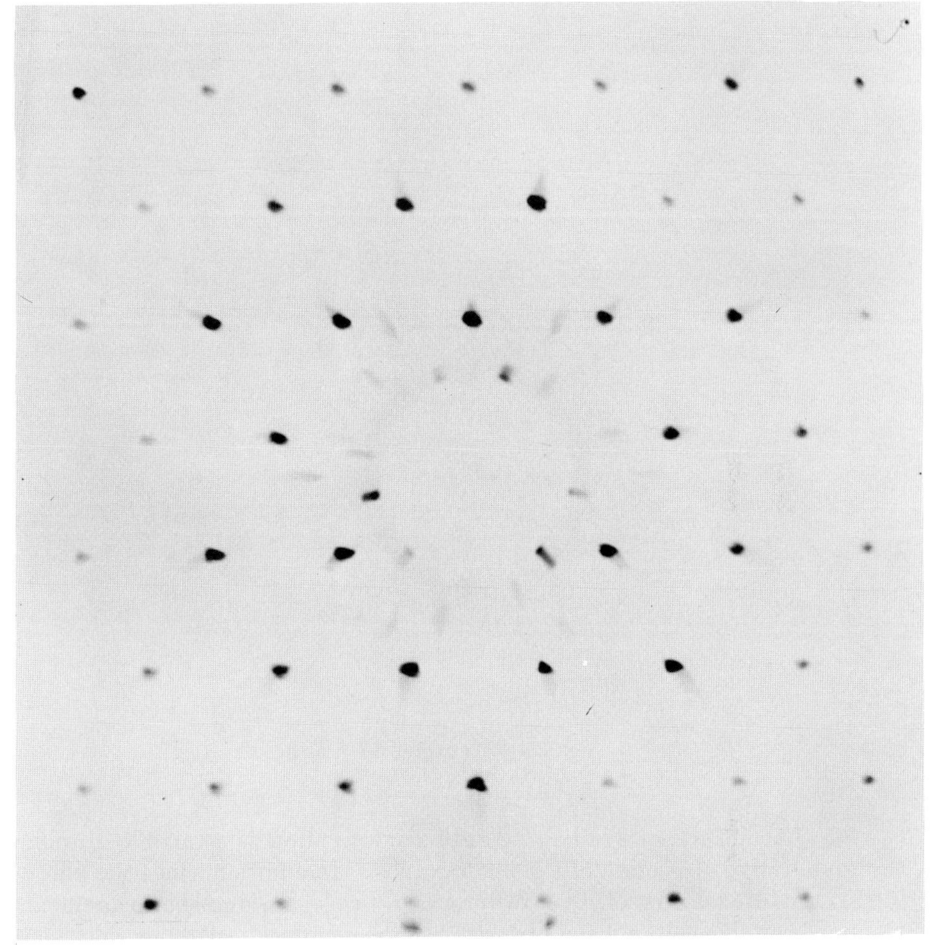

Fig. 24. Symmetry 3: dolomite.

Interpretation of precession photographs

As background for this section, it is suggested that the student again read Chapter 5, "Routine determination of symmetry by x-ray diffraction." The information required to determine the Friedel symmetry and the diffraction symbol is obtained with the greatest ease from a set of precession photographs because these constitute an undistorted picture, level by level, of the reciprocal lattice. In this enlarged duplicate of the reciprocal lattice, the intensity symmetry provides the Friedel symmetry, and the extinctions provide the reciprocal-lattice type and the remainder

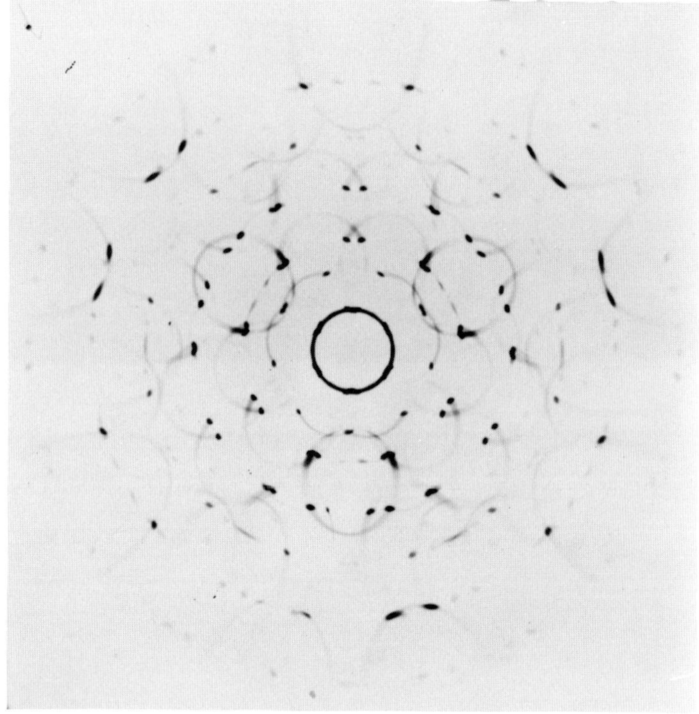

Fig. 25. Symmetry 3 m: calcite.

of the diffraction symbol. All this information is available through qualitative inspection of the photographs. In addition, since the photographs are scaled replicas of reciprocal-lattice levels, simple measurements made on them are sufficient to establish the reciprocal-cell dimensions and, by using standard transformations, the crystal-cell dimensions.

Friedel symmetry. To thoroughly establish the Friedel symmetry, an examination should be made of the stack of precession photographs perpendicular to each suspected symmetry element. For example, if the form development of a crystal is hexagonaloid, the crystal axis which is possibly a 6-fold axis should be oriented along the precessing axis of the instrument. If the crystal axis is truly 6-fold, then the upper-level rings of the cone-axis photograph and the upper-level precession photographs will show obvious 6-fold symmetry. If, in addition, the symmetry axis is the locus of intersecting mirrors, the plane symmetry 6 m m will be obvious in these photographs. By examining all axes which could reasonably be expected to yield symmetry information, the Friedel

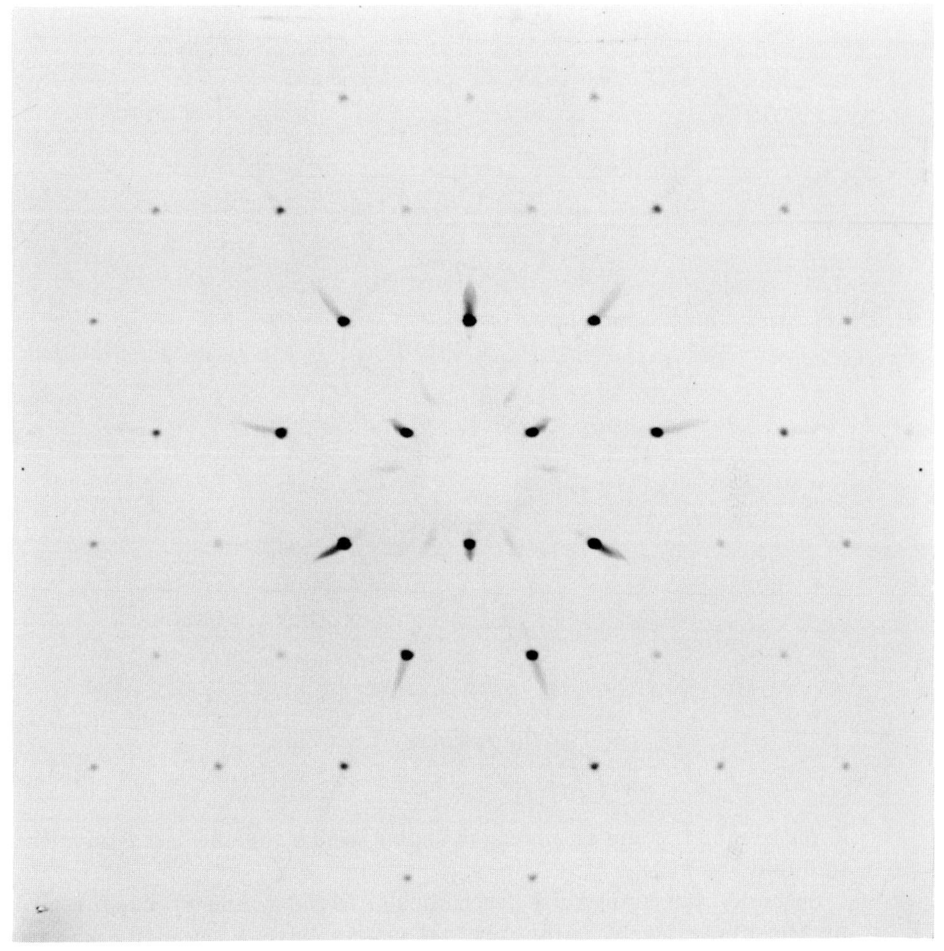

Fig. 26. Symmetry 3 *m*: calcite.

symmetry of the crystal can be quickly and certainly established by precession photographs. Examples are shown in even-numbered Figs. 16 to 34.

The zero-level cone-axis ring and the zero-level precession photographs often display more symmetry than the parallel upper-level photographs. There are two reasons for this.

In the first place, the zero level is the locus of the symmetry center which always occurs in the Friedel symmetry. Thus the zero level must show an even-fold symmetry. For example, if the upper-level photographs show 3-fold symmetry, the zero-level photograph must show

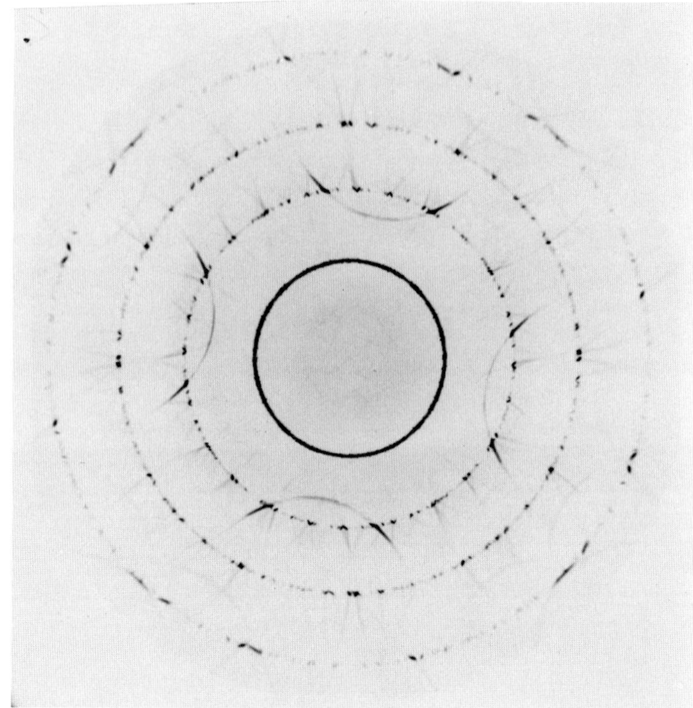

Fig. 27. Symmetry 4: scapolite.

3-fold symmetry and an inversion center, which together are equivalent to 6-fold symmetry.

Secondly, the zero level is perpendicular to the symmetry axis revealed by the upper levels. Many crystal classes have a set of 2-fold axes normal to the principal axis. These other axes accordingly lie in the zero level, and each contributes a symmetry which cannot be distinguished from a mirror. For example, if the upper levels have symmetry 6, the zero level may display symmetry 6 *m m*.

Lattice type. In Chapter 5 it was seen that the lattice types *P*, *I*, *A*, *B*, *C*, *F*, and *R* are distinguished by characteristic absent reflections, known as extinctions, in the general class of reflections *hkl*. The identification of the lattice type by lists of indices of observed reflections is required by many x-ray diffraction methods, such as the oscillating-crystal method, and can also be used with the precession method. But a simpler scheme is available for the precession method because it is possible to actually see a part of the reciprocal lattice in a sequence of two or more superposed photographs. A simple inspection, therefore,

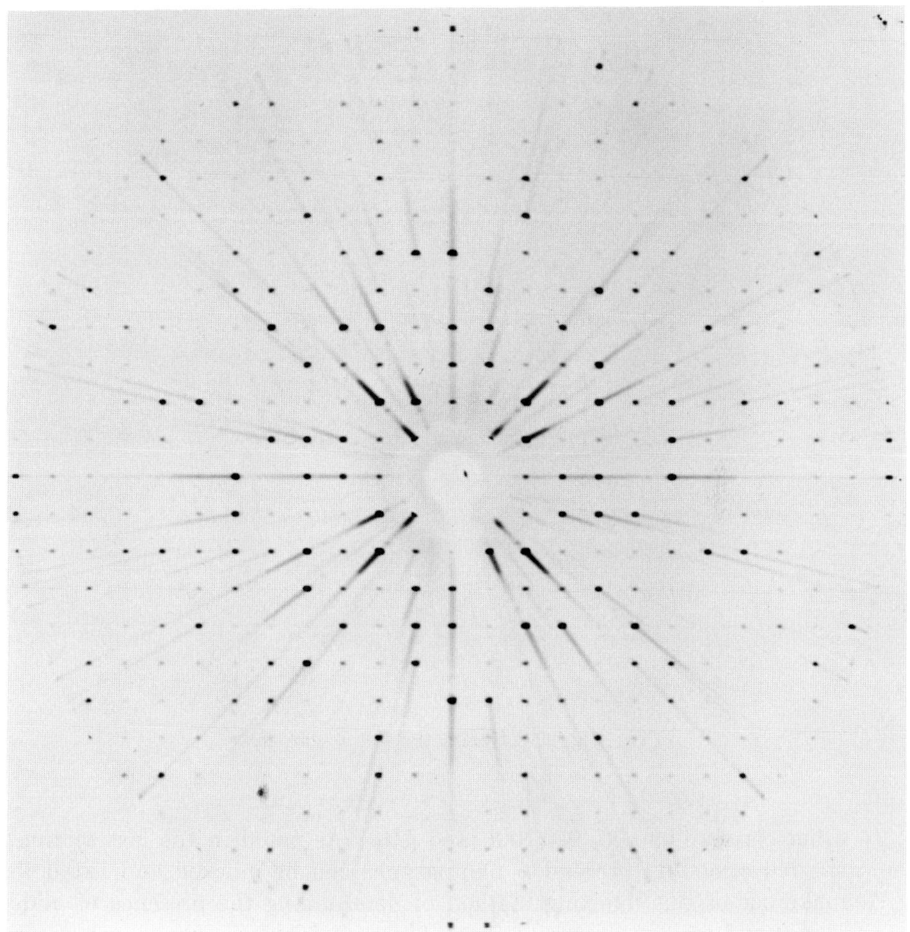

Fig. 28. Symmetry 4: scapolite.

reveals whether the reciprocal-lattice cell is P, I, A, B, C, F, or R. The crystal lattice is the reciprocal of these types. The crystal-lattice and the reciprocal-lattice cell types are the same for P, A, B, C, and R lattices; whereas for the two lattice types I and F, each is the reciprocal of the other, that is, $I^* \leftrightarrow F$ and $F^* \leftrightarrow I$. The lattice type can thus be determined rapidly by inspection when the precession method is used.

Diffraction symbol. In Chapter 5 it was shown that glide planes leave records of their existence by systematic absences of certain reflections in special classes like $hk0$, $h0l$, $0kl$, and hhl, while screw axes leave records of their existence by systematic absences of certain reflections in

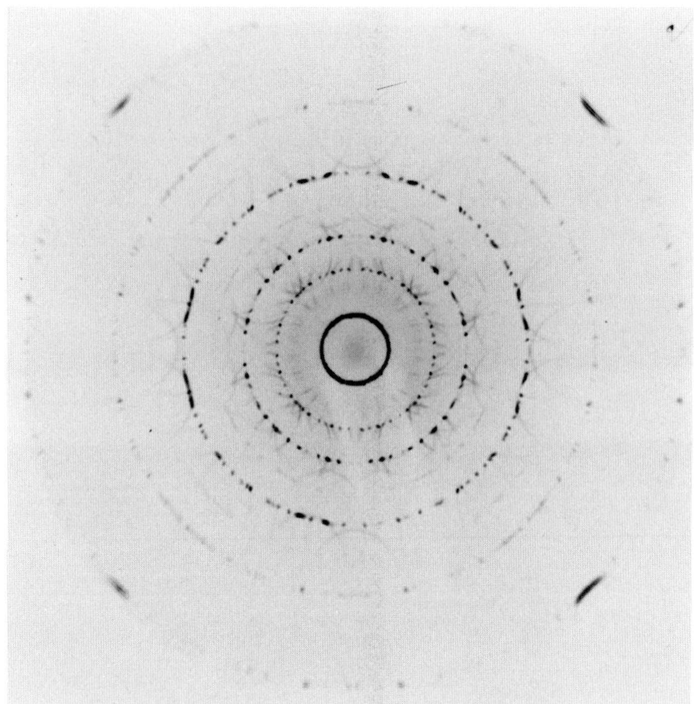

Fig. 29. Symmetry $4\,m\,m$: garnet.

other classes like $h00$, $0k0$, $00l$, and $hh0$. As noted in the last section, the reflections on precession photographs can be indexed and listed to make use of this indexing method of determining the presence of glide planes and screw axes in a crystal, but a simpler geometrical scheme is possible because the geometry of the reciprocal lattice is visible in precession photographs. All the special classes of reflections listed above occur in zero levels. If one such zero level and the corresponding second level (preferably) are superposed, a glide parallel to the level is indicated by a doubled period in the zero level (as compared with the second level); the direction of the glide is the direction of the period which is doubled.

Thus the presence of glide planes and screw axes can also be determined rapidly by inspection of precession photographs. The diffraction symbol is readily written (as discussed in Chapter 5) by adding this information to the Friedel symmetry and the lattice type.

Cell dimensions. The symmetry information just discussed is qualitative and can accordingly be derived from precession photographs by inspection. When the lattice type has been determined by inspection,

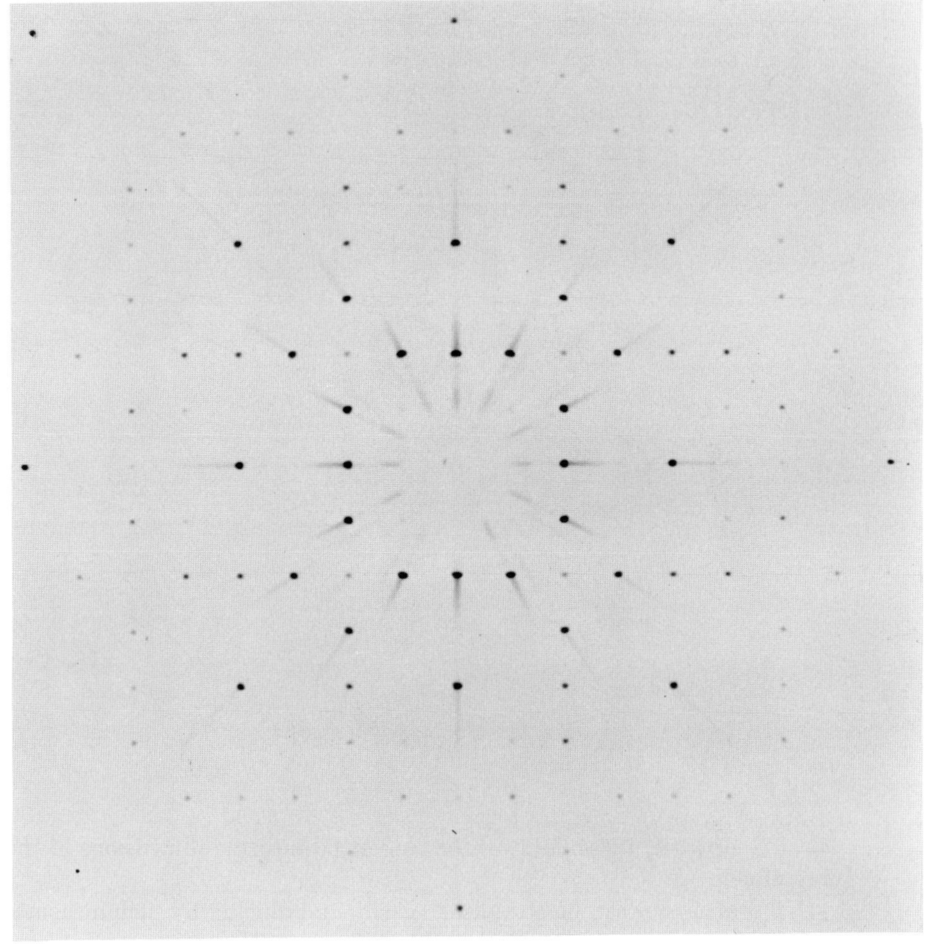

Fig. 30. Symmetry 4 *m m*: garnet.

it still remains to determine the dimensions of the cell. Again advantage is taken of the fact that each precession photograph is an accurately enlarged replica of the reciprocal lattice. It is an easy matter, with the aid of a special turntable measuring device shown in Fig. 35, to measure the angles between any desired rows of the reciprocal lattice which lie in the plane of the photograph, and to measure the spacings between any such rows. The angular measurements can be read to the nearest 5 minutes of arc, and the reciprocal spacings Md^* in the plane-lattice levels to the nearest 0.005 cm. From these measurements the reciprocal-cell dimensions can be computed, and with the aid of the relations in Table

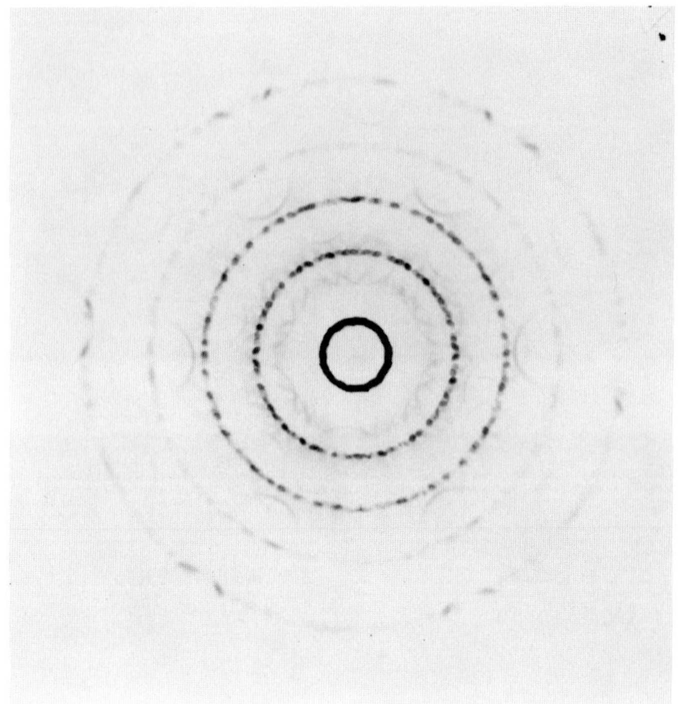

Fig. 31. Symmetry 6: apatite.

1*B* of Chapter 4, these data can be converted into the dimensions of the crystal cell.

Cell dimensions can be established easily and quickly by such measurements of precession photographs, chiefly because the measurements are made directly on scaled replicas of the reciprocal lattice. The way the measurements are made is advantageous because it is based upon setting a hairline on an entire row of spots, and this process involves an automatic averaging. As a result of both features, measurements of both lengths and angles can be made with a precision of about $\frac{1}{4}$ percent without special precautions.

(The rotating-crystal method, by contrast, permits the measurement of one translation at a time with modest precision, say 1 percent. At the other extreme, precision of the order of 0.01 percent can be attained by using the back-reflection Weissenberg method or the powder method, provided the approximate cell dimensions have already been determined.)

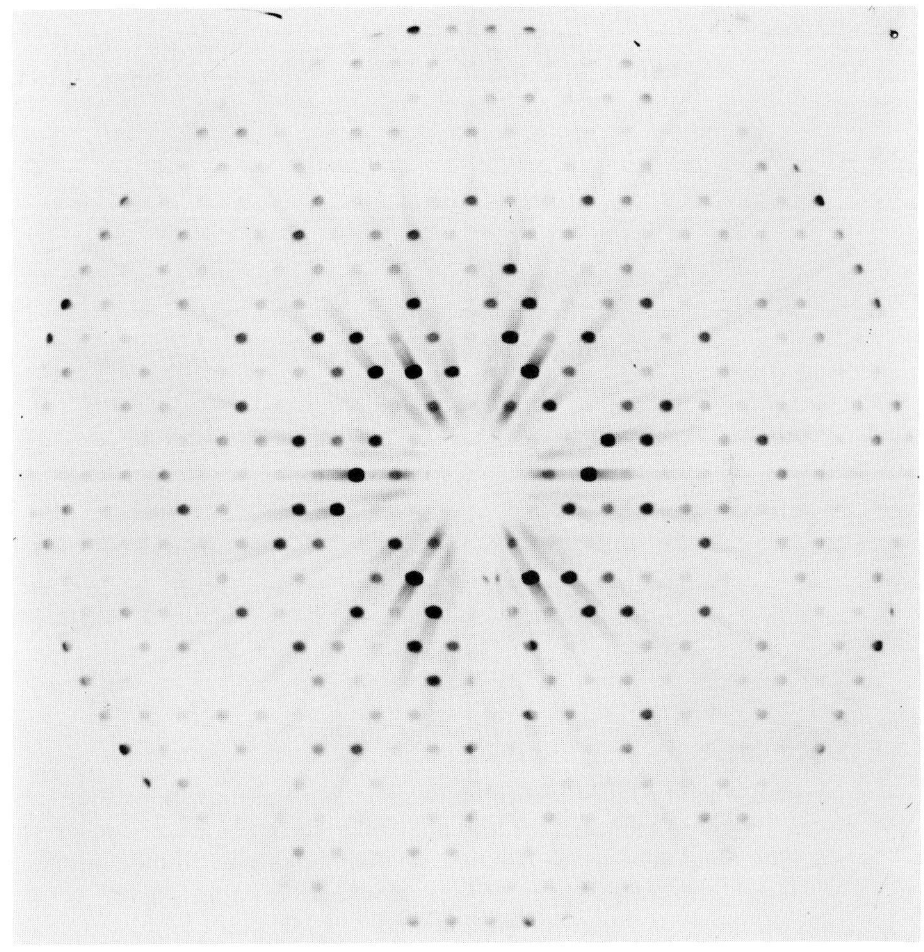

Fig. 32. Symmetry 6: apatite.

Summary

Any kind of change of orientation of the crystal with respect to the x-ray beam brings various crystallographic planes into the condition for producing Bragg reflections. One such way of changing orientation, which has many advantages, is to cause a rational direction of the crystal to precess about the direction of the x-ray beam, meanwhile preventing the crystal from merely rotating about the beam. This general motion is called *precession motion.* In precession instruments currently manu-

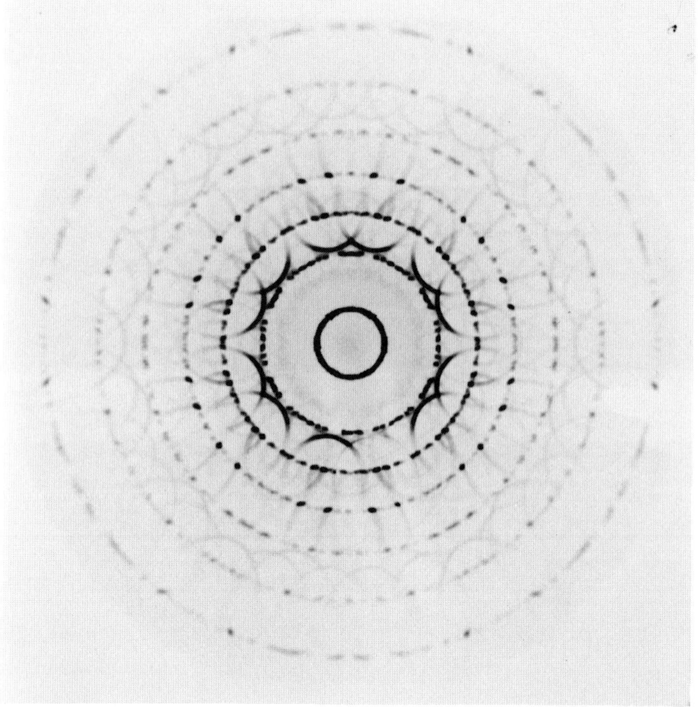

Fig. 33. Symmetry 6 *m m*: beryl.

factured, the crystal is prevented from merely rotating about the x-ray beam by being mounted on a gimbal (like an oar in an oarlock, with the long dimension of the oar representing the rational direction in the crystal). The precessing motion is caused by rotating a point on the rational direction about the x-ray beam.

The angle between the precessing rational direction and the x-ray beam is called the *precession angle* $\bar{\mu}$. This is held constant for a particular experiment, but can be set at any value from 0 to 30° in instruments presently available. The Laue condition for x-ray diffraction requires the directions of diffraction to be confined to the generators of a nest of Laue cones coaxial with each rational direction. The particular nest of cones coaxial with the precessing rational direction is of interest here. Since the precession angle $\bar{\mu}$ is held constant, the cones coaxial with the rational direction remain invariant during the precession cycle.

If a photographic film is placed so that its plane is normal to the rational axis of the crystal, and fixed to the crystal so that it also shares its precessing motion, the Laue cones intersect the film in a set of concentric circles, and these maintain a fixed location on the film during the

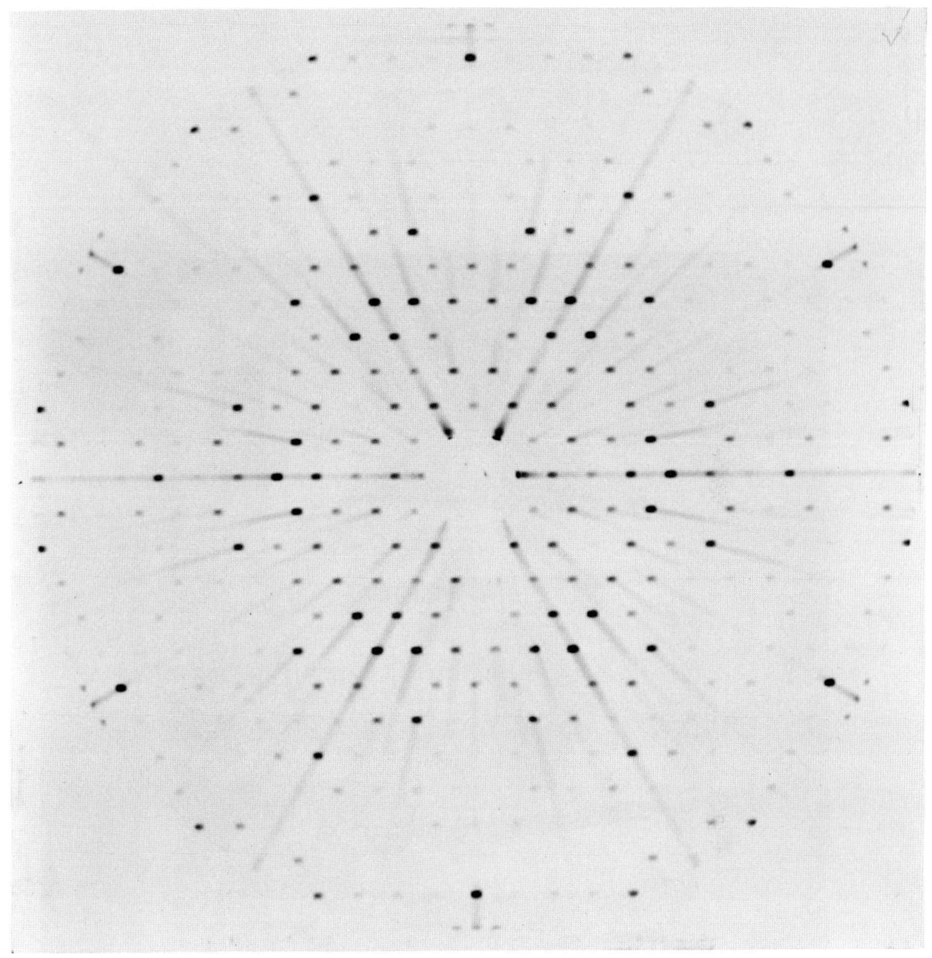

Fig. 34. Symmetry 6 *m m*: beryl.

precession cycle. The x-ray reflections are confined to these circles or rings, and the photograph made in this way is called a *cone-axis photograph.* Measurements of the diameters of the rings on a cone-axis photograph provide data from which the period along the rational axis can be computed. The symmetry of the reflections in a cone-axis ring is also the Friedel symmetry of the rational axis.

An alternative placement of the photographic film gives rise to a *precession photograph.* To produce this, the photographic film is mounted exactly as the crystal is mounted, with the centers of the two mounts located along the x-ray beam. The two mounts are also coupled together

Fig. 35. Device for measuring angles and distances on precession photographs. [*From M. J. Buerger: Am. Mineral.* **30** (1945) 553.]

mechanically so that any change of orientation of the crystal is duplicated by the film. The plane of the film is made perpendicular to the rational direction of the crystal, which therefore sweeps about the center of the film, marking out a circle. A flat screen with an annular opening permits only one Laue cone at a time to reach the film. The diffraction by the crystal recorded by this means turns out to be an undistorted picture of the points of the reciprocal lattice. On it, every linear feature of the wavelength-scaled reciprocal lattice is magnified by a factor M, which is the distance between the centers of the crystal and film mounts.

Precession photographs, being simply pictures of levels of the reciprocal lattice, are easy to interpret. The Friedel symmetry, lattice type, and diffraction symbol can be written down by inspection of sets of films. The cell dimensions, both linear and angular, can be determined by simple measurements with a precision of about $\frac{1}{4}$ percent without special precautions.

Precession photographs are also useful in orienting a crystal. Photographs for this purpose, usually made with unfiltered MoK radiation, are called *orientation photographs.*

Notes on history

The precession method was devised in the mid-1930s by Buerger. The motivation was the failure of the oscillating-crystal method to provide more than meager symmetry information. A precessing motion was substituted for the oscillation motion because this motion treats equally those planes which are symmetrically equivalent with respect to a symmetry element in the precessing axis and does not, therefore, degrade the symmetry of their reflections, as does the oscillating-crystal method.

In 1944 a monograph describing the method became available, and in 1945 two manufacturers began to supply precession instruments. These instruments are manufactured in Germany, Japan, The Netherlands, Switzerland, and the United States, so that they are now commonly available; most laboratories doing x-ray diffraction study of single crystals have at least one.

Additional reading

Martin J. Buerger. *The precession method in x-ray crystallography.* (Wiley, New York, 1964) 276 pages.

Significant literature

M. J. Buerger. *X-ray crystallography.* (Wiley, New York, 1942) 204–211.

M. J. Buerger. *The photography of the reciprocal lattice.* American Society for X-ray and Electron Diffraction, Monograph 1 (1944) 37 pages.

10

The cell contents and their general arrangement

Introduction

In Chapters 8 and 9 it was seen that the symmetry and cell of a crystal can be established provided that a homogeneous fragment is available. These results are independent of any accessory information about the crystal such as its chemical composition. But when the composition and density are also known, these additional data can be combined with the cell and symmetry data to produce some useful general information about the crystal structure, specifically, the number of atoms per cell and how the symmetry restricts their arrangement. These matters are discussed in this chapter.

As a preliminary to a complete crystal-structure determination the number of atoms per cell is computed as a standard routine, and the general distribution of this number of atoms is usually surveyed. The results are often sufficient to solve problems less extensive than the detailed arrangement of atoms; for example, they may be of aid in deciding the molecular weight or the configuration of a chemical molecule in the crystal. They are also useful in finding the location of a heavy atom in the structure, a bit of knowledge which may be vital to the determination of the crystal structure itself.

The density relation

Basic computation. The fact that the bulk density of a crystal and the density of one of its cells are the same permits equating these with useful consequences. The bulk density can be measured in many ways; a precision of 1 percent can be obtained without special precautions, and this can be improved by an order of magnitude if special care and precautions are observed. This experimental value G can be equated to the cell mass divided by the cell volume V. The cell mass is equal to the mass M of the atoms in the chemical formula, times the number Z of such formula units in the cell. The density relation is thus

$$G = \frac{\text{cell mass}}{\text{cell volume}} = \frac{ZM}{V}. \tag{1}$$

The cell volume, of course, is readily computed from the cell dimensions. As seen in Table 1B of Chapter 4, this is given, in the most general (triclinic) case, by

$$V = abc\sqrt{1 - \cos^2 \alpha - \cos^2 \beta - \cos^2 \gamma + 2 \cos \alpha \cos \beta \cos \gamma}, \tag{2}$$

which reduces considerably for other crystal systems.

It is customary to express density in grams per cubic centimeter. In using (1), therefore, other values should be expressed in cgs units. Specifically, the cell edges a, b, and c, ordinarily expressed in Ångström units, must be converted to centimeters by multiplying by 10^{-8}; the mass of the atoms in the chemical formula, ordinarily given in terms of a unit which is approximately the mass of a hydrogen atom (but more exactly one-twelfth the mass of the carbon nuclide ^{12}C), must be converted to grams by multiplying by the inverse of Avogadro's number (6.0225×10^{23} mol^{-1}), that is, by 1.6604×10^{-24}. With these conversions in mind the computation in (1) can be expressed as

$$G\ (\mathrm{g/cm^3}) = \frac{Z \times M\ (\text{atomic mass units}) \times 1.660 \times 10^{-24}\ (\text{g/atomic mass unit})}{V\ (\text{Å}^3) \times 10^{-24}\ (\mathrm{cm^3/Å^3})}. \tag{3}$$

Since G and V are experimentally determinable by simple physical measurements, (3) can be solved basically for $Z \times M$, the mass of atomic matter in the cell. This ordinarily is used in two ways, which are noted briefly below.

Number of formula units per cell. If the chemical formula is well established, (3) is commonly used to determine the number of formula units contained in the cell by solving for Z:

$$Z = \frac{G \times V \times 10^{-24}}{M \times 1.660 \times 10^{-24}}. \tag{4}$$

This is an integer which must be consistent with the symmetry of the cell in a manner considered later.

Cell mass and formula mass. For many crystals the exact chemical composition is not well established, either because of the difficulty of the analysis or because of complication of the ideal chemical formula due to solid solution. In such instances mass M in (3) has not been established, so that relation (4) cannot be used to determine Z. But in such cases (3) can still be solved for the total mass ZM in the cell:

$$ZM\ (\text{atomic mass units}) = \frac{G \times V \times 10^{-24}}{1.660 \times 10^{-24}}. \tag{5}$$

This information is often very useful. For example, mineral crystals commonly deviate from an ideal formula because of solid solution. There are many types of solid solution. The type frequently encountered in minerals and metallic systems is the substitution of some foreign atoms for a certain fraction of some of the atoms normally expected in the crystal structure. This is known as substitution solid solution. If this replacement occurs for several atoms, the chemical analysis deviates from that of the ideal compound and may not be easy to interpret. It is not unusual to find that the chemical analyses of the same mineral species from various localities differ so much that they do not suggest any single ideal formula for the species as a whole. But it is commonly noted that the number of large anions is constant for the whole set, that the total number of certain cations is constant for the whole set, and that both of these numbers are characteristic of, or permitted by, the symmetry of the space group, as discussed later. In such cases the locations of these sets of atoms with respect to the symmetry elements are partly or completely revealed.

A fairly complicated example of this is afforded by the mineral tourmaline, whose chemical analyses could not be completely and certainly interpreted until its structure was solved. The analysis of a very simple tourmaline, and the way this analysis was reduced, is outlined in Table 1. The space group of tourmaline is $R\,3\,m$. A computation using (5), based upon a cell volume of $V = 532$ Å^3 and a density of $G = 3.06$ g/cm^3, shows that the primitive cell contains a mass of 980 atomic mass units. If this mass is multiplied by the oxide percentages reported in the chemical analysis (listed in column 1 of Table 1), the results, listed in column 2, are the masses of each oxide contained in the cell. In column 4 these masses have been converted to the numbers of formula units of the oxides, which, in turn, are broken down into the numbers of metal atoms (listed

Table 1
Computation of cell content of the de Kalb tourmaline

	1	2	3	4	5	6
	Weight, %	Mass per cell, atomic mass units	Molecular mass of oxide	Number of formula units of oxide per cell	Number of metal atoms per cell	Number of oxygen atoms per cell
SiO_2	36.72	360	60.06	5.99	5.99	11.98
TiO_2	0.05	0.5	79.90	0.006	0.01	0.01
B_2O_3	10.81	106	69.64	1.53	3.05	4.58
Al_2O_3	29.68	291	101.94	2.85	5.70	8.55
FeO	0.22	2.2	71.84	0.03	0.03	0.03
MgO	14.92	146	40.32	3.62	3.62	3.62
CaO	3.49	34.2	56.08	0.61	0.61	0.61
Na_2O	1.26	12.35	61.97	0.20	0.40	0.20
K_2O	0.05	0.5	94.20	0.005	0.01	0.01
H_2O	2.98	29.2	18	1.62	3.24	1.62
F	0.93	9.1	19	0.48	0.48	
Σ	101.11					31.21
Less O equivalent to F	0.41					
	100.70					

in column 5) and oxygen atoms (listed in column 6) per cell. For reasons discussed later, the space group $R\,3\,m$ permits the primitive cell to contain atoms only in sets of 6, 3, and 1. Accordingly, the last two columns of Table 1 are to be examined to disclose such sets. For example, column 5 shows there are almost exactly 6 Si atoms, 3 B atoms, and about 1 large electropositive atom, which is partly Na and partly Ca. By means of this kind of analysis of the last two columns of Table 1, there arises a strong presumption that this tourmaline can be represented by the formula

$$\left|\begin{array}{r} Na_{0.40} \\ Ca_{0.61} \\ K_{0.01} \\ \hline 1.02 \end{array}\right| Mg_{3.05}B_{3.05} \left|\begin{array}{r} Al_{5.52} \\ Mg_{0.57} \\ Fe_{0.03} \\ \hline 6.12 \end{array}\right| \left|\begin{array}{r} Si_{5.99} \\ \\ Al_{0.18} \\ \hline 6.17 \end{array}\right| \left|\begin{array}{r} O_{31.21} \\ \\ F_{0.48} \\ \hline 31.69 \end{array}\right| H_{3.24}.$$

The fact that sets of atoms (listed in the vertical columns of this formula) are present in amounts which are approximately the integers 1, 3, and 6 is interpreted as meaning that the elements in a vertical column replace one another in substitutional solid solution. That these summations are about 2 percent higher than the integers 1, 3, 3, 6, 6, and 31 can be accounted for by assuming that the density 3.06, used in (5), was about 2 percent too high. A good conjecture would be that an ideal formula for tourmaline would be

$$NaMg_3B_3Al_6Si_6O_{27}(OH)_4.$$

This is borne out by a similar study of the analyses of tourmalines from other localities, and was confirmed by a complete crystal-structure analysis.

Coordinates of points

It is convenient to express the location of any point in the cell of a crystal structure by means of coordinates which are taken in the directions of the cell edges and whose magnitudes are fractions of these cell edges. Such coordinates are called *fractional coordinates* and are defined as follows: If X, Y, and Z are the absolute coordinates (say in Ångström units) parallel to the a, b, and c axes, respectively, then the fractional coordinates parallel to the same axes are

$$x = \frac{X}{a}, \qquad y = \frac{Y}{b}, \qquad z = \frac{Z}{c}. \tag{6}$$

These can be referred to any convenient origin, usually a cell corner. (In some of the illustrations which follow, however, the origin has been chosen at the cell center, which leads to more compact illustrations.)

Sets of equivalent positions

Generation of points by symmetry operations. If a point is placed in the neighborhood of one or more symmetry elements, other points must also be present in order that the meaning of the symmetry element be satisfied. It is convenient to regard the new points as generated from the original point by the operations of the symmetry group composed of these symmetry elements. For example, if a point is placed in an unspecialized location of space group $P\,m\,m\,2$, illustrated in Fig. 1, four points must be present in each cell to satisfy the space-group symmetry. It is convenient to consider this set of four points as generated from the original point by the four nonlattice operations of the space group $P\,m\,m\,2$, namely, 1, 2, $m_{\perp X}$, and $m_{\perp Y}$. The generation of the four points by the four operations is indicated in Fig. 1 by arrows. Note that operation 1 is regarded as generating the original point from itself. This information is also recorded in the first two columns of Table 2.

Reduction of multiplicity due to special locations. In Fig. 1, the point was placed in an unspecialized position; accordingly, the position of the point is without symmetry. There are, however, very specialized positions in the cell, namely, the locations of the symmetry elements themselves. If the point with general coordinates is shifted toward one of these symmetry elements, then both the point and its mate, generated by the operation of that symmetry element, approach one another until they eventually coincide on the symmetry element. When this occurs, these points are no longer distinct but become one point; instead of the symmetry element's relating the two original points, it requires the one point to have the symmetry it defines. Every distinct symmetry

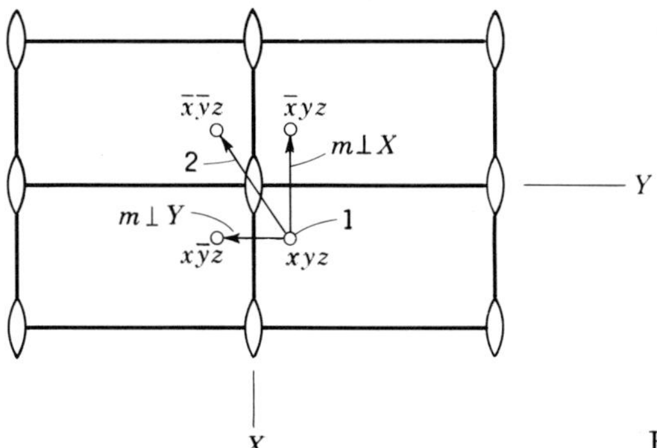

Fig. 1. $P\,m\,m\,2$

Table 2
Sets of equivalent positions of space group $P\ m\ m\ 2$

Operation generating the point	Specialization of location and its symmetry								
	1	$x = \frac{1}{2}$ $m_{\perp X}$	$x = 0$ $m_{\perp X}$	$y = \frac{1}{2}$ $m_{\perp Y}$	$y = 0$ $m_{\perp Y}$	$x = \frac{1}{2}$ $y = \frac{1}{2}$ $2\ m\ m$	$x = \frac{1}{2}$ $y = 0$ $2\ m\ m$	$x = 0$ $y = \frac{1}{2}$ $2\ m\ m$	$x = 0$ $y = 0$ $2\ m\ m$
1	xyz			$\rightarrow x\frac{1}{2}z$	$x0z$				
$m_{\perp Y}$	$x\bar{y}z$ $\rightarrow$	$\frac{1}{2}yz$	$0yz$			$\rightarrow \frac{1}{2}\frac{1}{2}z$	$\frac{1}{2}0z$	$0\frac{1}{2}z$	$00z$
$m_{\perp X}$	$\bar{x}yz$	$\rightarrow \frac{1}{2}\bar{y}z$	$0\bar{y}z$						
2	$\bar{x}\bar{y}z$			$\rightarrow \bar{x}\frac{1}{2}z$	$\bar{x}0z$				
Designation of set:	$4i$	$2h$	$2g$	$2f$	$2e$	$1d$	$1c$	$1b$	$1a$

element of this sort[†] presents a different location where points may coincide in this way. There are eight such locations for space group $P\ m\ m\ 2$, illustrated in Figs. 2 to 9.

A point and the points symmetrical with it constitute a set known as a *set of equivalent positions*. For brevity, these four words are sometimes replaced by the single word "equipoint" (singular). Equivalent positions are of two kinds: If the point is in an unspecialized, or general, location, the points are said to be in a *general position;* if the point is in a specialized location, the points are said to be in a *special position*. Together these two make up the several sets of equivalent positions (that is, the several equipoints) of the space group under discussion. There are nine sets of equivalent positions in space group $P\ m\ m\ 2$. Their locations are illustrated in Figs. 1 to 9, and their coordinates are listed in Table 2.

Whenever a point in a general position can be displaced toward a symmetry element with the result that symmetry mates approach one another and eventually coincide as one point on the symmetry element, that symmetry element constitutes a special position. Symmetry elements of this kind include rotation axes, rotoreflection axes, reflection planes, and certain specialized points on rotoreflection axes, specifically, those which have symmetries $\bar{6}$, $\bar{4}$, $\bar{3}$, and $\bar{1}$ (which is an inversion center). Special positions also occur on certain screw axes n_q, specifically, those for which n and q have a common factor, as noted in the next section.

† The meaning of "this sort" is explained later after the symmetry elements of space group $P\ 2_1\ c\ a$ have been discussed.

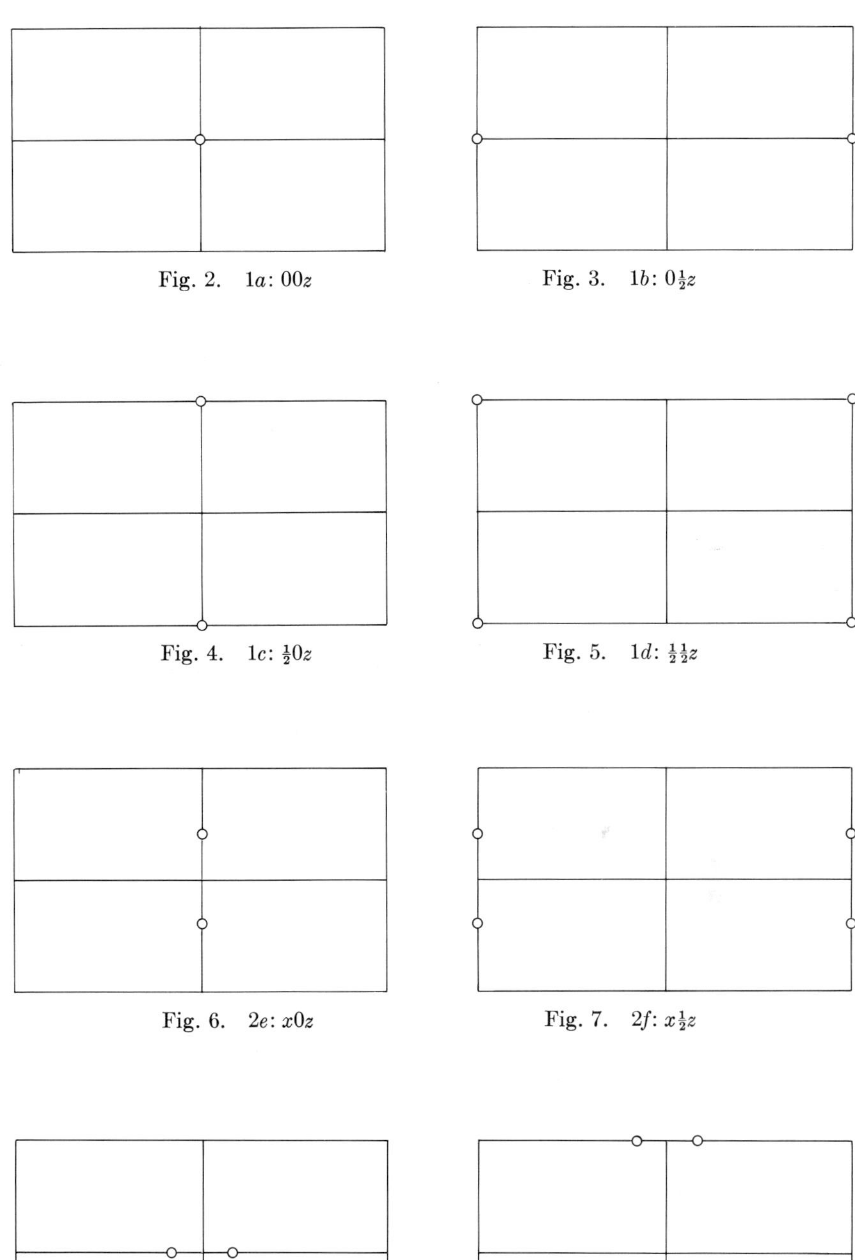

Fig. 2. 1a: $00z$

Fig. 3. 1b: $0\frac{1}{2}z$

Fig. 4. 1c: $\frac{1}{2}0z$

Fig. 5. 1d: $\frac{1}{2}\frac{1}{2}z$

Fig. 6. 2e: $x0z$

Fig. 7. 2f: $x\frac{1}{2}z$

Fig. 8. 2g: $0yz$

Fig. 9. 2h: $\frac{1}{2}yz$

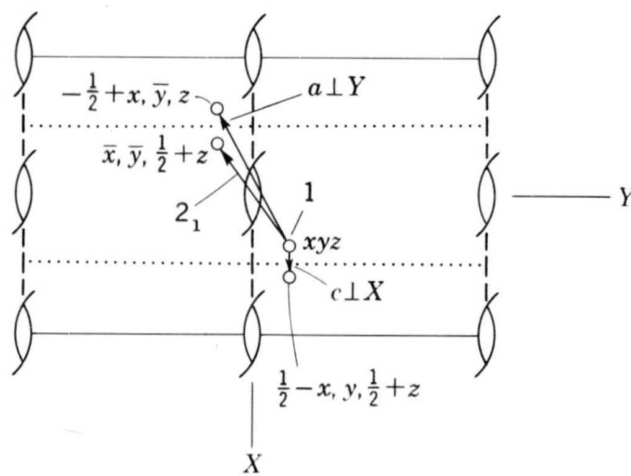

Fig. 10. $P\,c\,a\,2_1$

Symmetry elements not constituting special positions. The occurrence of special positions was illustrated by using space group $P\,m\,m\,2$. But in the isogonal space group $P\,c\,a\,2_1$, illustrated in Fig. 10, there are no special positions. For, if the initial point is placed on the 2_1 axis, its mate generated by the operation 2_1 is also on the axis; yet the mates do not coincide since they are separated by the translation component of the screw, namely, $\frac{1}{2}c$. If the initial point is placed on the glide plane c, the second point is removed from the first by the translation component of the glide, namely, $\frac{1}{2}c$; if it is placed on the glide plane a, the second point is removed from it by $\frac{1}{2}a$. In each case, the translation component of the symmetry element prevents the point and its symmetry

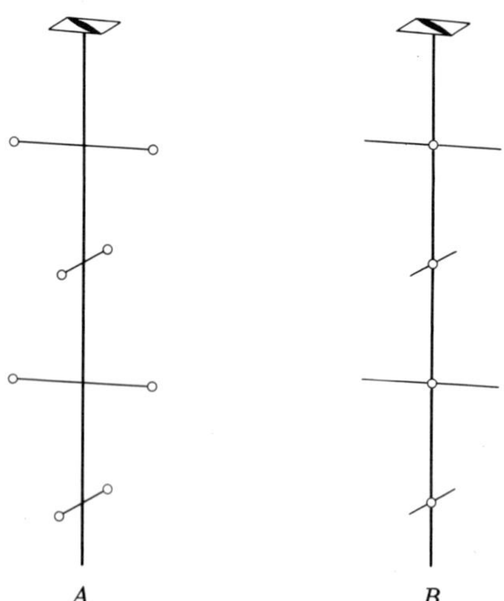

Fig. 11*A*. Equivalent points in the neighborhood of a 4_2 screw axis.
Fig. 11*B*. Equivalent points on a 4_2 screw axis.

mate from coinciding. This is true for any glide plane and also for any screw axis n_q provided that n and q have no common factor. When they do have a common factor, as in 4_2, 6_2, 6_3, and 6_4, the screw axis also functions as a special kind of rotation axis and therefore does behave as a special position. The example of 4_2 is illustrated in Fig. 11.

Sets of similar special positions. In the listing of special positions of $P\,m\,m\,2$ in Table 2 there are four different 1-fold equipoints, each of which occupies a different 2-fold axis of the cell, as seen in Figs. 2 to 5. Since these are distinct, a different atom could occupy each equipoint. Such sets of similar but distinct equipoints occur in many space groups. This is due to the fact that the combination of one symmetry element with the translations of the lattice produces other similar symmetry elements, as mentioned in Chapter 2. For primitive cells, the transla-

Table 3
Number of sets of equivalent positions having various multiplicities
Subscript ∘ indicates no variable parameters

Crystal class	Space group	Multiplicity					
		1	2	4	8	16	32
1	$P\,1$	1					
$\bar{1}$	$P\,\bar{1}$	$8_\circ$	1				
2	$P\,2$	4	1				
	$P\,2_1$		1				
	$A\,2$ $(B\,2,\ I\,2)$		2	1			
m	$P\,m$	2	1				
	$P\,a$ $(P\,b,\ P\,n)$		1				
	$A\,m$ $(B\,m,\ I\,m)$		1	1			
	$A\,b$ $(B\,a,\ I\,a)$			1			
$\frac{2}{m}$	$P\frac{2}{m}$	$8_\circ$	6	1			
	$P\frac{2}{a}$ $(P\frac{2}{b},\ P\frac{2}{n})$		$4_\circ + 2$	1			
	$P\frac{2_1}{m}$		$4_\circ + 1$	1			
	$P\frac{2_1}{a}$ $(P\frac{2_1}{b},\ P\frac{2_1}{n})$		$4_\circ$	1			
	$A\frac{2}{m}$ $(B\frac{2}{m},\ I\frac{2}{m})$		$4_\circ$	$2_\circ + 3$	1		
	$A\frac{2}{b}$ $(B\frac{2}{a},\ I\frac{2}{a})$			$4_\circ + 1$	1		

tions combined with a single symmetry element produce a "flock" of similar symmetry elements, so that they are present in the following numbers per cell:

8 inversion centers
2 parallel reflection planes
4 parallel 2-fold rotation axes
3 parallel 3-fold rotation axes
2 parallel 4-fold rotation axes
1 6-fold rotation axis

If no other operation relates these similar symmetry elements, then each constitutes a distinct special position. Often, however, some other symmetry operations within the cell cause one symmetry element of the similar set to be equivalent to another, in which case there are fewer distinct similar equipoints.

Since the various space groups have different sets of symmetry elements, they necessarily have different arrangements and numbers of equipoints. Some features of the distribution of the equipoints for the monoclinic and triclinic space groups can be seen in Table 3.

Distribution of atoms among the equipoints

It has just been seen that the presence of symmetry in a crystal places restrictions on the numbers of atoms which are permitted in a cell. If the space group contains no symmetry elements which can act as special positions, atoms can be present only in the general position, and their numbers in the cell are restricted to multiples of the multiplicity of that position. For example, the only set of equivalent positions of space group $P\,2_1\,2_1\,2_1$ has multiplicity 4; accordingly, atoms can occur only in sets of 4s in the cell of any crystal having this space group.

A crystal which has this symmetry is diglycine hydrobromide, $2(C_2H_5NO_2) \cdot HBr$. The number of atoms in its cell can be computed with relation (4), using the measured density $G = 1.94$ g/cm^3, the formula mass $M = 231$ atomic mass units, and a volume $V = 817$ Å^3 (based on an orthogonal cell whose edges are 8.21, 18.42, and 5.40 Å). The result is

$$Z = \frac{1.94 \times 817 \times 10^{-24}}{231 \times 1.660 \times 10^{-24}} = 4.13.$$

This value, which is about 3 percent higher than the integer 4, obviously indicates that there are 4 formula units of $2(C_2H_5NO_2) \cdot HBr$ per cell, but suggests that the measurements of density or cell edges, or both, are too high. In spite of this error it is apparent that the number of formula units per cell, as calculated by relation (4), is consistent with the multiplicity requirements of the space group, namely, that every atom must be present in multiples of 4.

When the space group has special positions in addition to the general position, there is more latitude to the numbers of each atom species which can be present in the cell. In the example just cited, the 4 Br atoms were required to be present in the general position of $P\,2_1\,2_1\,2_1$. Had the space group been $P\,2_1\,2_1\,2$, which has one 4-fold and two 2-fold positions, the 4 Br atoms could have been located again in the one 4-fold (general) position, but alternatively they could have been accommodated by occupying both of the 2-fold positions. But in neither space group could any odd number of Br atoms be accommodated.

For space groups with many special positions there are fewer restrictions. For example, $FeSb_2$ has symmetry $P\,2_1/n\,2_1/n\,2/m$, and use of relation (4) proves that there are 2 $FeSb_2$ per cell. This space group has one 8-fold, three 4-fold, and four 2-fold equipoints, these last all without degrees of freedom. The 2 Fe atoms must occupy one of the 2-fold sets. The 4 Sb atoms must be placed either in two of the three remaining 2-fold equipoints or in one of the three 4-fold equipoints, but cannot occupy the general, 8-fold equipoint.

In very symmetrical crystals having only a few atoms per cell, equipoint considerations alone may lead to a solution of the crystal structure, or to a small number of possible solutions. For example, if an isometric crystal whose composition is type AB has only four formula units per cell, its structure is fixed as either the NaCl type or the ZnS type. The first has symmetry $F\,4/m\,\bar{3}\,2/m$, the other $F\,\bar{4}\,3\,m$, so that the choice depends fundamentally on the symmetry. A decision between the two possibilities can also be made by comparing x-ray diffraction intensities expected from the two structure types with those actually observed.

Applications to crystals containing molecules or known groups

Molecular symmetry. The discussion just given was based upon consideration of the roles of the individual atoms in a crystal structure. This is usually necessary or desirable in the case of a crystal of inorganic composition and unknown structure. In most organic crystals, however, the building units are not the individual atoms but rather sets of atoms already organized into molecules. In such cases the whole chemical molecule must fit the equipoint restrictions as to both number and symmetry.

It is commonly, but not always, true that all the molecules in the crystal play the same role. When this is so, then all the molecules occupy the same equipoint; that is, they are *symmetrically equivalent* (or more briefly, simply *equivalent*). This implies that they are all related to each other by the various symmetry operations (and translations) of the space group. Of course this can be true only if the molecules occupy one equipoint, whether the general one or a special one. If the molecule occupies the general position, it need not have any symmetry, because the general position has symmetry 1. But if the molecule is on a special position, it must have (at least) the symmetry of that special position.† Accordingly, if the space group and the equipoint occupied by the molecule are known, the minimum symmetry which the molecule may have is determined. This information may be useful for a chemist who is trying to decide which, of several that a molecule might conceivably have, is the true configuration.

A molecule having a specific chemical symmetry often occupies an equipoint having less symmetry. Some simple examples are hexachlorobenzene, C_6Cl_6, whose chemical symmetry is at least $\overline{6}\ 2\ m$ (possibly as high as $6/m\ m\ m$), but whose crystal symmetry is only $\overline{1}$; naphthalene, $C_{10}H_8$, whose chemical symmetry is $2/m\ 2/m\ 2/m$, but whose crystal symmetry is only $\overline{1}$; and melamine, $C_3N_3(NH_2)_3$, whose chemical symmetry is at least 3, but whose crystal symmetry is only 1. In these cases the decrease in symmetry of the molecule when in a crystal structure depends on the lower symmetry of the packing and bonding.

On the other hand, there are rare instances in which a molecule occupies a position in a crystal which has a higher symmetry than its chemical symmetry. In all these instances the molecule is found to be rotating in the crystal and thus acquires the higher statistical symmetry of the rotating molecule. The compounds HCl, HBr, and HI each have a (relatively) high-temperature crystalline modification in which the space group is $F\ 4/m\ \overline{3}\ 2/m$, with 4 molecules per cell. This small number of molecules per cell requires each molecule to occupy an equipoint having symmetry $4/m\ \overline{3}\ 2/m$, and this high-symmetry location for a rather unsymmetrical molecule is interpreted as evidence that the symmetry is statistical and due to rotation of the molecule.

Coordination groups. The number and geometrical arrangement of neighbors immediately surrounding an atom are called the *coordination* of that atom. Certain atoms can assume different coordinations under various circumstances. For example, Al is sometimes surrounded by four other atoms (such as oxygen) which lie in the directions of the corners of a tetrahedron; this is called tetrahedral coordination. Alternatively, Al is sometimes surrounded by six other atoms which lie in the direction of the corners of an octahedron; this is called octahedral coordination.

† The symmetries of the several special positions of a space group are listed under each space group in *International tables for x-ray crystallography*, vol. 1.

The information provided by the symmetry of an equipoint is often useful in predicting the coordinations of some atoms in crystals. When an atom and its surrounding coordinating atoms are considered, the entire group must conform to the symmetry of the equipoint on which it is placed. The symmetry of the octahedron is $4/m\,\bar{3}\,2/m$. It can be located on any equipoint having this symmetry or that of one of its subgroups. The tetrahedron has an inferior symmetry, specifically, $\bar{4}\,3\,m$, and it can occupy any position having this symmetry or that of a subgroup. It *cannot*, however, occupy an inversion center, because $\bar{1}$ is not a subgroup of $\bar{4}\,3\,m$. In the case of $FeSb_2$, mentioned in a foregoing section, the 2 Fe atoms per cell were required to occupy one of the four sets of 2-fold equipoints of space group $P\,2_1/n\,2_1/n\,2/m$. These are all on inversion centers, symmetry $\bar{1}$, so that the Fe atoms cannot have tetrahedral coordination. When the structure was solved, it was, in fact, found that the Fe atoms are surrounded by Sb atoms in octahedral coordination. This was permissible because $\bar{1}$ is a subgroup of $4/m\,\bar{3}\,2/m$.

Summary

The symmetry and cell can be established for any crystal provided a homogeneous fragment of the crystal is available. If the density of the crystalline material is known, the mass of the atoms in the cell can be computed by making use of the fact that the density of the bulk crystalline material and that of the crystal cell are the same. The resulting knowledge of the total chemical mass in each cell may be useful in deciding the general magnitude to assign to the molecular weight of a large molecule. It is also useful as noted below.

If, in addition to the density of the crystalline material, its chemical analysis is available, the analysis can usually be interpreted in terms of chemical composition (even when the interpretation is complicated because of solid solution) by making use of the symmetry requirements of the space group. Like atoms can satisfy symmetry requirements of the crystal only by being present in certain integral numbers which are consistent with that symmetry.

The discussion of the numbers of atoms in a cell is simplified by the use of the concept of *sets of equivalent positions* (sometimes shortened to *equipoints*): Since each space group is inherently a different symmetry, each provides different permissible ways of accommodating points (or atoms) in the cell. The number of such distinct ways is limited for each space group, and each way constitutes a set of equivalent positions, with a characteristic number of points, coordinates, and symmetry.

From a knowledge of the number of atoms (which the computations based upon cell volume, density, and chemical composition have shown must be present in the cell) it is often possible to assign one or more of the species of atoms to specific equipoints. When this is possible, the conclusion is useful in several possible ways. For example, if a small number of heavy atoms can be definitely located, the whole crystal structure can be solved by the heavy-atom method (discussed in Chapter 12).

In the event that the crystal is composed of molecules, the molecules themselves must be accommodated on the available equipoints. Since there are ordinarily only a few molecules per cell, they usually occupy only one or, at most, a few equipoints; therefore the minimum symmetry the molecule can have is often limited or fixed. This symmetry information is frequently sufficient to permit the chemist to make a decision between alternative possible chemical linkages in the molecule.

Notes on history

The determination of the number of chemical formula units in a cell has been a common part of the procedure in investigating a crystal structure since the earliest days. Indeed, this was introduced by the Braggs themselves, and was described in 1915 in their book *X-rays and crystal structure.*

The earliest crystal-structure investigations did not make specific use of space groups and consequently made no use of the multiplicities and symmetries of special positions. These and other properties of the 230 space groups were drawn to the attention of crystallographers by Niggli, who tabulated the special positions of all space groups and their multiplicities in his 1919 book *Geometrische Kristallographie des Diskontinuums.* The coordinates and multiplicities of all equipoints of all space groups were also tabulated by Wyckoff in 1922, while the symmetries of these positions were emphasized by Astbury and Yardley in their extensive 1924 paper published in the Proceedings of the Royal Society of London.

Additional reading

Martin J. Buerger. *Elementary crystallography.* (Wiley, New York, 1956, 1963) 460–470.

Martin J. Buerger. *Crystal-structure analysis.* (Wiley, New York, 1960) 242–258.

Significant literature

Calculation of number of formula units per cell

W. H. Bragg and W. L. Bragg. *X-rays and crystal structure.* (G. Bell, London, 1915) 110–111, 114, 164.

Properties of equivalent positions

Paul Niggli. *Geometrische Kristallographie des Diskontinuums.* (Gebrüder Borntraeger, Leipzig, 1919).

Ralph W. G. Wyckoff. *The analytical expression of the results of the theory of space-groups.* (Carnegie Institution, Washington, 1922) 180 pages; 2d edition (1930) 239 pages.

W. T. Astbury and Kathleen Yardley. *Tabulated data for the examination of the 230 space-groups by homogeneous x-rays.* Proc. Roy. Soc. **224** (1924) 221–257.

C. Hermann. *International tables for crystal structure determination,* vol. 1. (Gebrüder Borntraeger, Berlin, 1935) 84–377.

Norman F. M. Henry and Kathleen Lonsdale. *International tables for x-ray crystallography,* vol. 1. (Kynoch Press, Birmingham, 1952) 74–352.

11

Investigation of the arrangement of atoms in the cell

Introduction
Diffraction amplitudes
    Amplitude scattered by an atom
    Scattering by the cell contents
    Phase of scattering by an atom
    Alternative derivation of the phase
    Condensation due to symmetry
    Temperature factor
Test of the correctness of a crystal structure
Measurement of intensities
    Relation of measurement to $|F|^2$
    Film methods
    Diffractometer methods
Correction of measured intensities
    The Lorentz factor
    The polarization factor
    Absorption
    Extinction
Summary
Notes on history
Additional reading
Significant literature

Introduction

In Chapter 10 it was seen that by combining the data of the cell dimensions and symmetry of the crystal with a knowledge of density and chemical analysis, a limited amount of information about the arrangement of atoms in a crystal can be gained. Except in rare instances, however, the result does not determine the arrangement of atoms in the crystal. This detailed arrangement, known as the *crystal structure* of that crystal, can be deduced, in general, only from a consideration of the diffraction intensities. This conclusion is consistent with the theorem enunciated in Chapter 3; from the point of view of this chapter the theorem may be stated in reversed form, as follows:

Theorem: The cell of a crystal is determined by the geometrical features of the diffraction by the crystal, but the arrangement of atoms in the cell depends on the intensities of the diffraction maxima.

The way the arrangement of atoms affects the diffraction intensities was considered briefly in Chapter 3. The matter is discussed again, below. Later in the chapter some of the ways of measuring the required intensities and of transforming them to diffraction amplitudes are outlined.

Diffraction amplitudes

Amplitude scattered by an atom. The individual electrons of an atom are the loci of scattering of x-rays. Since they are distributed through the volume of the atom, whose linear dimensions are comparable to x-ray wavelengths, the wavelets they scatter are not generally in phase with each other. In the direction of the primary beam, however, all Z electrons of the atom scatter in phase, so that the combined wave has an amplitude whose magnitude is Z times that scattered by one electron. This direction corresponds to a Bragg glancing angle of $\theta = 0$. At larger values of θ the various wavelets interfere with one another, and the amplitude of the resultant scattering, generally designated by the symbol f, is less than Z and declines gradually with $\sin \theta$.

Scattering by the cell contents. Some of the basic features of the diffraction by a crystal can be brought out by considering the contributions of the various atoms in the cell to a Bragg reflection from a stack of axial planes. The geometrical features of this case are illustrated in Fig. 1, which shows a cell with two bounding axial planes in a horizontal attitude. Radiation scattered by matter in these two terminal planes, which are separated by spacing d, differs in phase by 2π for the first-order

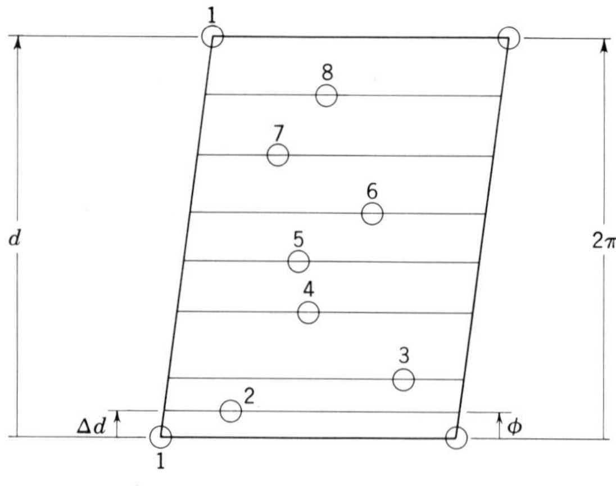

Fig. 1

Bragg reflection, as shown in Chapter 3. If the x-rays reach the upper plane of the cell first, the scattering by that plane is 2π in advance of that scattered by the bottom plane. The phase scattered by matter in any intermediate plane is proportional to the height of that plane above the lower plane and is equal to

$$\phi = \frac{\Delta d}{d} 2\pi. \tag{1}$$

Each atom in the cell scatters a wavelet whose amplitude is f and whose phase is a value of ϕ as determined by (1). As pointed out in Chapter 3, this individual wave can be represented graphically by a vector on the Argand diagram, the vector having a magnitude f and a direction making an angle ϕ with the horizontal axis (Fig. 2). This vector determines a

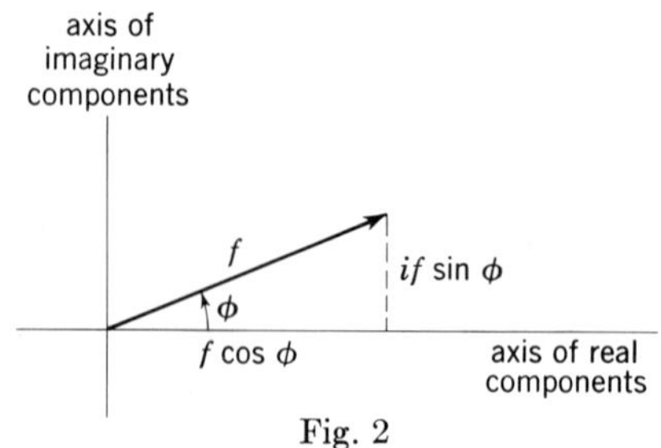

Fig. 2

point in the complex plane known as a complex number. Either the point or the vector to it can be represented in two ways:

(*a*) It can be represented by the vector sum of its two components, one parallel to the real axis, the other parallel to the imaginary axis, namely, $f \cos \phi + if \sin \phi$.

(*b*) It can be represented by the exponential $fe^{i\phi}$.

These representations are identical because of Euler's formula

$$e^{i\phi} = \cos \phi + i \sin \phi. \tag{2}$$

Thus either side of (2) can be taken to represent a unit vector in the complex plane which makes an angle ϕ with the real axis, while

$$fe^{i\phi} = f(\cos \phi + i \sin \phi) \tag{3}$$

can be taken to represent a corresponding vector of magnitude f.

In the application of interest here, this vector embodies the two important parameters of a wave: its amplitude f and its phase ϕ. The result of combining several waves can be found by adding the vectors which represent them. In plotting the several vectors representing the wavelets scattered by the several atoms in the cell it is worthwhile taking them in the order of increasing distance Δd from the lower plane of Fig. 1. Then the phase ϕ is larger for each succeeding atom, and the result is a systematic plot like Fig. 3. Each vector has a length proportional to the scattering power f of the atom it represents, and a direction ϕ fixed by the height of the atom in the cell. This sequence of vectors

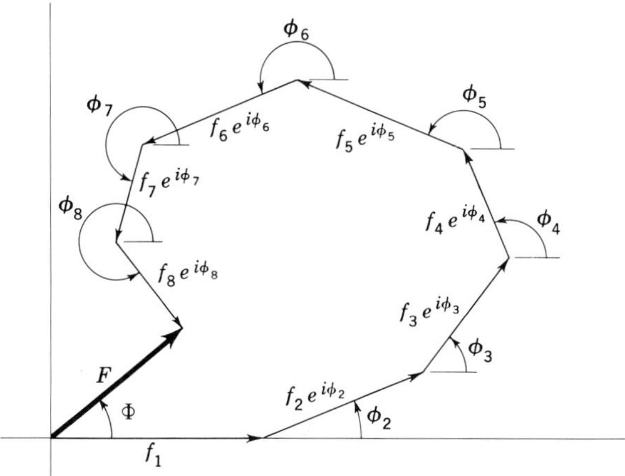

Fig. 3

forms a chain winding counterclockwise. Its resultant, designated F, is a vector representing the amplitude of the composite wave scattered by all the atoms in the cell.

The sum of the vectors, expressed graphically in Fig. 3, can also be expressed analytically by summing terms like those on either the left or the right of (3), one term for each of the N atoms in the cell. If the left-hand side of (3) is used, the summation is written

$$F = \sum_{j=1}^{N} f_j e^{i\phi_j}, \tag{4}$$

whereas if the right-hand side of (3) is used, the summation is written

$$F = \sum_{j=1}^{N} f_j(\cos \phi_j + i \sin \phi_j). \tag{5}$$

Both the graphical representation in Fig. 3 and the presence of the imaginary i on the right of (4) and (5) show that, in general, F is itself a complex quantity. Accordingly, it can also be expressed as a vector sum of its components along the real and imaginary axes, a procedure which is convenient for various computations. Figure 4 illustrates that the expression of F in this way takes the form

$$F = A + iB, \tag{6}$$

where

$$A = |F| \cos \Phi \quad \text{(its real component)}, \tag{7}$$

$$iB = i|F| \sin \Phi \quad \text{(its imaginary component)}, \tag{8}$$

and

$$\tan \Phi = \frac{A}{B}. \tag{9}$$

The diagram (Fig. 4) shows that the magnitude of F is given by

$$|F| = \sqrt{A^2 + B^2}. \tag{10}$$

A simplification of (2) always results if the crystal is centrosymmetrical, provided that the origin is taken at a symmetry center. This requires that the plane for which ϕ is taken as zero contain the symmetry center.

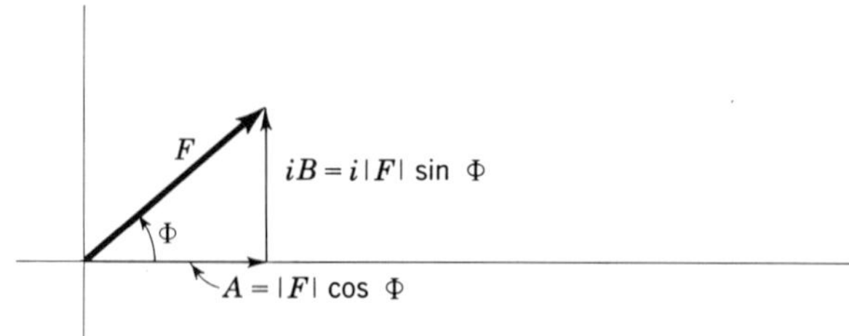

Fig. 4

Under these circumstances the boundaries of the cell are chosen so that the plane which determines the zero phase is located through the cell's center. When this is done, an atom above the plane is always matched by an identical atom below it, and these atoms contribute waves of identical amplitude but opposite phases. The contribution of one such pair is

$$\begin{aligned} fe^{i\phi} + fe^{-i\phi} &= f(e^{i\phi} + e^{-i\phi}) \\ &= f(2\cos\phi). \end{aligned} \tag{11}$$

When the atoms are taken as $N/2$ matched pairs, in this way, the complex aspect of (4) and (5) disappears, and the resultant of the summation becomes a real number, specifically,

$$(4),\ (5)\rightarrow \qquad F = 2\sum_{j=1}^{N/2} f_j \cos\phi_j. \tag{12}$$

The graphical representation of this simplification is seen in Fig. 5. Each member of a matched pair is part of one of two chains of vectors, one of which winds clockwise, the other counterclockwise. Since the two sequences appear as mirror images across the real axis, their resultant must lie along the real axis.

The conclusions just reached were based upon a Bragg reflection from atoms in a series of planes parallel to one of the faces of the cell. This situation appears to be a special case; actually, all the results derived for it are generally valid for the following reason: Any crystal can be referred

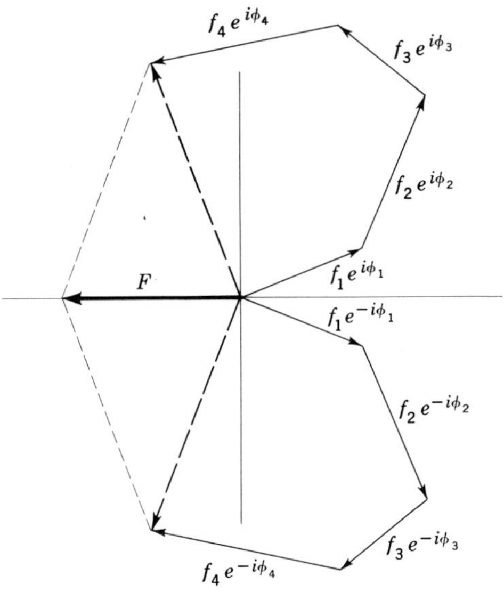

Fig. 5

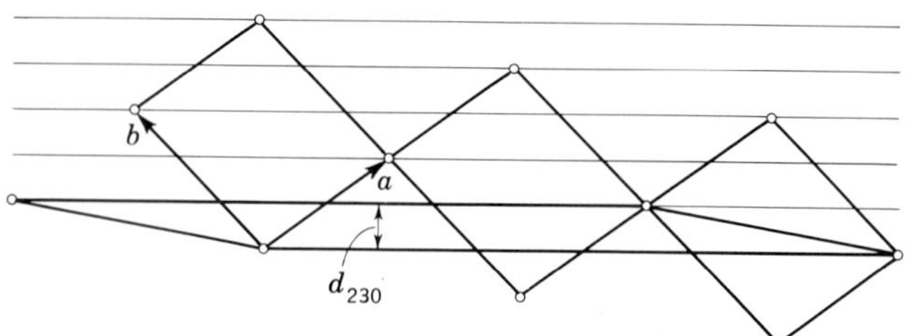

Fig. 6. The lattice points shown would normally be referred to the cell determined by translations a and b. An alternative primitive cell can be chosen which is bounded by neighboring planes having any index, such as the planes (230) shown here.

to a cell, two of whose parallel enclosing faces are neighboring planes of any stack of the lattice. An example, illustrated in two dimensions, is shown in Fig. 6. For the lattice points shown, the obvious choice of cell is the nearly rectangular one outlined there, but a legitimate alternative cell is the very thin, very oblique one whose upper and lower surfaces are the neighboring horizontal planes of the stack in question. Any relation derived for the oblique cell can be stated for the nearly rectangular cell by making use of a transformation of axes from the oblique cell to the nearly rectangular one. In such a transformation only the specific values of the reflection indices are involved for the relations already derived. The general conclusions are not affected by these specific values.

Phase of scattering by an atom. To derive the values of the phases of the waves scattered by the atoms it is, perhaps, most direct to return to the conventional cell, shown in more detail in Fig. 7. The phase shift caused by the displacement of an atom by a shift of coordinates of amount $x = X/a$ can be calibrated by the total phase shift caused by its displacement so that it lies in a plane containing the end of vector **a**. In Fig. 7 this is the phase shift corresponding to two spacings of magnitude d, or more generally, to h spacings. The following proportion can therefore be written:

$$\frac{X}{a} = \frac{x/a}{a} = \frac{x}{1} = \frac{\phi_x}{h \cdot 2\pi}, \tag{13}$$

so that

$$\phi_x = 2\pi hx. \tag{14}$$

Similarly, it can be shown that

$$\phi_y = 2\pi ky \tag{15}$$

and

$$\phi_z = 2\pi lz. \tag{16}$$

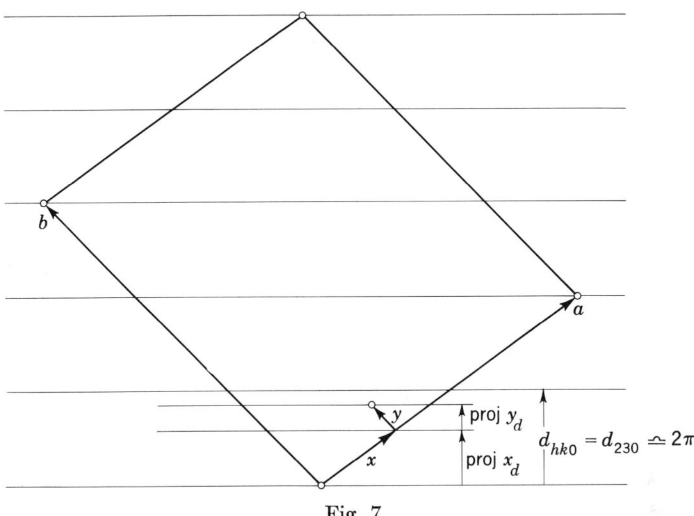

Fig. 7

The total phase shift caused by displacing an atom from the origin to a point in the cell having fractional coordinates xyz is the sum of the separate phase shifts of (14) to (16), namely,

$$\begin{aligned} \phi &= \phi_x + \phi_y + \phi_z \\ &= 2\pi(hx + ky + lz). \end{aligned} \tag{17}$$

The wave contributed by this atom is represented, therefore, by substituting this in the left or right of (3) to give

$$fe^{i2\pi(hx+ky+lz)} = f[\cos 2\pi(hx + ky + lz) + i \sin 2\pi(hx + ky + lz)]. \tag{18}$$

The final forms of the amplitude of the wave scattered by the cell, from (4) and (5), are therefore

$$F_{hkl} = \sum_{j=1}^{N} f_j e^{i2\pi(hx_j+ky_j+lz_j)} \tag{19}$$

and

$$F_{hkl} = \sum_{j=1}^{N} f_j[\cos 2\pi(hx_j + ky_j + lz_j) + i \sin 2\pi(hx_j + ky_j + lz_j)]. \tag{20}$$

In the last form the sine terms vanish for centrosymmetrical crystals, provided that the origin of coordinates is taken at a symmetry center.

The resultant F_{hkl} in (19) and (20) is known as the *structure factor* of the crystal. This quantity, which is generally a complex number, represents the amplitude of the wave scattered in the orders hkl by the contents of the crystal cell.

Alternative derivation of the phase. The derivation of the phase of scattering by an atom can be reformulated so that it becomes part of the generalization of reciprocal space, which will be discussed in Chapter 12. It was noted in (1) that the phase scattered by a point in a plane at a level Δd above the plane whose scattering phase is taken as zero is proportional to $\Delta d/d$. For present purposes it is convenient to rewrite (1) as

$$\frac{\phi}{2\pi} = \Delta d \, \frac{1}{d}. \tag{21}$$

This relation can be expressed in another, more elegant, way by making appropriate use of vectors. With the aid of relation (20) of Chapter 4, the term $1/d_{hkl}$ can be recognized as the length of a vector $\mathbf{t}^*_{hkl}$ which extends from the origin to the reciprocal-lattice point hkl. This vector has the same direction as the spacing vector $\mathbf{d}_{hkl}$. The other important vector in the problem is the one which extends from the origin O to the scattering point P (Fig. 8) and which can be labeled $\mathbf{r}_{xyz}$. So long as the end of $\mathbf{r}$ terminates at a level Δd above the origin, it determines the same relative phase $\phi/2\pi$ as in (21). This condition can be described by saying that the projection of $\mathbf{r}$ on the direction of vector $\mathbf{d}$ (or $\mathbf{t}^*$) is the length Δd. A relation equivalent to (21), but expressed in terms of vectors $\mathbf{r}$ and $\mathbf{t}^*$, can be written as follows:

$$\frac{\phi}{2\pi} = |\mathbf{r}_{\text{proj on }\mathbf{t}^*}|\,|\mathbf{t}^*| \tag{22}$$

$$= |\mathbf{r}|\;|\mathbf{t}^*| \cos(\angle \mathbf{r}\mathbf{t}^*). \tag{23}$$

In vector algebra the terms on the right of (23) are called the scalar product of the vectors $\mathbf{r}$ and $\mathbf{t}$; in the shorthand of vector algebra this is written as a *dot product* $\mathbf{r} \cdot \mathbf{t}$. In the notation of vector algebra,

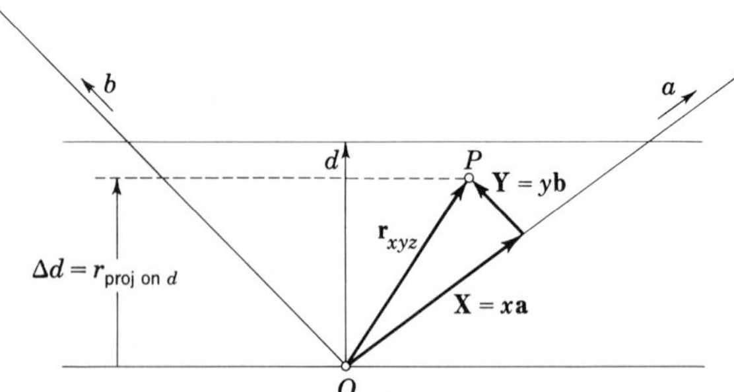

Fig. 8. Part of the lower portion of Fig. 7 shown in more detail.

therefore, (22) and (23) are expressed simply by

$$\frac{\phi}{2\pi} = \mathbf{r} \cdot \mathbf{t}^*. \tag{24}$$

To be able to compute the phase with the aid of a relation like (24) it is convenient to express both vectors in terms of their natural components. The vector $\mathbf{r}$ is the sum of its components along the crystal axes a, b, and c:

$$\mathbf{r} = x\mathbf{a} + y\mathbf{b} + z\mathbf{c}. \tag{25}$$

The vector $\mathbf{t}^*$, on the other hand, is easily expressed as the sum of its components along the axes of the reciprocal cell:

$$\mathbf{t}^*_{hkl} = h\mathbf{a}^* + k\mathbf{b}^* + l\mathbf{c}^*. \tag{26}$$

If the component forms in (25) and (26) are substituted into (24), it becomes

$$\frac{\phi}{2\pi} = (x\mathbf{a} + y\mathbf{b} + z\mathbf{c}) \cdot (h\mathbf{a}^* + k\mathbf{b}^* + l\mathbf{c}^*). \tag{27}$$

The scalar product on the right can be readily simplified by multiplying out the individual terms and making use of two basic properties of sets of reciprocal vectors. The first property was given in (22) of Chapter 4, specifically, that

$$\begin{aligned} \mathbf{a} \cdot \mathbf{a}^* &= 1, \\ \mathbf{b} \cdot \mathbf{b}^* &= 1, \\ \mathbf{c} \cdot \mathbf{c}^* &= 1. \end{aligned}$$

The second property, which arises because pairs of reciprocal axes represented by different letters are perpendicular to each other (and hence have zero projection on each other), is that dot products of such pairs vanish:

$$\begin{aligned} \mathbf{a} \cdot \mathbf{b}^* &= \mathbf{a} \cdot \mathbf{c}^* = 0, \\ \mathbf{b} \cdot \mathbf{c}^* &= \mathbf{b} \cdot \mathbf{a}^* = 0, \\ \mathbf{c} \cdot \mathbf{a}^* &= \mathbf{c} \cdot \mathbf{b}^* = 0. \end{aligned}$$

With the aid of these properties of sets of reciprocal axes, (27) reduces to

$$\frac{\phi}{2\pi} = hx + ky + lz. \tag{28}$$

This result is the same as that in (17).

The derivation of the phase makes use of a scalar product of a vector in crystal space with a vector in reciprocal space, as specifically expressed in (24). This is a general feature of a Fourier transformation, discussed in Chapter 12.

Condensation due to symmetry. It has been seen that the structure factor becomes simpler, and the sum need be computed for only half the atoms in the cell, when the crystal has a center of symmetry. A condensation of terms can also be effected if the crystal has any other symmetry. Every symmetry element of the cell serves to relate symmetrical sets of atoms, as a consequence of which the scattering powers of all atoms of the set are the same and their coordinates are related. The condensation reduces the number of atoms N over which the final summation must be made.

Temperature factor. The discussion of amplitude given so far has tacitly assumed that the crystal is composed of a pattern of static atoms. Actually the thermal energies of these atoms require them to vibrate about mean positions. The general effect of this is to reduce somewhat the amplitudes of all reflections.

The general manner in which this comes about can be understood in the following way: Consider a large set of atoms in a crystal which lie in some rational plane (hkl) because they are related to one another by the lattice translations in that plane. If no motions of the atoms occurred, all atoms would contribute in phase to the reflection hkl. But the thermal agitation disturbs this perfect translation equivalence, for the atoms each execute a vibrational motion about their ideal mean positions. At any instant of time each atom occupies a position which departs from its mean position by some small displacement. At that instant the atoms of the collection do not occupy exactly the same plane, and so the amplitudes of the wavelets they scatter are not quite in phase. Accordingly, the resultant amplitude they contribute to reflection hkl is not the simple arithmetic sum of all their f's, so that a correction which reduces this ideal amplitude must be applied to such relations as (4).

This correction is a function of the mean-square displacement $\langle u^2 \rangle$ of the atoms from the plane in question, and has the general form e^{-M}. The exponential M is related to the mean-square displacement $\langle u_H^2 \rangle$ normal to the plane (hkl), which is abbreviated H in the subscript, in the following way:

$$M = (8\pi^2 \langle u_H^2 \rangle) \frac{\sin^2 \theta}{\lambda^2}. \tag{29}$$

It is customary to bulk together the terms in the parentheses as a temperature coefficient B_H, so that the correction of an amplitude reflected by a plane H is usually written

$$e^{-M} = e^{-B_H (\sin^2 \theta)/\lambda^2}. \tag{30}$$

The dimensions of B are in square Ångströms since, according to (29), it is proportional to the mean-square displacement of the atom. Not only does its value vary from one kind of atom to another, but even for a particular atom in the structure it varies with the direction of the normal to the plane whose reflection amplitude is being corrected. The variation must, of course, conform to the symmetry of the equivalent position of the atom in the cell. The various values of B tend to lie in the general neighborhood of 0.5 Å^2 for many common inorganic crystals at room temperature. In the initial stages of computing the amplitudes to be expected from a model, values of B which had been found in related crystal structures are usually used. More exact values of B for each atom, and its variation with direction, are ordinarily determined as a result of a least-squares refinement of the parameters of the crystal structure, as discussed in Chapter 14.

Test of the correctness of a crystal structure

For a particular arrangement of atoms, such as suggested by Fig. 1, there is a definite resultant amplitude F_{hkl} corresponding to each x-ray reflection *hkl*. The collection of all such resultants is characteristic of a particular arrangement of atoms. Indeed, as will be apparent from the discussion in the next chapter, to every arrangement of atoms there is a unique set of F_{hkl}'s (each F has not only a phase component but also a magnitude component); and reciprocally, to every set of F_{hkl}'s there is a unique pattern of atoms. It is this relation, of course, which offers the possibility of discovering the arrangement of atoms from the diffraction effects it produces.

There are a number of ways of deducing the arrangement of atoms from its diffraction effects; the ones most commonly used are outlined in the next two chapters. When some pattern of atoms has been deduced by one of these methods, it is evident that if the deduced pattern (often called the *model*, or the *proposed structure*) is correct, the set of F_{hkl}'s calculated for it with the aid of (20), for every value of *hkl*, should match the set of $|F_{hkl}|$'s experimentally observed, within the precision of the measurements of the $|F_{hkl}|$'s. The calculation yields both the magnitudes of the F's (10) and their phases (9) for the model, but the only experimental observables are the magnitudes $|F_{hkl}|$. Thus, only the magnitudes of the F_{hkl}'s can be examined for suitable matching. These are ordinarily designated

$|F_{\text{obs}}|$: the magnitude of an amplitude observed experimentally,
$|F_{\text{cal}}|$: the magnitude of an amplitude calculated for the model.

If the model is correct, the set $|F_{\text{cal}}|$ for every *hkl* should match the set $|F_{\text{obs}}|$ for corresponding diffraction indices *hkl*.

Of course, the observed value of any $|F_{hkl}|$ depends on the precision with which it can be measured, and also on how well it can be corrected for various experimental effects, the most important of which are discussed later in this chapter. On the other hand, the calculated value of the $|F_{hkl}|$ depends on how well the theory upon which its calculation is based represents the true state of affairs (for example, on how well the thermal agitation has been taken into account). For any particular reflection *hkl*, a measure of the failure of $|F_{\text{cal}}|$ to match $|F_{\text{obs}}|$ is the ratio of the difference of these to the measured value, neglecting the sign of the difference:

$$\frac{|F_{\text{obs}}| - |F_{\text{cal}}|}{|F_{\text{obs}}|}.$$

It is common practice to compute, for any model, a fraction like this, but with the numerator and denominator summed over all reflection indices, namely,

$$R = \frac{\Sigma \big| |F_{\text{obs}}| - |F_{\text{cal}}| \big|}{\Sigma |F_{\text{obs}}|}. \tag{31}$$

The value of R is expressed either as a decimal fraction or as a percentage. This provides a measure of the mean *discrepancy*, or residual,† which exists between the match of the set of $|F_{\text{cal}}|$'s and $|F_{\text{obs}}|$'s for the model.

It is interesting to consider what values R is likely to assume. A. J. C. Wilson showed that, for structures composed of atoms with equal scattering powers, the most probable value of R for a completely wrong structure proposal is 82.8 percent for a centrosymmetrical crystal, or 58.6 percent for a noncentrosymmetrical one. For structures composed of atoms with equal scattering powers, a model with R in the neighborhood of 45 percent can be regarded as having an arrangement of atoms which is at least partly correct and worth attempting to refine; if R falls below 20 percent, the structure is probably correct. If good intensities are available, a correct model can be refined to a value of R in the range of 5 to 10 percent and even lower.

On the other hand, the meaning of R should be appraised with due consideration to the nature of the structure and the nature of the intensity measurements. For example, if the structure is composed of just heavy atoms and light atoms, such as the heavy-metal oxides, carbides, nitrides, or borides, a low value of R simply means that the heavy atoms are probably located correctly with respect to each other. In the case of a complicated structure with many atoms per cell, many weak x-ray

† The symbol R in (31) does *not* stand for "reliability," and it is not a measure of the reliability of the model, but rather of its unreliability. The symbol can be interpreted as a fractional "residue," and carries the significance of "disagreement" or "discrepancy."

reflections can be expected, and some of these will be too small to be measurable. If these are omitted in the calculation of R, the many errors in these reflections are omitted from the summation, and the resulting value is usually considerably too low. On the other hand, if these are included as if they had zero intensity, many abnormally large errors for these reflections are contributed to the summation, and so the resulting value of R is always too high, perhaps double what it would be if these were omitted from the summation.

The value of R is often used as a measure of the lack of complete adjustment of a model to the experimental observables. The crystal-structure analyst accordingly attempts to modify a model so as to reduce the value of R as much as possible, for example, by adjusting the coordinates of the atoms or adjusting the factors which take account of their thermal vibrations. It is often found, however, that for a particular model, R cannot be reduced below some intermediate value, say 25 or 30 percent. This can usually be taken to mean that some fairly large aspect of the structure is correct, but that it is, perhaps, misplaced in the cell or that its relation to another part of the structure is incorrect. The reason that the R value is not very large in such cases is that, for most reflections, the correct structure of the fragment dominates the Argand diagrams, such as Fig. 3, yet the perfection of the matches is spoiled because of the incorrect relation of this fragment to the rest of the structure.

Measurement of intensities

X-ray diffraction intensities are measured in one of two ways: measuring or estimating the blackenings of spots on a photographic film, or receiving the diffracted beams in a photon detector and counting the photons in the diffracted beams. The second way, which is becoming common, makes use of an instrument known as a *diffractometer*.

Relation of measurement to $|F|^2$. Both methods require characteristic radiation accompanied by as little general (polychromatic) radiation as can be arranged. Under these circumstances the crystal orientation with respect to the primary x-ray beam must be varied in order to produce Bragg reflections. As the crystal thus passes through the condition of reflection, an elementary volume element diffracts radiation whose energy in a single pass is given by

$$E_{hkl} = KLp\,|F_{hkl}|^2, \tag{32}$$

where L and p are the Lorentz and polarization factors, to be discussed later, and K is a constant made up of the universal constants e, m, and c, as well as other constants of an experimental nature, as follows:

$$K = \frac{e^4}{m^2c^4}\left[\frac{I_0\lambda^3N^2\Delta V}{\omega}\right]. \tag{33}$$

Here I_0 is the intensity of the primary x-ray beam reaching the crystal, λ its wavelength, N the number of cells per unit volume, ΔV the volume of the crystal (assumed to be so small that it does not give rise to appreciable absorption or extinction, as discussed later), and ω the rate of rotation of the crystal.

Another formulation of the relation between F and E is

$$\begin{aligned}\frac{E\omega}{I_0} &= \frac{e^4}{m^2c^4}[\lambda^3N^2\Delta V]Lp|F|^2 \\ &= K'Lp|F|^2.\end{aligned} \tag{34}$$

The quantity on the left is known as the *integrated reflection*. Its numerator has the dimensions of energy · radians/time = power · radians and represents the power of the reflection integrated over the angular range in which the reflection occurs. This is independent of the gross imperfections of the crystal. The ratio $E\omega$ to I_0, on the left of (34), is therefore a characteristic of a particular reflection hkl and is directly proportional to $|F_{hkl}|^2$.

Film methods. Although the earliest x-ray intensity measurements were made with the aid of what is now known as a diffractometer, the detector then in use proved unreliable, so that for many years thereafter measurements were made with the aid of photographic films. Such methods suffer from many complications, but the general features are as follows:

Under favorable conditions, the optical density on the developed film is directly proportional to the exposure. The density can be measured readily, so that the use of film methods appears to be simple and easy. They have, however, two inherent disadvantages:

In the first place, an x-ray reflection takes place over an angular range, varies from zero to a peak intensity, and then declines again to zero. This records on the film as a spot of variable blackness. The peak intensity could be readily measured by measuring the density of the blackest part of the spot. But this is not what is wanted; rather, according to (34) and the subsequent discussion, the energy integrated over the range of the reflection is required. Unfortunately, there is no simple or easy way of measuring the corresponding integrated density of the spot. It is common practice, therefore, to estimate or measure the peak density and to assume that the result is proportional to the integrated reflection.

To avoid this problem, an integrating cassette can be used to obtain more accurate integrated reflections by film methods. The theory of this device is illustrated in Fig. 9. The upper part shows how the power received by the film varies with the position on the film. If this distribution curve is duplicated at small regular intervals, as seen in the middle part of Fig. 9, the sum of all the curves, which is shown in the lower part of Fig. 9, has a plateau whose height is equal to the sum of the vertical lines in the upper part of Fig. 9. This geometry can be arranged in recording the x-ray spots on the film of a Weissenberg or a precession

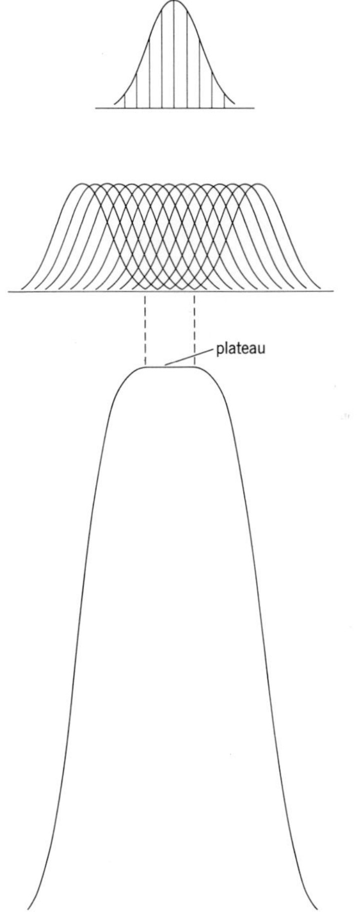

Fig. 9. [*From Martin J. Buerger: Crystal-structure analysis.* (*Wiley, New York,* 1960) 104.]

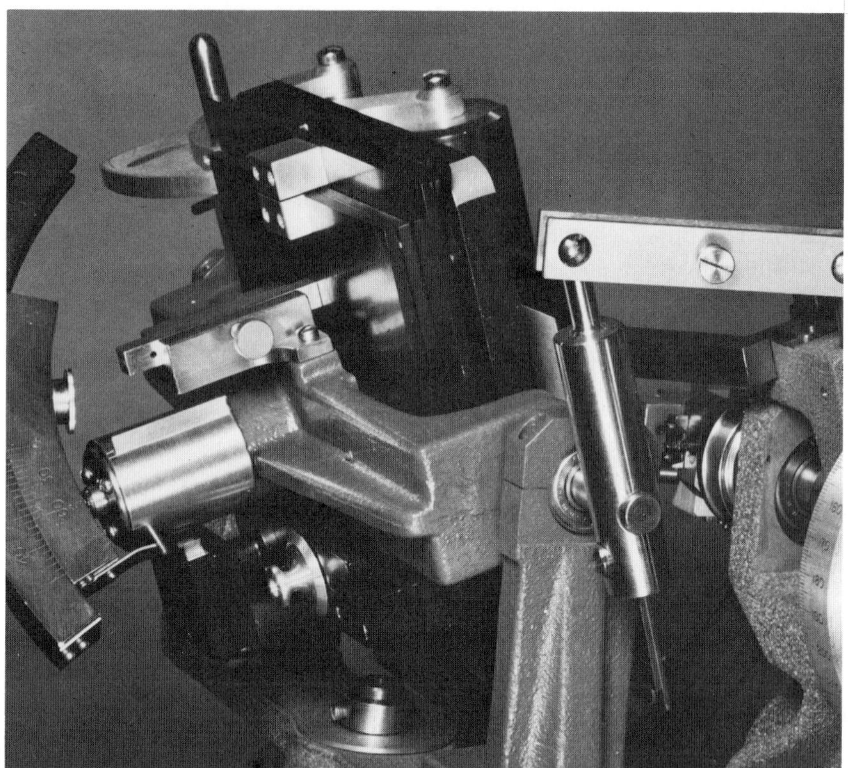

Fig. 10. Integrating mechanism for the precession instrument. [*The Charles Supper Company.*]

apparatus by causing the cassette to be displaced regularly and periodically to different positions in two directions. The mechanism for the precession instrument is shown in Fig. 10, and a photograph made with it is seen in Fig. 11. In such a photograph the densities of the plateaus are proportional to the integrated reflections of the spots.

Another difficulty with film methods is that a photographic film can record only a limited range of densities. Furthermore, the linear relationship between density and exposure deteriorates above density values of about 1.4. To extend the range of $|F|^2$'s which can be measured, differently timed exposures of the same photograph can be made. By this means the strong reflections can be measured on films which have had short exposures, and the weak reflections on those which have had long exposures. This is usually accomplished systematically by a scheme known as the *multiple-film method,* in which several films (usually three to six) are placed in the same cassette and exposed at once. The films

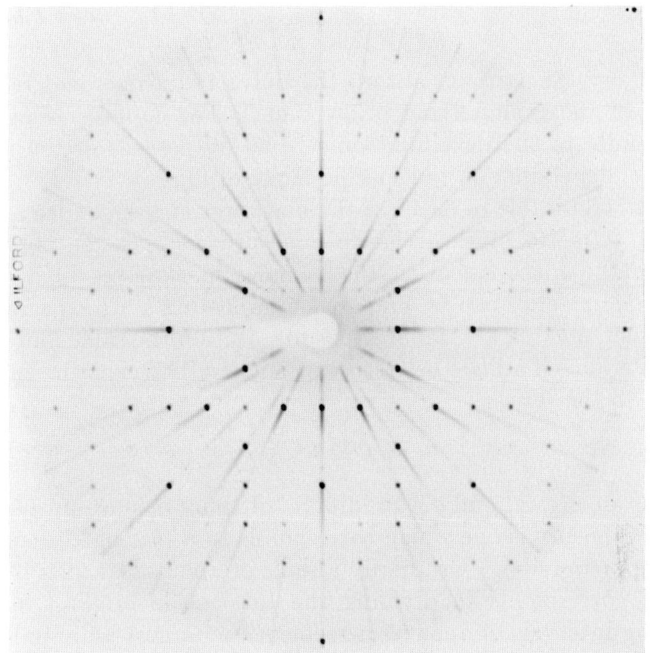

A

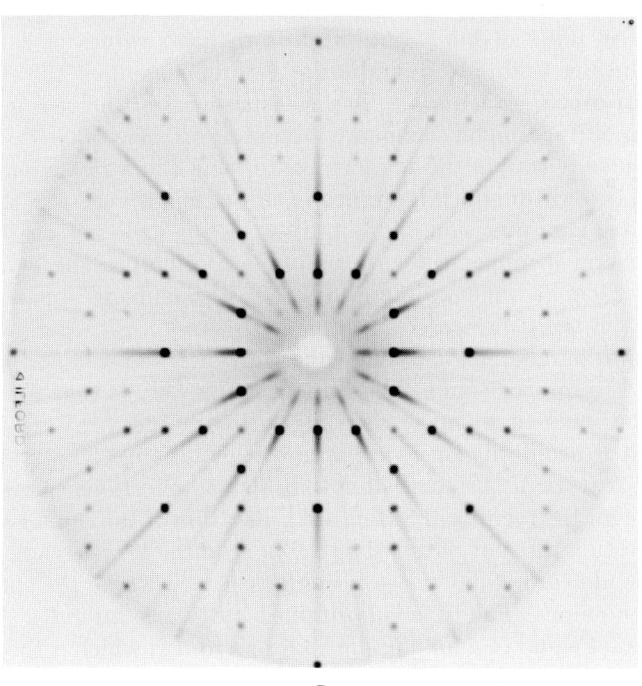

B

Fig. 11. *A*: Ordinary precession photograph (garnet, *a* axis, zero level, $\mu = 30°$; Mo$K\alpha$, 35 kV, 17 mA, 8 hr). *B*: Precession photograph made with eight cycles of the integrating mechanism shown in Fig. 10 (same data as *A*, except 13 mA, 19 hr). [*From Martin J. Buerger: The precession method in x-ray crystallography.* (*Wiley, New York,* 1964) 189.]

nearer the crystal partially absorb the diffracted x-rays and reduce the exposures of those films lying below them. For copper radiation each film transmits to the next film some 25 to 50 percent of the radiation reaching it, depending on the specific make of film.

While it is desirable to measure the blackness of a spot with a densitometer, an estimate of the relative exposure can also be obtained by comparing the blackness with the varying blacknesses in a standard series. Such a series can be prepared by selecting a strong x-ray reflection and then making a series of exposures of it on the same film but with the cassette moved a little between exposures. The exposures are timed according to the series

$$a,\ at,\ at^2,\ at^3,\ \ldots$$

(where a is a constant and t is an interval of time) in such a way that the minimum exposure is one which corresponds to a just barely perceptible blackening of the film. Ordinarily a scale of about 20 values is convenient; this corresponds to dividing the total scale of $|F_{obs}|$ into parts differing by intervals of about 5 percent of the largest amplitude.

One of the disadvantages of film methods is that the threshold level, below which an intensity cannot be measured, is relatively high. As a result, many $|F|^2$'s of low values appear to have values of zero. The actual values of these are desirable for the refinement of the structure.

Diffractometer methods. The measurement of diffraction intensities with a diffractometer has many advantages. Basically, the method is direct since it counts the photons comprising the x-ray intensity. It is capable of recording a large range of $|F|^2$'s, has a very low threshold, and provides high precision.

Two general arrangements are in use for orienting the crystal and detector. One has the geometry of the Weissenberg apparatus; the other, that of the Eulerian cradle.

In the diffractometer with Weissenberg geometry (Fig. 12), the crystal is arranged to rotate, as in a Weissenberg apparatus, so that the rotation axis can be inclined to make any angle $\bar{\mu}$ with the x-ray beam. The instrument is ordinarily operated at equi-inclination (Chapter 8). The important feature of this method is that all the reflections corresponding to one level of the reciprocal lattice are generators of the same Laue cone. These reflections are not recorded on a film as in the Weissenberg method, but enter a detector tube whose opening is placed where the spot would occur on the film. The tube is positioned by two coordinates (Fig. 12): ν, which is set equal to μ for equi-inclination,† and Υ, which is the position of the generator of the Laue cone and corresponds to the distance of a spot above the centerline on the Weissenberg film. For a specific

† The complement of ν, designated $\bar{\nu}$, is the half-opening angle of the Laue cone.

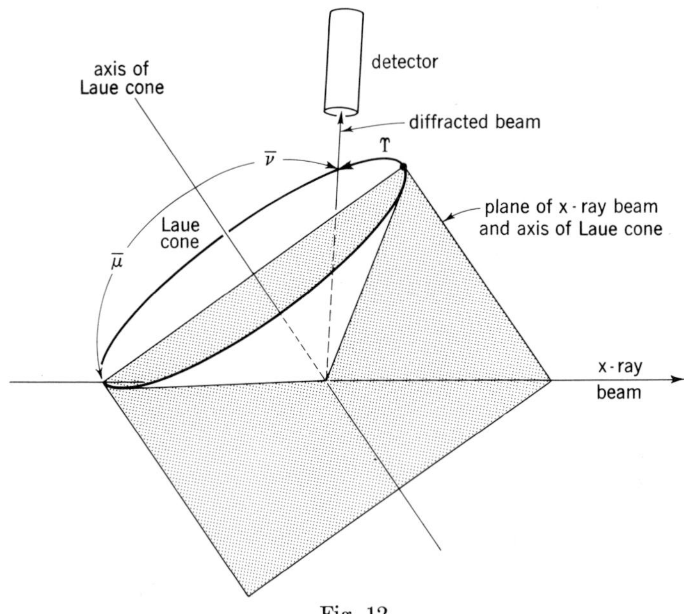

Fig. 12

reflection to occur, the crystal must be turned about its rotation axis to a setting ω, which corresponds with the horizontal coordinate of a spot on a Weissenberg film.

In apparatus based upon the Eulerian cradle (Fig. 13), the crystal is adjusted by means of the three Eulerian angles ϕ, χ, and ω until it is oriented so as to cause a desired reflection *hkl* to occur in a horizontal plane. The detector tube, which can only be adjusted so as to point in various directions in this plane, is then set at the deviation angle $2\theta_{hkl}$ and so receives the desired reflection.

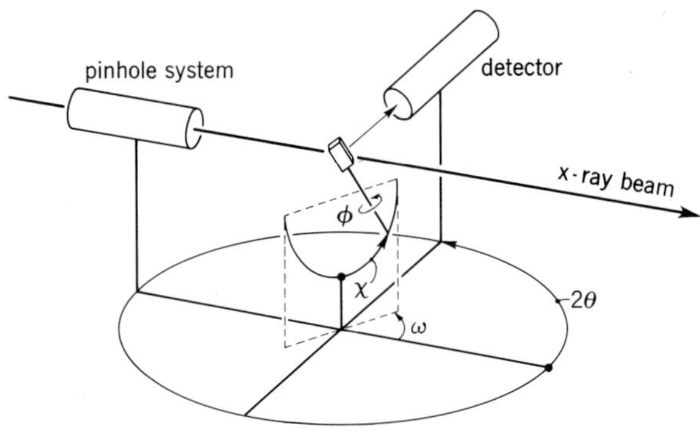

Fig. 13

Instruments based upon both of these geometrical arrangements are available commercially. Either may be set and operated manually, and both are also available in automated models. The type with Weissenberg geometry is simpler to operate, both manually and by automation, because only two settings need be made to record all reflections on the same reciprocal-lattice level: the crystal position with coordinate ω, and the detector position with coordinate Υ. When an entire reciprocal-lattice level of reflections has been measured, then the μ settings of both rotation axis and detector must be changed in order to record the next level. This manual change need be made only as many times as there are levels, which is a small number. The Eulerian-cradle variety of diffractometer must be operated by a different strategy. For example, each reflection can be regarded as occurring on a line, called a central line, which radiates from the origin of the reciprocal lattice. Lines to reciprocal-lattice points having simple indices contain many points whose indices are multiples of those of the first point, but many other lines with complicated indices each contain only one point. Every central line must be brought into initial position by adjusting two coordinate angles. To obtain reflections from other points in the line requires making one new setting for the crystal and one for the detector, and these may be coupled.

In both kinds of diffractometer, the detector tube is connected with electronic circuitry which amplifies and counts the photon pulses received by it. The detector itself may be a Geiger, proportional counter, or scintillation detector. The first is simplest and has simple circuitry, but the linearity of the ratio of its count response to the intensity received is limited to a few hundred counts per second. The proportional counter and scintillation counter have much greater linearity ranges and have other advantages, but require complicated and expensive circuitry.

The diffractometer is usually operated so that the detector collects the reflected beam while the crystal is rotated through the Bragg reflecting condition; in this way the total number of counts received during the waxing and waning of the reflection is recorded. When a suitable value of the background, as measured just before the reflection begins and just after it ends, is subtracted from this value, the result is the total photon count proportional to $E\omega/I_0$ in (34), which is the integrated power of the desired reflection. This value need only be corrected, as noted in the next section, to provide $|F_{hkl}|^2$. If reasonable precautions are observed, this value can be relied upon to a few percent.

Correction of measured intensities

Relation (34) shows that the measured integrated power of a reflection must be corrected for Lorentz and polarization factors before a quantity is obtained which is proportional to the desired $|F|^2$. In addition, a correction must be made for the effect of the absorption of x-rays by the crystal, and for another phenomenon known as *extinction*.† These additional corrections are required because (34) specifically applies to a volume so small that these two additional effects are negligible.

It will be seen that the Lorentz and polarization effects have a simple geometrical basis and are relatively easy to calculate. The effects of absorption are much more tedious to compute. The correction of these several effects, however, must be made for each of the many x-ray reflections, a number likely to be of the order of several thousand for an ordinary inorganic crystal. Applying these corrections was therefore a laborious procedure until the advent of high-speed electronic computers. With the aid of these new computing facilities, however, all these corrections can be easily and quickly calculated.

The Lorentz factor. It was seen in Chapter 4 that the various x-ray reflections can be understood in terms of reciprocal-lattice points making contact with the sphere of reflection. If the crystal is rotated in an x-ray beam, some of the reciprocal-lattice points pass through the sphere, and as each does so, a reflection occurs. These occurrences are not instantaneous, a fact which can be appreciated if the reciprocal-lattice points are not regarded as infinitesimal points but are each assigned small equal volumes. It is then evident that the time taken for a point to pass through its reflecting condition varies with the position of the point and its motion. The Lorentz factor was devised to take care of this varying time opportunity for reflection. This is inversely proportional to the component of the velocity of the reciprocal-lattice point normal to the surface of the sphere. For the zero level of a crystal rotating about an axis perpendicular to the x-ray beam, this has a simple form, namely,

$$L = \frac{1}{\sin 2\theta}. \tag{35}$$

For other levels, with a crystal rotating about an axis inclined to the x-ray beam, it has a more complicated form,

$$L = \frac{1}{\cos \mu \cos \nu \sin \Upsilon}, \tag{36}$$

where the angles are instrumental coordinates commonly employed in the Weissenberg method (Chapter 8). Since the form of the Lorentz factor depends on the geometry of the motion of the reciprocal-lattice points, it differs from one method to another and is most complicated for the precession method.

† Unfortunately, this is the same word used to describe classes of absent reflections (Chapter 5).

When the Lorentz factor has been computed for the various reflections, it can be removed from the measured integrated power of the reflections, according to (34), by multiplying the experimental values by $1/L_{hkl}$.

The polarization factor. When x-rays are reflected by a crystal plane, they become partly plane-polarized. As the electromagnetic wave of x-radiation is reflected by a plane, the component of the wave whose vibration is parallel to the reflecting plane does not suffer reduction, but the component in the plane of the incident and diffracted rays is reduced in intensity by an amount proportional to $\cos^2 2\theta$. As a consequence, the intensity of the diffracted wave is reduced by a factor, called the *polarization factor*, amounting to

$$p = \tfrac{1}{2} + \tfrac{1}{2} \cos^2 2\theta. \tag{37}$$

This reduction is independent of the method of experimentation and depends only on the angle of deviation, 2θ. The polarization factor can be removed from the measured values of the integrated power, according to (34), by multiplying the experimental value by $1/p_{hkl}$.

Absorption. As x-rays traverse a crystal, they are partly absorbed by the atoms along their path and so are reduced in intensity. A beam whose original intensity is I_0 suffers a reduction in intensity while it traverses a path of length t, to

$$I = I_0 e^{-\mu t}, \tag{38}$$

where μ, the linear-absorption coefficient, is a characteristic of the composition of the material and can be computed from available tables.

When an x-ray reflection occurs, this reduction in intensity also occurs for each possible ray of the kind shown in Fig. 14. The total path consists of two parts: t_1, the primary beam from the surface of the crystal to the volume element dV, and t_2, the reflected beam from the volume element to the surface of the crystal. The total transmission consists of the sum of such rays to and from all volume elements. The fraction of

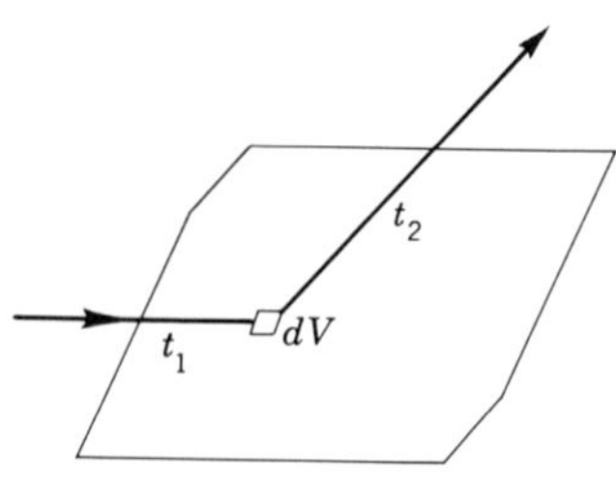

Fig. 14

the total intensity transmitted is given by

$$T = \frac{1}{V} \int^{V} e^{-\mu(t_1+t_2)}\, dV. \tag{39}$$

To compute this by hand for even one reflection is tedious, and prohibitive for many thousand reflections. Since the value of (39) for spheres of various radii and various absorption coefficients was easily computed and had been tabulated, it was usual to grind the crystal into a spherical specimen when possible. But with the availability of electronic computing it is now an easy matter to compute quickly the transmission factors for all reflections for a crystal of any shape which can be properly measured.

Extinction. When the primary beam makes the proper glancing angle θ with the planes of a stack to produce a reflection, the reflected ray also makes this angle θ with the planes and is therefore reflected back parallel to the primary beam. Each reflection changes the phase of the wave by $\pi/2$, so the twice-reflected ray is out of phase by π with the primary wave which it joins. Thus this multiple reflection weakens the primary ray, and attenuates it as it penetrates farther into the crystal. This effect is called *primary extinction;* it is proportional to $(\tanh sq)/sq$, where s is the number of planes in perfect sequence, and q is the reflection efficiency of the plane. Primary extinction is large for reflections having large structure factors (because of large q) and large spacings d_{hkl}, and is appreciable in volume units greater than 10^{-4} cm across. Because of this last feature, it is very serious in perfect crystals, but considerably alleviated in slightly imperfect crystals. Formula (34), which relates the integrated reflection to $|F|^2$, is actually valid only when the crystal has a suitable imperfection such that it is composed of a "mosaic" of small volumes within which extinction is not serious. Although primary extinction can be measured experimentally in large slabs, it is very difficult to allow for in the tiny crystal specimens used in crystal-structure analysis.

Another effect is called *secondary extinction.* This is the reduction of intensity with depth of penetration of the primary beam due to diversion of some of the energy into the reflected beam. This kind of extinction behaves somewhat like absorption.

Summary

While the dimensions and shape of the unit cell can be determined from the geometrical relations between the directions of the x-ray reflections, the determination of the relative position of the atoms in the cell is a function of the amplitudes of these reflections. For a correct structure the magnitudes of the amplitudes calculated for the model must match the corresponding magnitudes of the observed amplitudes. A common measure of the mean discrepancy of the match is the fraction

$$R = \frac{\Sigma\big||F_{obs}| - |F_{cal}|\big|}{\Sigma|F_{obs}|}.$$

Refinement procedures are available by which this residual can be reduced for a correct structure to the range of 5 to 10 percent and even lower, provided good values of $|F_{obs}|$ are available.

The experimental amplitudes $|F_{obs}|$ are proportional to quantities called the integrated reflections which can be measured as the plane (hkl) passes through the Bragg reflecting condition. The values of the integrated reflections can be determined by measuring the blackenings of the spots on a photographic film made by any of the standard methods; alternatively, they can be determined by counting the photons in the reflected x-ray beams with the aid of an instrument called a diffractometer.

To transform the measured integrated reflections into $|F_{obs}|$'s they must be corrected for background, Lorentz factor, polarization factor, and transmission factor. The corrections can be rapidly applied by high-speed electronic computing. Some corrections ought also to be applied to a limited number of reflections for an effect known as extinction, but the extent to which this effect is present cannot be easily assessed for the tiny specimens used in crystal-structure analysis. Fortunately, this circumstance is not serious enough to interfere with the analysis. A correction must also be applied to the calculated amplitude for temperature effects.

Notes on history

Most of the theory concerning the relation of the intensities of x-ray reflections to the arrangement of atoms in the crystal was developed in the earliest days of x-ray diffraction. The correction of calculated intensities for thermal agitation of the atoms was presented by Debye in 1913, and subsequently improved by Waller in 1923. As a result of these investigations it turned out that the reduction of intensity of a reflection due to thermal agitation can be computed by multiplying the structure factor by a term which is now known as the Debye-Waller temperature factor.

The use of $E\omega/I_0$ as the value of the reflecting power of a crystal plane was introduced by Bragg, James, and Bosanquet in 1921 as an appropriate measurement for the spectrometer. This instrument was an early form of diffractometer, but because its ionization chamber presented experimental difficulties, its use in measuring diffracted intensities gradually gave way to photographic methods. The latter were then used extensively until after the Second World War, when good photon counters were perfected. Now the best intensity measurements are made with diffractometers equipped with proportional counters or scintillation counters.

H. A. Lorentz explained to the students in his classes at the University of Leiden that the diffracted intensities of increasing order n were proportional to n^{-2}, a term which later became known as the Lorentz factor. This also appeared in the form of $1/\sin 2\theta$ in the formulae given for integrated reflections in 1914 by C. G. Darwin. He also included a factor, first deduced by J. J. Thompson, requiring a decline in intensity due to the polarization of x-rays scattered by electrons.

In his detailed study of 1922 on the amount of x-radiation reflected by a stack of crystal planes, Darwin discovered two effects which seriously reduced the intensity of the reflected beam. These he called primary and secondary extinction. The latter had already been encountered experimentally by W. H. Bragg.

The reduction of diffracted intensities due to absorption of x-rays by the specimen had long been recognized, but a scheme of routine correction for it did not become available until 1930, when Claassen showed how to allow for absorption in cylindrical samples. Very exact corrections are now easily made for specimens of any arbitrary shape with the aid of high-speed electronic computers.

Additional reading

H. Lipson and W. Cochran. *The determination of crystal structures*, vol. III, *The crystalline state.* (G. Bell, London, 1953) 54–63.

Martin J. Buerger. *Crystal-structure analysis.* (Wiley, New York, 1960) 21–48, 77–238, 259–282.

U. W. Arndt and B. T. M. Willis. *Single crystal diffractometry.* (University Press, Cambridge, England, 1966) 331 pages.

Significant literature

Integrated reflections

C. G. Darwin. *The theory of x-ray reflexion.* Phil. Mag. **27** (1914) 315–333, 675–690.

W. Lawrence Bragg, R. W. James, and C. H. Bosanquet. *The intensity of reflexion of x-rays by rock-salt.* Phil. Mag. **41** (1921) 309–337.

Temperature factor

P. Debije. *Über den Einfluss der Wärmebewegung auf die Interferenzerscheinungen bei Röntgenstrahlen.* Verh. deut. phys. Ges. **15** (1913) 678–689.

P. Debye. *Interferenz von Röntgenstrahlen und Wärmebewegung.* Ann. Phys. **43** (1914) 49–95.

Ivar Waller. *Zur Frage der Einwerkung der Wärmebewegung auf die Interferenz von Röntgenstrahlen.* Z. Phys. **17** (1923) 398–408.

Extinction

C. G. Darwin. *The reflexion of x-rays from imperfect crystals.* Phil. Mag. **43** (1922) 800–829.

Absorption

A. Claassen. *The calculation of absorption in x-ray powder-photographs and the scattering power of tungsten.* Phil. Mag. (7) **9** (1930) 57–65.

12

Fourier synthesis and the phase problem

The Fourier transform

One of the most important theoretical tools for dealing with problems in crystal-structure analysis is known as the *Fourier transform*. This is a mathematical function which provides a direct relation between the crystal structure and the diffraction effects it produces.

To introduce the Fourier transform as a purely mathematical device tends to obscure the relatively simple relationship which it involves. Fortunately, the focusing action of a lens furnishes a physical phenomenon corresponding to Fourier transformation, which can be appreciated in simple geometrical terms. An introduction of this kind follows.

A lens as a Fourier transformer

A lens is a natural Fourier transformer. By tracing rays through a lens in the formation of the image of an object, the Fourier transform of the object is encountered in a very natural way.

Figure 1 illustrates some of the basic features of the formation of an image by a lens. Consider some particular rays with easily predictable courses as they diverge from two points of the object through the optical system. One point is on the lens axis, at O, which may be regarded as the origin for the object, while the other is a more general point H, off the axis. From H, a ray through the center C of the lens does not suffer deviation by the lens and reaches the plane of the image at H', so that the ray HCH' is a single straight line. Similarly, OCO' is a straight line.

Another ray of importance from H is the ray parallel to the axis, namely, HA. It is the characteristic of a lens to refract any ray whose direction is parallel to the axis so that it crosses the axis at the focal distance D beyond the lens. The lens also refracts the ray OB, which is

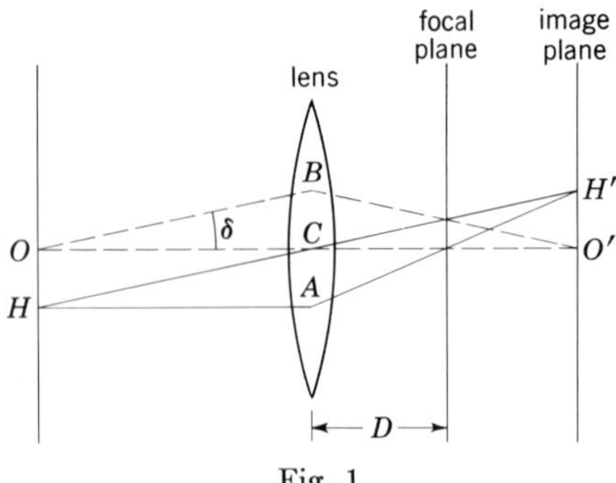

Fig. 1

parallel to HC, so that rays OB and HC cross each other at a distance D beyond the lens.

There are, in fact, two unique positions behind the lens where rays cross. The obvious one is in the image plane, where the convergence of the rays reaching each point there, such as H', builds up the image of the original object by their summation. But it has just been seen that the rays also converge in a different combination in the back focal plane of the lens. In that plane, rays cross which do not originate in the same point of the object but, rather, have the same direction (defined by angle δ) in leaving the object. The convergence of the various sets of rays, each set having a different direction parameter δ, builds up in the back focal plane another kind of image which is the *diffraction image* of the object. This image is related to the object by being, mathematically, its Fourier transform.

In Fig. 2 (page 233) the relations between the object plane and the diffraction plane are examined in more detail. The rays that are converged by the lens to some arbitrary point Q in the diffraction plane are parallel before reaching the lens. Their common direction is defined by an angle δ lying in the plane determined by the optical axis OO^* and the vector $\mathbf{r}^*$ which locates the arbitrary point Q. This is the plane $J_1J_2J_3J_4$ of Fig. 2, shown also in section in Fig. 3. For simplicity, suppose that the object plane is illuminated by a plane wave advancing in the direction of the optical axis, and that each point in the object scatters this wave as a spherical wavelet. Then the phases of the wavelets scattered in the direction defined by angle δ, by points of the object on the line OJ_1, decrease with the distance of the point from the origin. The phase of a point at a distance x from O is given by the proportion

$$\begin{aligned}\frac{\phi}{2\pi} &= \frac{OG}{\lambda}\\ &= \frac{x \sin \delta}{\lambda}. \end{aligned} \tag{1}$$

The angle δ is easily evaluated from the right side of Fig. 3 as

$$\delta = \tan^{-1} \frac{|\mathbf{r}^*|}{D}, \tag{2}$$

where D is the focal length of the lens. If (1) and (2) are compared, δ can be eliminated so that the phase is provided by

$$\phi = \frac{2\pi x}{\lambda} \sin \tan^{-1} \frac{|\mathbf{r}^*|}{D}. \tag{3}$$

For small angles the sine and tangent are nearly equal, so that for ray directions making small angles δ with the optic axis, a good approximation of (3) is

$$\phi = \frac{2\pi x|\mathbf{r}^*|}{\lambda D}. \tag{4}$$

The rays considered so far are those scattered by points which lie on the line J_1J_4 in Fig. 2. But all points on a line normal to the line J_1J_4, such as the line EE' in Fig. 2, give rise to rays whose paths to the lens are the same; these rays are all scattered with the same phase in the direction defined by δ. Accordingly, if a point in the object is located by a vector $\mathbf{r}$ (Fig. 2), the length of the projection of this vector on J_1J_4 (which has the same direction as vector $\mathbf{r}^*$) determines the phase scattered by that point. A generalization of (4), therefore, is that the phase scattered by a general point in the object plane is given by

$$\phi = \frac{2\pi\{|\mathbf{r}_{\text{proj on } r^*}|\,|\mathbf{r}^*|\}}{\lambda D}. \tag{5}$$

The term in braces has a value

$$\begin{aligned} |\mathbf{r}_{\text{proj on } r^*}|\,|\mathbf{r}^*| &= [|\mathbf{r}| \cos \phi]\,|\mathbf{r}^*| \\ &= |\mathbf{r}|\,|\mathbf{r}^*| \cos \angle\mathbf{rr}^*. \end{aligned} \tag{6}$$

In the language of vector algebra $|\mathbf{r}|\,|\mathbf{r}^*| \cos \angle\mathbf{rr}^*$ is called the scalar product of the two vectors $\mathbf{r}$ and $\mathbf{r}^*$ and is written in standard vector notation as the dot product $\mathbf{r} \cdot \mathbf{r}^*$. Thus, (5) can be shortened to

$$\phi = \frac{2\pi\mathbf{r} \cdot \mathbf{r}^*}{\lambda D}. \tag{7}$$

The terms λ and D are experimental constants. Suppose that these are chosen so that $\lambda D = 1$. Then (7) reduces to

$$\phi = 2\pi\mathbf{r} \cdot \mathbf{r}^*. \tag{8}$$

This presents the phase relation in its simplest form, and one which appears in the purely mathematical expression of the Fourier transform.

Of course, the points in the object plane do not all scatter light with the same efficiency; otherwise the entire object plane would appear uniformly illuminated, so that no object could be detected. Suppose that the object is regarded as made up of very many small parts which can be numbered from 1 to N. Suppose part 1 scatters a wavelet whose amplitude is f_1, while part 2 scatters a wavelet whose amplitude is f_2, etc. Then, since the lens converges the rays from all these parts to a point like Q in Fig. 2, a summation of waves represented by these rays occurs there. The resultant amplitude, which occurs at the end of

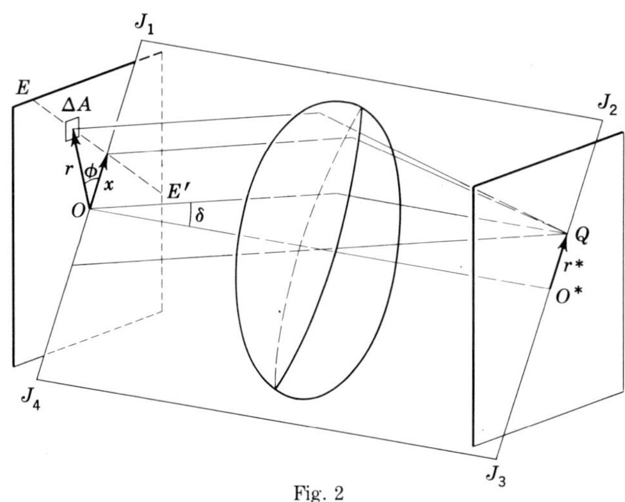

Fig. 2

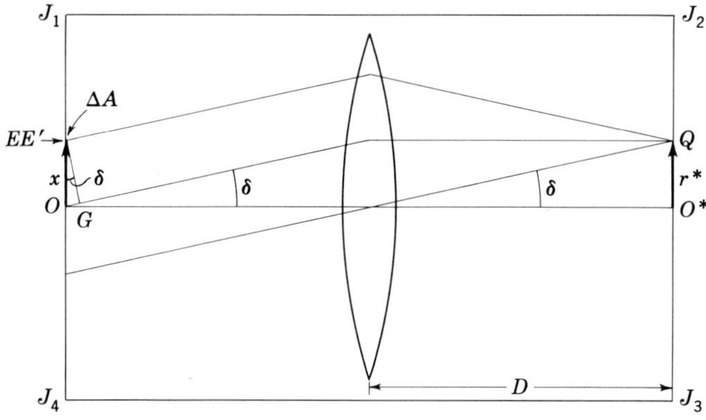

Fig. 3

vector $\mathbf{r}^*$, can be represented by summing these individual amplitudes, with proper phases, in the following way:

$$\begin{aligned} R(\mathbf{r}^*) &= f_1 e^{i2\pi \mathbf{r}_1 \cdot \mathbf{r}^*} + f_2 e^{i2\pi \mathbf{r}_2 \cdot \mathbf{r}^*} + \cdots + f_N e^{i2\pi \mathbf{r}_N \cdot \mathbf{r}^*} \\ &= \sum_{j=1}^{N} f_j e^{i2\pi \mathbf{r}_j \cdot \mathbf{r}^*}. \end{aligned} \tag{9}$$

Relation (9) holds for the resultant amplitude at any point in the diffraction plane located by a vector $\mathbf{r}^*$, provided the object is a set of N small discrete parts. The relation can be readily generalized to scattering by a continuous object by imagining the object plane to be subdivided into small elementary areas ΔA within which the scattering amplitude per unit of area is ρ. This amplitude, in general, varies from one elementary area to another, and so the amplitude scattered by the elementary area located at the end of the vector **r** is $\rho_{\mathrm{r}}\Delta A$. If this is substituted for f_j in (9), the result is

$$R(\mathbf{r}^*) = \sum_{\mathbf{r}} \rho_{\mathrm{r}}\, \Delta A\, e^{i2\pi\mathbf{r}\cdot\mathbf{r}^*}. \tag{10}$$

If the elementary area is reduced in size, in the limit (10) becomes

$$R(\mathbf{r}^*) = \int_{\substack{\text{area}\\ \text{of the}\\ \text{object}\\ \text{plane}}} \rho(\mathbf{r}) e^{i2\pi\mathbf{r}\cdot\mathbf{r}^*}\, dA. \tag{11}$$

If the resultant in (11) is to be actually computed, it is convenient to express the vectors **r** and **r*** in coordinate form. To do this it is customary to choose unit axial vectors **a** and **b** in the object plane, and to express **r** in terms of a component $x\mathbf{a}$ along the direction of **a** and $y\mathbf{b}$ along the direction of **b**. A corresponding resolution is carried out for **r*** in the diffraction plane. Thus the vectors **r** and **r*** are expressed in component form as

$$\begin{aligned} \mathbf{r} &= x\mathbf{a} + y\mathbf{b} \\ \text{and} \qquad \mathbf{r}^* &= x^*\mathbf{a}^* + y^*\mathbf{b}^*. \end{aligned} \tag{12}$$

Then the product $\mathbf{r} \cdot \mathbf{r}^*$ in (11) is expressed in component form as

$$\begin{aligned} \mathbf{r} \cdot \mathbf{r}^* = {} & xx^*\mathbf{a} \cdot \mathbf{a}^* + xy^*\mathbf{a} \cdot \mathbf{b}^* \\ & + yx^*\mathbf{b} \cdot \mathbf{a}^* + yy^*\mathbf{b} \cdot \mathbf{b}^*. \end{aligned} \tag{13}$$

This expression can be simplified if the axial pair $\mathbf{a}^*$, $\mathbf{b}^*$ is chosen as the set reciprocal to the axial pair **a**, **b**; in this case the pairs have the properties given in Chapter 11. Under these circumstances the cross terms in (13) vanish, but symmetrical products like $\mathbf{a} \cdot \mathbf{a}^*$ are unity, and so (13) reduces to the simple form

$$\mathbf{r} \cdot \mathbf{r}^* = x \cdot x^* + y \cdot y^*. \tag{14}$$

In component form, then, (11) is written

$$R(x^*y^*) = \int_{\substack{\text{area}\\ \text{of the}\\ \text{object}\\ \text{plane}}} \rho(xy) e^{i2\pi(x\cdot x^* + y\cdot y^*)}\, dx\, dy. \tag{15}$$

Relations (9), (11), and (15) are merely quantitative descriptions of the diffraction amplitudes which occur in the diffraction image at the desired location. In (9) and (11) the point at which the summation is sampled is at the end of the vector $\mathbf{r}^*$; in (15) this is expressed in terms of coordinates x^* and y^*. In both cases it should be noted that the resultant amplitude, in general, is complex, that is, that it involves both a magnitude and a phase, since the summation involves the exponential phase component.

It is illuminating to apply these results to diffraction by a lattice array. Considering first a one-dimensional lattice and a two-dimensional lens, as suggested in Fig. 4, suppose that the scattering is produced by a series of small holes in an opaque screen, equally spaced at interval d. The diffraction can be understood directly from Fig. 4. According to (8), the scattering phase in the direction defined by δ is $\phi = 2\pi\mathbf{r} \cdot \mathbf{r}^*$. In the one-dimensional grating and its sequence of diffraction spectra on a parallel line, the vector $\mathbf{r}$ is replaced by a scalar r. The nth hole of the sequence is located at an origin distance $r = nd$. Then (9) becomes $R(r^*) = \Sigma f_j e^{i2\pi n[dr^*]}$. Unless the term in brackets is an integer, this sum essentially vanishes (the more nearly so the larger the number n of the holes), because when the exponent is not zero, each exponential represents a vector (whose length is f) pointing in a different direction on the Argand diagram. These tend to annul each other unless the term in brackets is an integer. If it is an integer, then each exponential has the form $e^{i2\pi p}$, which implies a phase rotated away from the real axis by an angle which is an integral number p of 2π. This defines the same phase as e^0, so that all vectors point the same way, and the summation becomes an arithmetic addition. This occurs at positions r^* in the diffraction plane for which dr^* is an integer N, that is, at positions $r^* = N(1/d)$. Thus the diffracted rays arrive in phase at the diffraction plane at positions which are multiples of a fundamental distance $1/d$. Only these points of the diffraction plane are illuminated; all others are dark. The sequence of illuminated positions occupies the points of a one-dimensional lattice; this lattice has a period $1/d$, and so it is the reciprocal of the lattice of the holes which produced the diffraction.

Each lattice is the reciprocal of the other. If the direction of the initial wave is reversed, the scattering from points along the reciprocal

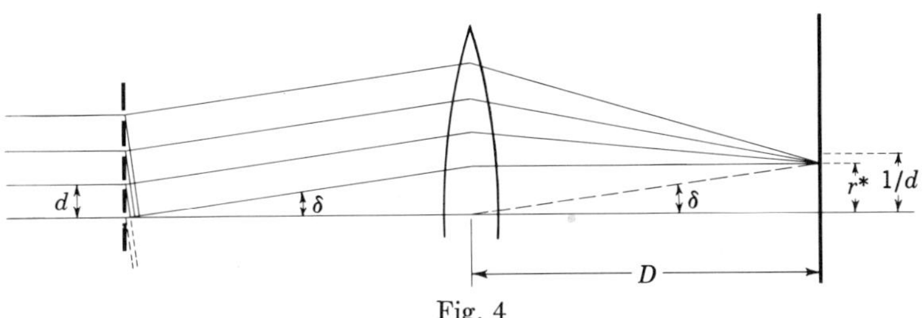

Fig. 4

lattice produces illumination at points of the original lattice. This is easy to see if the image plane and diffraction plane are both placed at the equal distance D from the lens, as shown in Fig. 5.

Figure 6 shows the corresponding application to a two-dimensional array of scattering points. Suppose that the plane on the left side of the figure is the object plane. The lattice there is arranged so that some spacing, labeled d, is in the vertical plane. The rows corresponding to this spacing produce a diffraction which is in phase at certain points along the vertical line in the diffraction plane at positions t^* such that $dt^* = N$, or $t^* = N(1/d)$, as in the discussion of the one-dimensional example. This diffraction result holds for every set of parallel rows of the lattice array on the left, regardless of the direction of the row. For every such set of rows there is a d which determines, in the diffraction plane, a central line of translations parallel to that d, and whose translation interval is $t^* = 1/d$. In Chapter 4 it was shown that $d \cdot t^* = 1$ characterizes a pair of reciprocal lattices. Therefore the diffraction from a lattice array is seen to be an array of illuminated spots at the points of the reciprocal lattice. These two lattice arrays are mutually reciprocal,

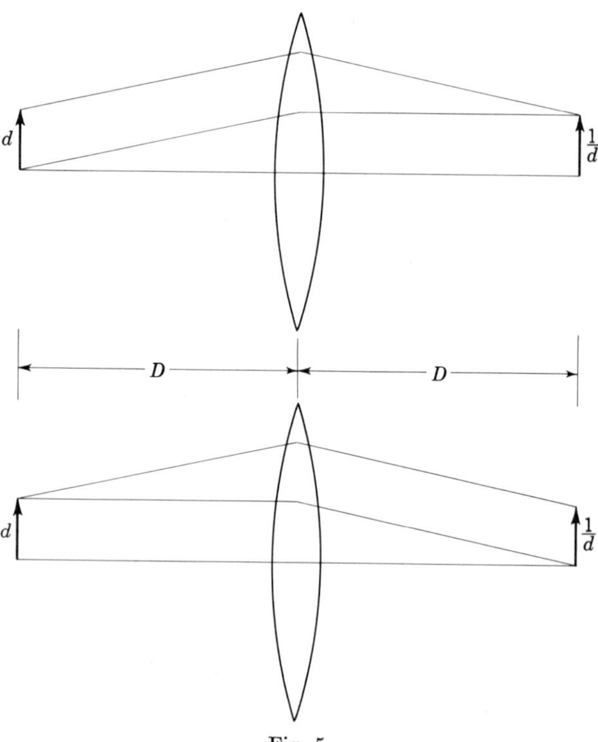

Fig. 5

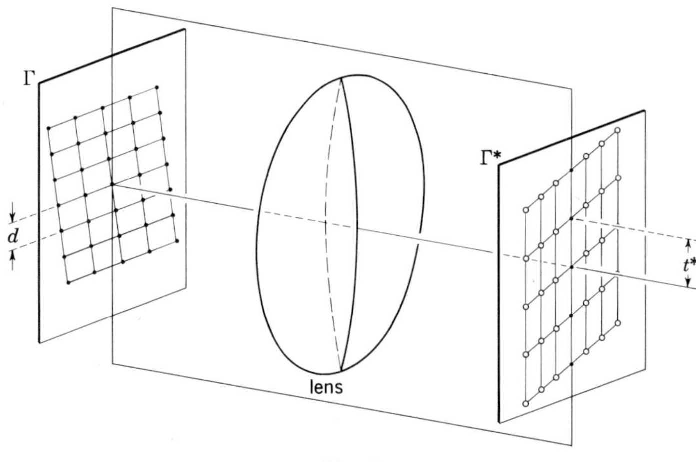

Fig. 6

for reversing the direction of the light wave produces the left array from the right one.

For a somewhat more sophisticated treatment, relation (9) can be used. In this case all the vectors **r** have the form

$$\mathbf{r} = m\mathbf{a} + n\mathbf{b}, \tag{16}$$

where m and n are integers. With this restriction, (9) becomes

$$\begin{aligned} R(\mathbf{r}^*) &= \Sigma f e^{i2\pi(m\mathbf{a}+n\mathbf{b})\cdot\mathbf{r}^*} \\ &= \Sigma f e^{i2\pi(m\mathbf{a}\cdot\mathbf{r}^*+n\mathbf{b}\cdot\mathbf{r}^*)} \\ &= \Sigma f e^{i2\pi m\mathbf{a}\cdot\mathbf{r}^*} e^{i2\pi n\mathbf{b}\cdot\mathbf{r}^*}. \end{aligned} \tag{17}$$

Whenever both the dot products in (17) have integral values, specifically,

$$\begin{aligned} &\mathbf{a}\cdot\mathbf{r}^* = h \qquad \text{(an integer)} \\ \text{and} \qquad &\mathbf{b}\cdot\mathbf{r}^* = k \qquad \text{(an integer)}, \end{aligned} \tag{18}$$

then both exponents vanish simultaneously, the exponentials both become unity, and the resultant summation becomes the arithmetic sum of all the f's. Whenever either part of (18) (or both) is not satisfied, the sum corresponds to vectors pointing more or less uniformly in all directions in the complex plane, and these tend to annul one another, leaving a resultant which is zero for an infinite lattice array and nearly zero for a reasonably large number of lattice points. The illuminated points thus occur at the ends of vectors which conform to (18). The condition that is satisfied is

$$\mathbf{r}^* = h\mathbf{a}^* + k\mathbf{b}^*, \tag{19}$$

where $\mathbf{a}^*$, $\mathbf{b}^*$ is a set of vectors reciprocal to the set $\mathbf{a}$, $\mathbf{b}$, that is, where

$$\mathbf{a} \cdot \mathbf{a}^* = 1, \qquad \mathbf{a} \cdot \mathbf{b}^* = 0,$$
$$\mathbf{b} \cdot \mathbf{b}^* = 1, \qquad \mathbf{b} \cdot \mathbf{a}^* = 0. \tag{20}$$

That this is the requirement on $\mathbf{r}^*$ can be shown by direct substitution of (19) in (18). Thus the diffraction of a plane-lattice array of points is a set of maxima at points defined by (19) and (20), that is, at points of the reciprocal lattice. If the direction of light is reversed in Fig. 6, the diffraction by a set of scattering points located on points of the reciprocal lattice produces maxima at the points of the original lattice.

Fourier transformations

In the last section it was seen that the formation of a diffraction image from a plane object by a lens can be represented as an integration, over the area of the object, of the density at a point in the object, modified by an exponential which supplies the phase of the wave scattered by that point. The general form of this representation was given by (11), and this was expressed by (15) in coordinate form suitable for computation. It was also seen that the diffraction from a lattice array composed of equal scattering points is a set of equally illuminated points located at the points of its reciprocal lattice. The mathematical form of (11) is known as the Fourier transform of the function, namely, $\rho(x)$, which describes the object. The designations "diffraction image of," "reciprocal of," and "Fourier transform of" have similar meanings. The first actually designates a physical result obtained from an object, while the last two are mathematical functions of the object which describe, and permit one to compute, the physical result.

Since matter has a refractive index with respect to x-rays which is substantially 1, the x-rays diffracted by a crystal cannot be converged by a lens to a set of points, as in the physical example discussed in the last section. Nevertheless, the general mathematical form describing the diffraction of x-rays by a crystal structure is the same as that describing the diffraction of light by a periodic object. This can be seen by comparing the diffraction of x-rays by a set of discrete atoms, given by (4) of Chapter 11, with that for optical diffraction by discrete points, given by (8) and (9) of this chapter. The general form of a Fourier transform computed for a point at the end of vector $\mathbf{r}^*$ in "Fourier space" is a generalization of the two-dimensional form of (11), specifically,

$$R(\mathbf{r}^*) = \int_{\substack{\text{volume} \\ \text{of crystal} \\ \text{space}}} \rho(\mathbf{r}) e^{i2\pi \mathbf{r}\cdot\mathbf{r}^*}\, dV. \tag{21}$$

The corresponding coordinate form is a generalization of the two-dimensional form of (15), namely,

$$R(x^*y^*z^*) = \int_{\substack{\text{volume}\\ \text{of crystal}\\ \text{space}}} \rho(xyz)e^{i2\pi(x\cdot x^*+y\cdot y^*+z\cdot z^*)}\,dx\,dy\,dz. \tag{22}$$

In the actual diffraction by a crystal, the triperiodic translational repetition causes a limitation of x^*, y^*, and z^* to the integers h, k, and l, respectively; these correspond to the integers in (19). Thus this transform becomes the complex amplitude of diffraction in the Laue orders hkl. Since the scattering atoms are regarded as occupying the discrete points $x_jy_jz_j$, the integration of (22) is replaced by the summation

$$F(hkl) = \sum_j f_j(x_jy_jz_j)e^{i2\pi(hx_j+ky_j+lz_j)}. \tag{23}$$

The phase problem

Because the focusing of light diffracted by a grating provides a simple analog of diffraction by a crystal, it can be used to provide a striking demonstration of what is known as the *phase problem of x-ray crystallography*. The focusing of light coming from an object to form an image is illustrated in Fig. 7*A* by using a symmetrical lens. If the lens is separated into two halves, and the right half L_2 is displaced to the right by twice its focal length D, the focusing by the resulting lens pair is as shown in Fig. 7*B*. The main result of splitting the lens into two equal parts is to lengthen the entire optical system; the cone of rays diverging from any point in the object becomes a set of parallel rays in the space between the lenses, and these rays are then converged by the second half of the lens to form the image at the same distance from the right side of the lens system as before. With this separated arrangement the location at which parallel rays from the object cross (that is, the diffraction plane) is *within* the lens system. Figure 7*C* shows the location of the diffraction image. It can be seen that the second lens L_2 plays the same role in receiving the parallel rays which reach it from the diffraction image and focusing them to produce the final image as the first lens L_1 does in receiving the parallel rays from the original object and focusing them to form the diffraction image. Thus the formation of an image from an object by a lens system may be separated into two stages; the sequence in the process can be described as the *diffraction of the diffraction of the original object*. If the operation of diffraction, or of forming the Fourier transform, is represented by R, then the process of image formation is represented by the two stages

Stage 1: R (object) = diffraction, (24)

Stage 2: R (diffraction) = image. (25)

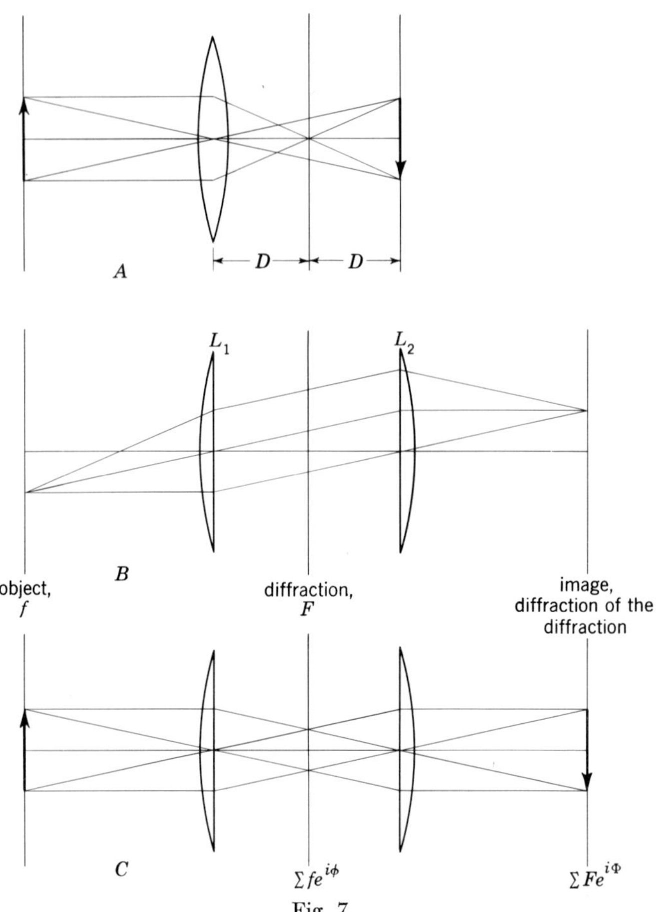

Fig. 7

The combination of these two is

$$R\ \{R\ (\text{object})\} = \text{image}, \tag{26}$$

or, briefly,†

$$R^2\ (\text{object}) = \text{image}. \tag{27}$$

† An interesting sidelight is that the image has the same appearance as the object except that all coordinates have opposite signs; that is, the image is the inverse of the object. This can be represented as

$$\text{Image} = -\text{object}.$$

If this is combined with (27), there emerges

$$R^2(\text{object}) = -\text{object}.$$

This states that two Fourier transforms in sequence merely invert the object. It follows that four Fourier transforms in sequence are required to restore the original object.

Now, these two optical stages correspond to the two stages in the solution of a crystal structure. The first stage is analogous to the experimental diffraction of x-rays by the crystal. This stage provides the diffraction maxima whose intensities can be measured experimentally. The second stage corresponds to converting these intensity data by means of a Fourier transformation to produce the image of the structure. This pair of successive stages would succeed in solving any crystal structure except for one flaw: The fact that the whole process is interrupted in the middle of Fig. 7*C* requires that the amplitudes $F(hkl)$ available there, which are complex quantities, each be measured in magnitude and in phase. It is easy to measure the magnitude, but no experimental method of determining the phase is known. It is impossible, therefore, to carry out the Fourier transformation represented by the right-hand half of Fig. 7*C*, because only half the required data are actually available, namely, the magnitudes of the required coefficients of the Fourier summation; the phases have been lost.

The performing of a crystal-structure analysis in two stages instead of using a continuous flow-through process, as represented by the continuous flow of rays through the optical system of Fig. 7*C*, thus carries the penalty that the phases are lost. Unfortunately, this penalty cannot be avoided by making a continuous optical system like Fig. 7*C* for x-rays, for two reasons. In the first place, matter has such a small refractive index (substantially 1) for x-rays that they cannot be controlled by lenses. But even if they could be focused by lenses, an optical system like that of Fig. 7*C* could not be devised unless all diffraction maxima $F(hkl)$ could be produced simultaneously. But in Chapter 3 it was seen that, with monochromatic radiation, each maximum occurs only when the three Laue conditions are simultaneously satisfied, and each maximum therefore occurs when the crystal has a different orientation. Accordingly, the full set of such maxima cannot be generated simultaneously in three dimensions, and the scheme of Fig. 7*C* cannot be used in a continuous flow.†

Solving a crystal structure from x-ray diffraction data, therefore, is not a routine matter of substituting experimental data in a general equation and then solving this for the electron density, because half the required data are not available. As a result, the solution of a crystal structure requires an indirect approach, and a solution may not even be possible.

There are two general routes which can be taken toward a solution. A

† The continuous flow is possible for a diperiodic pattern by an optical system in three dimensions, or for a monoperiodic pattern by an optical system in two dimensions. In general, the Laue conditions for all maxima can be satisfied simultaneously for a pattern with n independent translations by an optical system in a space of $n + 1$ dimensions.

Fourier summation can be prepared in which the coefficients make use of magnitudes only. This results in a map of a function defined in crystal space which bears a relation to the image of the crystal structure that is actually sought. This kind of approach in avoiding the phase problem is discussed in Chapter 13.

A second kind of effort to circumvent the phase problem can be made in Fourier space by attempting to assign phases to the $F(hkl)$'s in order that the mathematical summation corresponding to stage 2 can be carried out. This approach is discussed in this chapter.

Fourier synthesis of the electron density

Fourier transformations are extensively used in the study of crystal structures. The computation of the amplitude F of the reflection hkl, which was introduced in (22) of Chapter 3 and again in (4) of Chapter 11, is a Fourier transformation from crystal space to Fourier space. This corresponds to the focusing action of the lens in the upper part of Fig. 5. The transformation from Fourier space back to crystal space, which corresponds to the focusing action of the lens in the lower part of Fig. 5, is also a Fourier transformation. These two transformations are reverses of one another; if the first is represented by the symbol R, the second is often designated R^{-1}. If the phase of a ray connecting points located by vectors $\mathbf{r}$ and $\mathbf{r}^*$ in the one transformation is represented by ϕ, then it is $-\phi$ in the reverse transformation.

The reverse transformation, specifically transforming the diffraction image back into the electron density, is the general theme of this chapter. This transformation can be approached in a more sophisticated manner than in the last section, in the following way, which makes use of some additional mathematics.

The electron density, that is, the number of electrons per unit volume, is the same in all cells; therefore it can be represented by a periodic function. A periodic function can be expressed as a Fourier series, in which the series is a sum of terms representing a wave and its harmonics, having various amplitudes. A simple form of this is

$$\begin{aligned} f(\phi) &= A_0 + A_1 \cos \phi + A_2 \cos 2\phi + A_3 \cos 3\phi + \cdots \\ &= \sum_{n=0}^{\infty} A_n \cos n\phi. \qquad (28) \end{aligned}$$

The corresponding form for representing an electron density which is periodic in one dimension, with period a, is

$$\rho(X) = A_0 + A_1 \cos 2\pi \frac{X}{a} + A_2 \cos 2\pi \cdot 2 \frac{X}{a} + A_3 \cos 2\pi \cdot 3 \frac{X}{a} + \cdots$$

$$= \sum_{h=0}^{\infty} A_h \cos 2\pi \cdot h \frac{X}{a}. \tag{29}$$

Here X is the absolute coordinate in the one dimension, so that the fractional coordinate in the period is $x = X/a$. The electron-density function used in (29) is centrosymmetrical in the origin because the cosine is centrosymmetrical in the origin. In order to represent a more general series, an exponential can be substituted for the cosine, and the summation extended to include negative as well as positive values of h:

$$\rho(X) = \sum_{h=-\infty}^{\infty} A_h e^{-i2\pi hX/a}. \tag{30}$$

The coefficients A_h of these terms are called Fourier coefficients. They are unknown but can be found by a standard routine. This is accomplished by multiplying each side of (30) by $e^{i2\pi h'X/a}$ and integrating from 0 to a:

$$\int_0^a \rho(X) e^{i2\pi h'X/a}\, dX = \int_0^a e^{i2\pi h'X/a} \sum_{h=-\infty}^{\infty} A_h e^{-i2\pi hX/a}\, dX \tag{31}$$

$$= \int_0^a \sum_{h=-\infty}^{\infty} A_h e^{i2\pi(h'-h)X/a}\, dX. \tag{32}$$

The integral on the right of (32) vanishes, except for the term for which $h' = h$, when it is equal to Aa. Thus (32) reduces to

$$\int_0^a \rho(X) e^{i2\pi hX/a}\, dX = aA_h. \tag{33}$$

If the same routine is followed for an electron-density distribution which is triperiodic, then the three-dimensional equivalent of (30) is

(30):
$$\rho(XYZ) = \sum_h \sum_k \sum_l A_{hkl} e^{-i2\pi(hX/a+kY/b+lZ/c)}. \tag{34}$$

The routine solution of this for the Fourier coefficients A_{hkl} yields a result like (33). The equivalent of (33) for three dimensions, but with right-hand and left-hand sides interchanged, is

$$VA_{hkl} = \int_0^a \int_0^b \int_0^c \rho(XYZ) e^{i2\pi(hX/a+kY/b+lZ/c)}\, dX\, dY\, dZ, \tag{35}$$

where V is the volume of the cell. This can be compared with the diffraction amplitude (or structure factor) F_{hkl}, given in (22) of Chapter 3:

$$F_{hkl} = \Sigma f_j e^{i2\pi(hx_j+ky_j+lz_j)}. \tag{36}$$

The right-hand sides of these two equations mean the same thing, but they are differently expressed. In (35) the coordinates are absolute, while in (36) they are fractional, so that the exponentials are identical. Relation (36) formulates the amplitude scattered by the contents of a cell expressed as a set of discrete atoms, while (35) formulates the amplitude from the same atoms expressed as the variation they produce in electron density. Since these are different formulations of the same thing, their left-hand sides can be equated to give

$$A_{hkl} = \frac{1}{V} F_{hkl}; \tag{37}$$

that is, the Fourier coefficients required for (34) are just the amplitudes per unit volume. If this is substituted back into (34), and if fractional coordinates are used, the result is

$$\rho(xyz) = \frac{1}{V} \sum_h \sum_k \sum_l F_{hkl} e^{-i2\pi(hx+ky+lz)}. \tag{38}$$

This is the reverse Fourier transformation. It permits recovering the electron density $\rho(xyz)$ of the crystal from the diffraction amplitudes F_{hkl} which the crystal produces.

The form of (38) is that of a summation. Each term in the summation is a wave of electron density whose amplitude is provided by its Fourier coefficient F_{hkl}/V (whose dimensions are electrons per unit volume, the standard measure of electron density). This density wave has a wavefront whose orientation is given by the plane of constant phase; this is provided by the exponent in (38). This phase is constant for

$$hx + ky + lz = n = \text{constant}. \tag{39}$$

In Chapter 3 it was shown that this is the equation of the nth plane of the stack whose indices are (hkl). Thus the wavefront of the component wave of electron density corresponding to F_{hkl}/V is parallel to the planes (hkl) of the lattice and has the spacing of these planes. It does not necessarily have its maximum density at the origin, however; rather, the location of maximum density is removed from the origin by the phase component of F_{hkl}.

The direct and reverse Fourier transforms which are used in crystal-structure analysis are accordingly the following:

From crystal space to Fourier space:

(36): $$F_{hkl} = \sum_j f_j e^{i2\pi(hx+ky+lz)}.$$

From Fourier space to crystal space:

(38): $$\rho(xyz) = \frac{1}{V} \sum_h \sum_k \sum_l F_{hkl} e^{-i2\pi(hx+ky+lz)}.$$

Both of these are summations which are customarily computed by modern electronic digital computers when these are available.

The heavy-atom method

Theory. In Chapter 11 it was shown that the complex amplitude F of a reflection hkl is the resultant of the summation of the vectors in the complex plane, each representing the magnitude and the phase of the wavelet scattered by an atom of the cell. The form of this, given in (4) of Chapter 11, is

$$F_{hkl} = \sum_{j=1}^{N} f_j e^{i\phi_j}. \tag{40}$$

The phase scattered by each atom depends on its position in the cell, as given in (21) of Chapter 11, namely,

$$\phi_j = \frac{\Delta d_j}{d_{hkl}} 2\pi. \tag{41}$$

A graphic illustration of the summation of wavelets in (40) is given in Figs. 8 and 9. In Fig. 8 the unit cell of a fictitious two-dimensional crystal is shown. The crystal structure is specialized in that

(*a*) there is a heavy atom H at the cell's corner,
(*b*) the other atoms in the cell are light atoms; for simplicity, the light atoms are shown located on uniformly spaced levels parallel to the reflecting plane (hkl).

The amplitude of the wavelet scattered by an atom depends on the number of electrons in the atom, so that the vector representing the magnitude of this amplitude is large for heavy atoms and small for light atoms. Since the heavy atom is shown located at the origin, the wavelet it scatters is represented on the Argand diagram (Fig. 9) by a long vector f_H; its direction is such that $\phi = 2\pi\ (\Delta d/d) = 2\pi(0/d) = 0$. The lighter atoms are represented by shorter vectors in Fig. 9. These atoms are distributed through the cell in Fig. 8 on levels separated by $\Delta\phi =$

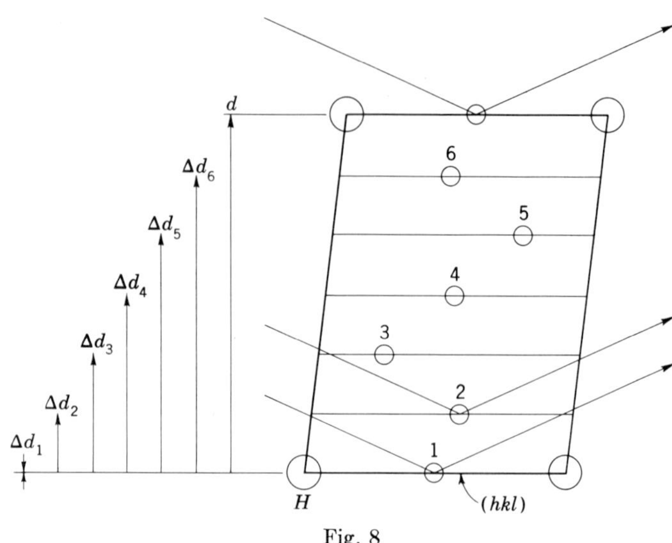

Fig. 8

$2\pi(\Delta d/d) = 2\pi(\frac{1}{6}) \approx 60°$. Accordingly, the six short vectors f_1, f_2, f_3, f_4, f_5, and f_6, when placed tail-to-head, outline a hexagon; the head of the last vector f_6 lies at the origin, and so the resultant of these six wavelets is zero. The net wave scattered by the contents of the crystal cell, therefore, is due only to the heavy atom. The second order of the reflection from the plane (hkl), namely, $2h\ 2k\ 2l$ (Fig. 10), gives the same general result, as also does $3h\ 3k\ 3l$. On the other hand, for reflection $6h\ 6k\ 6l$, all light atoms contribute wavelets whose phases are multiples of 2π, and so their vectors in Fig. 10 add arithmetically.

This example, though quite specialized, nevertheless brings out an interesting feature of x-ray scattering by the atoms of a crystal. If the cell contained a uniform electron density, the intensity of the wave scattered by this matter would be zero for all reflections except the unique reflection $hkl = 000$, for which all electrons would scatter in phase. A rough approximation to this condition occurs for crystals with large cells composed of atoms containing approximately equal numbers of electrons. If such atoms are distributed uniformly or are distributed at random throughout the cell, all reflections except 000 tend to be built up of small wavelet contributions whose magnitudes are nearly equal and whose phases are uniformly distributed in direction; such reflections have small magnitudes. For a cell composed of atoms with similar scattering powers, most reflections are likely to have large magnitudes only if the atoms have a rather nonuniform distribution in the cell.

Suppose the cell contains a more or less uniform distribution of light atoms, and also one heavy atom whose location is known (in one of the ways noted below). Then for many reflections the wavelet contributed by the heavy atom is the major contribution to the resultant wave scattered by the cell. For those reflections, the wave scattered by the cell is clearly dominated by the contribution of the heavy atom. In the case of a centrosymmetrical crystal, the phase is limited to 0 or π, which amounts to assigning a sign + or − to F_{hkl}. In this case the sign of the Fourier coefficient F_{hkl} tends to be the same as the sign of the contribution to the reflection by the heavy atom alone. If the crystal is noncentrosymmetrical, then the phase of the reflection tends to be roughly approximated by the phase scattered by the heavy atom.

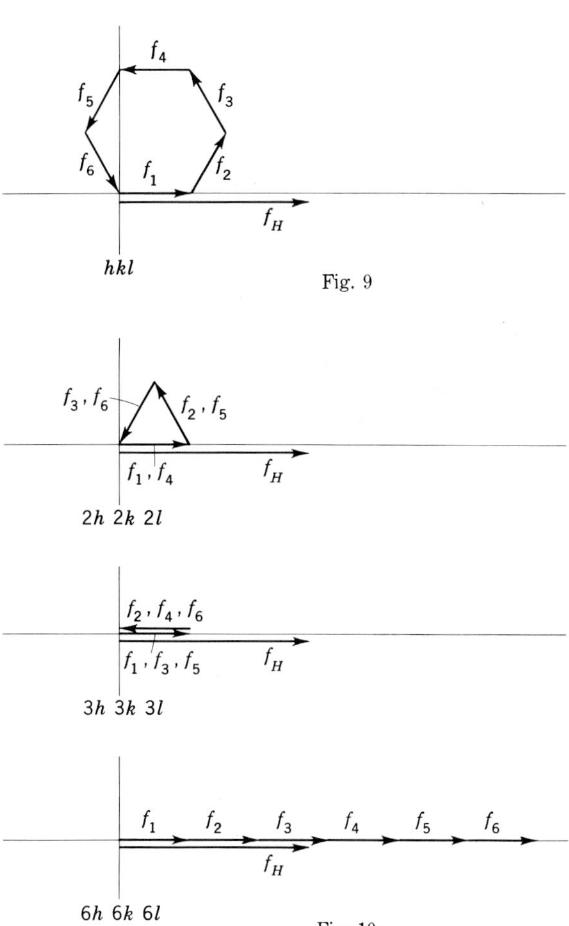

Fig. 9

Fig. 10

The *heavy-atom method*, which is based upon this relation, has been extensively used to solve crystal structures, especially those of crystals which can be grown artificially and whose compositions, accordingly, can be controlled. Many organic crystal structures have been determined by this strategy; indeed, it is now common practice to determine the shape of an organic molecule by attaching to it a heavy atom such as S, Se, Br, or I, crystallizing the resulting compound, and then solving the crystal structure by the heavy-atom method. This, incidentally, reveals the arrangement of atoms in the molecule. The heavy-atom method can also be used in the determination of the structures of naturally occurring compounds, that is, of minerals, provided the composition normally contains a heavy atom.

If the atomic number of the heavy atom is designated as Z_H and that of a light atom as Z_L, then the ratio $Z_H/\Sigma Z_L$ provides a measure of the probable success of assigning phases by using the heavy-atom method. In favorable cases a ratio of 0.3 is not too small to allow a one-stage solution of the structure.

Locating the heavy atom. There are three general methods of finding the locations of the heavy atom. The most general one is use of the Patterson function, which is discussed in Chapter 13.

A simpler method, which was used before the discovery of the Patterson function and is still in use, is based upon strategic use of equipoint information, which was discussed in Chapter 10. The heavy-atom method is most simply applied when the cell contains a small number of heavy atoms and a considerably larger number of light atoms. In such cases the number of heavy atoms is so small that they must occupy a set of equivalent positions of small multiplicity. Such positions, as shown in Table 3 of Chapter 10, commonly have no variable parameters.† When several alternative sets having the same multiplicity are available, it is sometimes immaterial which is chosen; sometimes the correct alternative must be determined along the general lines detailed in the next paragraph.

The several heavy atoms of a symmetrical set can often be located, at least approximately, even if they do have variable parameters. This calls for computing the intensities expected from the heavy-atom contribution as their coordinates are varied, allowance being made for the maximum (or probable) increase or decrease in magnitude which can be expected from the light atoms.

† Even in cases which do have variable parameters, these can sometimes be fixed arbitrarily. For example, the two equipoints of multiplicity 1 in space group Pm have coordinates $xy0$ and $xy\frac{1}{2}$. These can be arbitrarily assigned $x = 0$, $y = 0$, since they are the coordinates of a point on the only symmetry element, namely, a mirror. This places the heavy atom at the origin in the first case, or $00\frac{1}{2}$ in the second.

Examples. The neatest example of structure determination by the heavy-atom method is Robertson and Woodward's solution of the structure of platinum phthalocyanine, $PtC_{32}H_{16}N_8$. The ratio $Z_H/\Sigma Z_L$ for this compound is 0.29. The symmetry of this structure is $P2_1/a$, and the cell contains $2PtC_{32}H_{16}N_8$. The only equipoints having multiplicity 2 in this space group, and therefore available for the two Pt atoms, are four different sets with no variable parameters (Table 3, Chapter 10). These are all inversion centers and can be converted into one another by permissible changes of origin. The set represented by 000 was chosen as the site for platinum. With this choice, the platinum scatters with phase zero, except in extinguished reflections. When these phases are attributed to the scattering by the entire cell, the Fourier synthesis of its electron-density projection on a plane normal to the unique axis is as shown in Fig. 11. This summation required 302 Fourier coefficients F_{h0l}. All phases proved to be correct when a new set was computed with the contribution of the C and N atoms which were revealed in the first Fourier synthesis of Fig. 11.

A more general case is afforded by diglycine hydrobromide, $2(C_2H_5NO_2)$ HBr. The ratio $Z_H/\Sigma Z_L$ for this compound is 0.43. The symmetry of this structure is $P\,2_1\,2_1\,2_1$, and the cell contains four formula units of $2(C_2H_5NO_2)HBr$. Space group $P\,2_1\,2_1\,2_1$ has only one equivalent position whose multiplicity is 4, which is exactly the number of Br atoms. In this example, therefore, the heavy atom Br is in an unspecialized position in the cell and has three undetermined parameters, x, y, and z. These were found with the help of the Patterson function, discussed in Chapter 13. Although space group $P\,2_1\,2_1\,2_1$ is noncentrosymmetrical, its projections in the three axial directions are centrosymmetrical, so that a Fourier summation for a projection requires a knowledge of phases in terms of the signs of the Fourier coefficients only. A Fourier synthesis for projection of the electron density along c was computed with signs

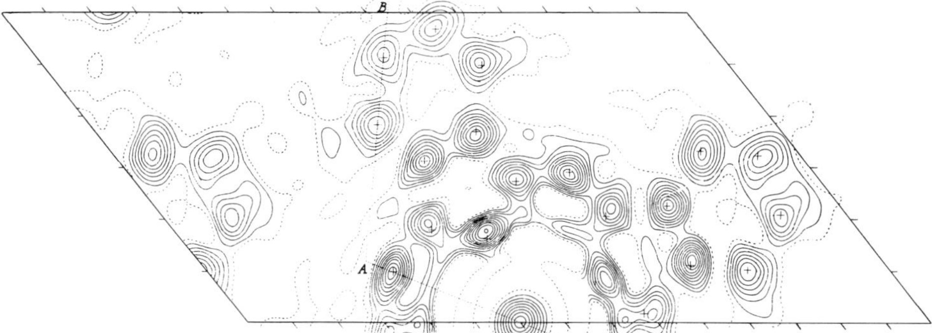

Fig. 11. Platinum phthalocyanine, $PtC_{32}H_{16}N_8$: electron-density projection $\rho(xz)$. [*From J. Monteath Robertson and Ida Woodward: J. Chem. Soc.* (1940) 44.]

determined by the Br atoms, with the result shown in Fig. 12*A*. The peaks of this projection show not only the Br atoms but lighter atoms as well. When the contribution of these additional atoms was added to a new Fourier summation, 5 of the original 137 signs were found to have been incorrectly given by Br alone. The new electron-density projection, improved by correcting these signs, is shown in Fig. 12*B*. The improvement in resolving power due to correcting the five signs is obvious. Yet it is evident that, even with these several incorrect phases, the contribution of the heavy atom was sufficient to provide a preliminary result in the form of a rough picture of the desired structure which, by repeating the Fourier synthesis, could be perfected.

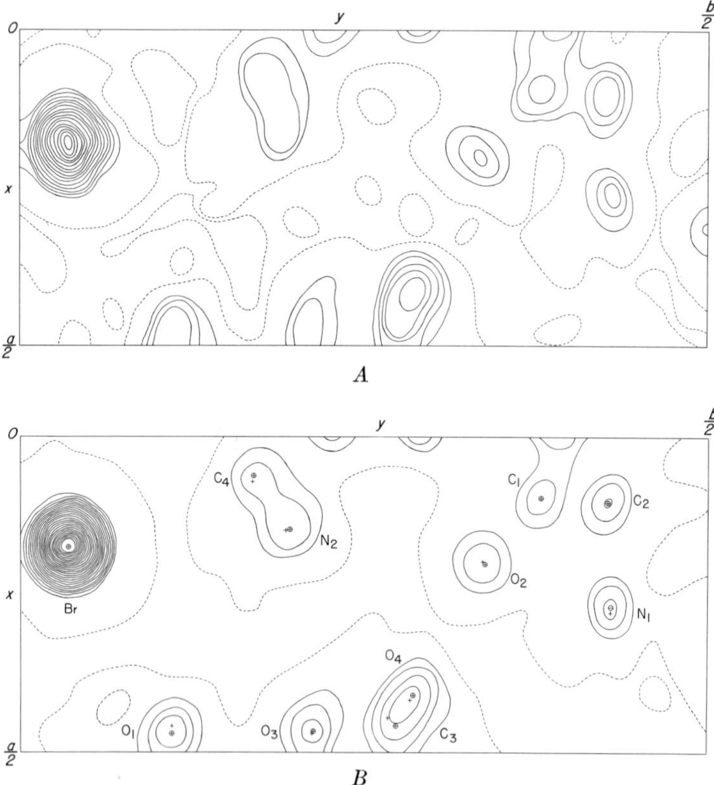

Fig. 12. Diglycine hydrobromide: electron density $\rho(xy)$, projected parallel to c. *A*: Fourier synthesis with Fourier coefficients based upon signs due to Br atoms alone. *B*: Fourier synthesis with Fourier coefficients based upon signs due to all atoms found in first Fourier synthesis above. [*From Martin J. Buerger: Crystal-structure analysis.* (*Wiley, New York*, 1960) 518, 519.]

The replaceable-atom method

Replaceable atoms. Many crystals are available in a variety of compositions each having the same arrangement of atoms. An especially simple case is a pair of compositions which are the same except for one atom, which is atom species M in one crystal and N in the other. Let the common part of the composition be represented by C. Then the two compositions are MC and NC. The atom M (or N) is called the replaceable atom in the structure. When such a pair of crystals is available, a comparison of the magnitude of each reflection *hkl* of one crystal with the magnitude of the corresponding reflection of the other may lead to a knowledge of the phases of the reflection for both crystals. It should be emphasized that this requires that the two structures have identical coordinates for all atoms.

Basic theory of phase determination. The basis of this method of phase determination is as follows: The wave scattered by the crystal can be resolved into a wavelet due to the replaceable atom M and the resultant of the wavelets scattered by the atoms of the common part C of the structure. For the two similar crystals this relation can be expressed as

$$F_{\mathrm{CM}} = F_{\mathrm{C}} + F_{\mathrm{M}} \tag{42}$$

and

$$F_{\mathrm{CN}} = F_{\mathrm{C}} + F_{\mathrm{N}}. \tag{43}$$

If these are subtracted, the result is

$$\begin{aligned} F_{\mathrm{CM}} - F_{\mathrm{CN}} &= F_{\mathrm{M}} - F_{\mathrm{N}} \\ &= \Delta F_{\mathrm{M-N}}. \end{aligned} \tag{44}$$

If the location of the replaceable atom in the structure can be determined, for example because it occupies a known set of equivalent positions or because it can be found by Patterson synthesis as discussed in Chapter 13, then the right side of (44) can be computed with respect to both magnitude and phase. On the other hand, only the experimental magnitudes of each of the two terms on the left side of (44) are known, but if (44) is properly dealt with, the phases of these terms can also be learned. To do this the magnitudes of the reflections on the left must be placed on an absolute scale. There are adequate methods of accomplishing this which need not be discussed here.

Centrosymmetrical crystals. If the crystal is centrosymmetrical, the phases reduce to the signs of the F's. The magnitude and sign of the right of (44) are easily computed. On the other hand, the magnitude of each term on the left is known, but not its sign. Each term can have either a positive or a negative sign, so there are four sign combinations.

Table 1
Determination of phases for chlorine camphor and bromine camphor
[After E. H. Wiebenga and C. J. Krom. Recl. Trav. Chim. **65** (1946) 673.]

1	2		3				4	5	
$h0l$	$\|F_{BrC}\|$	$\|F_{ClC}\|$	$F_{BrC} - F_{ClC}$ (+ +)	(+ −)	(− +)	(− −)	Computed $F_{Br} - F_{Cl}$	Deduced signs F_{BrC}	F_{ClC}
001	45	36	+ 9	+81	−81	− 9	+18	+	+
100	13	10	+ 3	+23	−23	− 3	− 3	−	−
$10\bar{1}$	15	35	−20	+40	−40	+20	+27	−	−
103	32	34	− 2	+66	−66	+ 2	+ 1		
300	7	<4	$>+3$	$<+11$	>-11	<-3	+ 7	+	
$20\bar{3}$	19	11	+ 8	+30	−30	− 8	+25	+	−
$30\bar{1}$	27	10	+17	+37	−37	−17	−18	−	−
301	40	19	+21	+59	−59	−21	+24	+	+
004	22	12	+10	+34	−34	−10	−14	−	−
$60\bar{7}$	11	7	+ 4	+18	−18	− 4	− 6	−	
606	9	7	+ 2	+16	−16	− 2	− 6	−	
$80\bar{1}$	9	7	+ 2	+16	−16	− 2	+ 5	+	+
800	8	4	+ 4	+12	−12	− 4	+ 4	+	+

The correct pair should match the computed value on the right of (44) in both magnitude and sign.

An example of how this scheme is applied is afforded by the solution of Br camphor and Cl camphor by Wiebenga and Krom. The sign determination for certain reflections is carried out in Table 1. Column 2 lists the measured magnitudes on an absolute scale. Column 3 shows their difference for the four possible sign combinations. Each of these possibilities is compared with the computed value of the right side of (44), which is listed in column 4. The resulting comparison justifies assigning the phases listed in column 5. This procedure determined all but 44 of the 160 F_{h0l}'s. These permitted computing the projected electron densities shown in Fig. 13. From these preliminary results, the arrangement of atoms in these halogen camphors was evident, and more perfect electron-density projections could be synthesized by another iteration.

Noncentrosymmetrical crystals. If the crystal lacks a center of symmetry, then the F's of (42) to (44) are complex, and so they must be treated as vectors in the complex plane. This requires that the vector F_{CM} be spanned by the vectors F_{CN} and ΔF_{M-N}. The magnitude and phase of ΔF_{M-N} are known, but for F_{CM} and F_{CN} only the magnitudes are available. The last two are accordingly represented as circles of radii

$|F_{\text{CM}}|$ and $|F_{\text{CN}}|$ on the Argand diagram (Fig. 14), since the directions of the vectors determined by their phases are unknown. Whenever the vector $\Delta F_{\text{M-N}}$ spans these circles, a solution of (44) occurs. There are, in general, two such solutions symmetrically located about the direction of $\Delta F_{\text{M-N}}$, so that (44) fails to provide a unique determination of phase.

A way out of this dilemma, which has been extensively used in establishing the structures of proteins, was pointed out by Harker. Large biological molecules commonly form crystals of small density within whose open spaces there is sometimes room for attaching various heavy

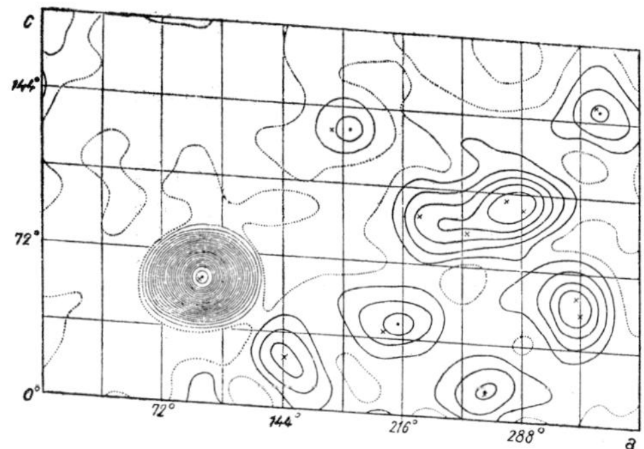

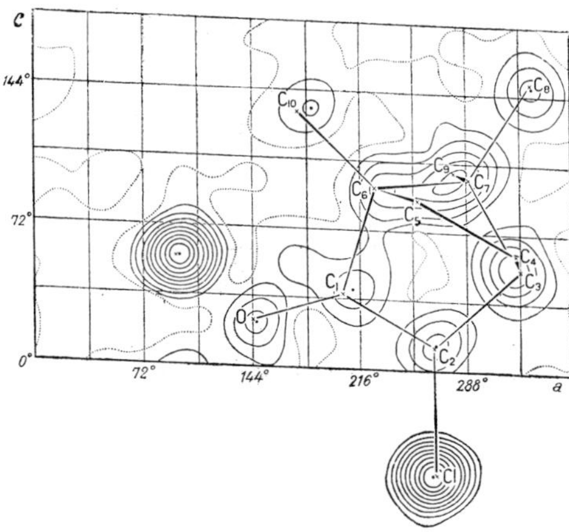

Fig. 13. Preliminary Fourier synthesis $\rho(xz)$ for Br camphor (*above*) and Cl camphor (*below*) prepared by using the signs predicted in Table 1. [*From E. H. Wiebenga and C. J. Krom: Recl. Trav. Chim.* **65** (1946) 673.]

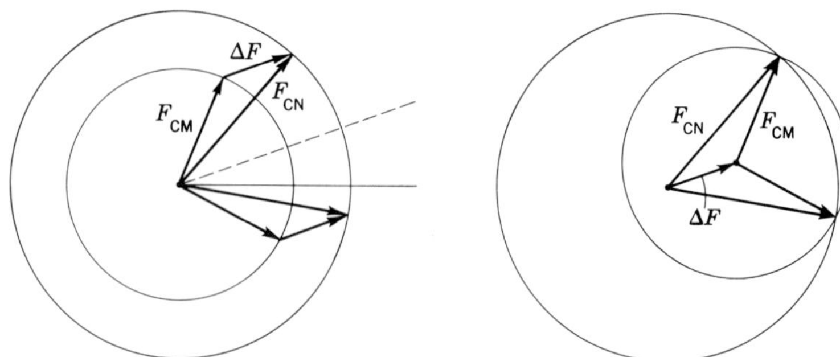

Fig. 14

atoms, or even dye molecules containing heavy atoms. Thus there are often available crystals having compositions C, CM, and CQ, where C is the protein itself, M is the heavy atom in one location in the cell, and Q is another heavy atom in a different location in the cell. The heavy atoms M and Q can be located; therefore their contributions F_M and F_Q are known in both magnitude and phase for every reflection. On the other hand, F_C, F_{CM}, and F_{CQ} are known in magnitude only from experimental measurement, and so must be represented on an Argand only as circles. The relation of F_C, F_M, and F_{CM} can be treated in the following way:

$$F_{CM} = F_C + F_M; \tag{45}$$

$$\therefore\ F_C = -F_M + F_{CM}. \tag{46}$$

The form of (46) is suitable for plotting on an Argand diagram; it can be interpreted as follows: F_C, whose solution is sought, is the sum of $-F_M$ and F_{CM}. The first is fully known and can be plotted on the Argand diagram as a vector, as seen in Fig. 15. On the other hand, F_C and F_{CM} are known only as magnitudes and can be plotted only as circles. F_C is a circle centered at the origin, while F_{CM} is seen from (46) to be a circle centered at the end of vector $-F_M$. These two circles, centered at the head and tail of the vector F_M, are shown in Fig. 15; they intersect, in general, in two places and thus provide two solutions, F_C' and F_C'', for the desired F_C of (46). Relations corresponding to (45) and (46) can also be written by substituting another atom Q for M. Again there arise two solutions for F_C of (46), as seen in Fig. 16. If $-F_M$ and $-F_Q$ lie along the same line, the pairs of solutions based on them coincide (because they are symmetrically directed about the vectors F_M and F_Q). But if $-F_M$ and $-F_Q$ do not lie along the same line in the Argand diagram, only one solution of the pair of solutions based upon $-F_M$ can coincide with one of the pair based upon $-F_Q$. In this case the common solution is the one sought. This is indicated in Fig. 17, where the com-

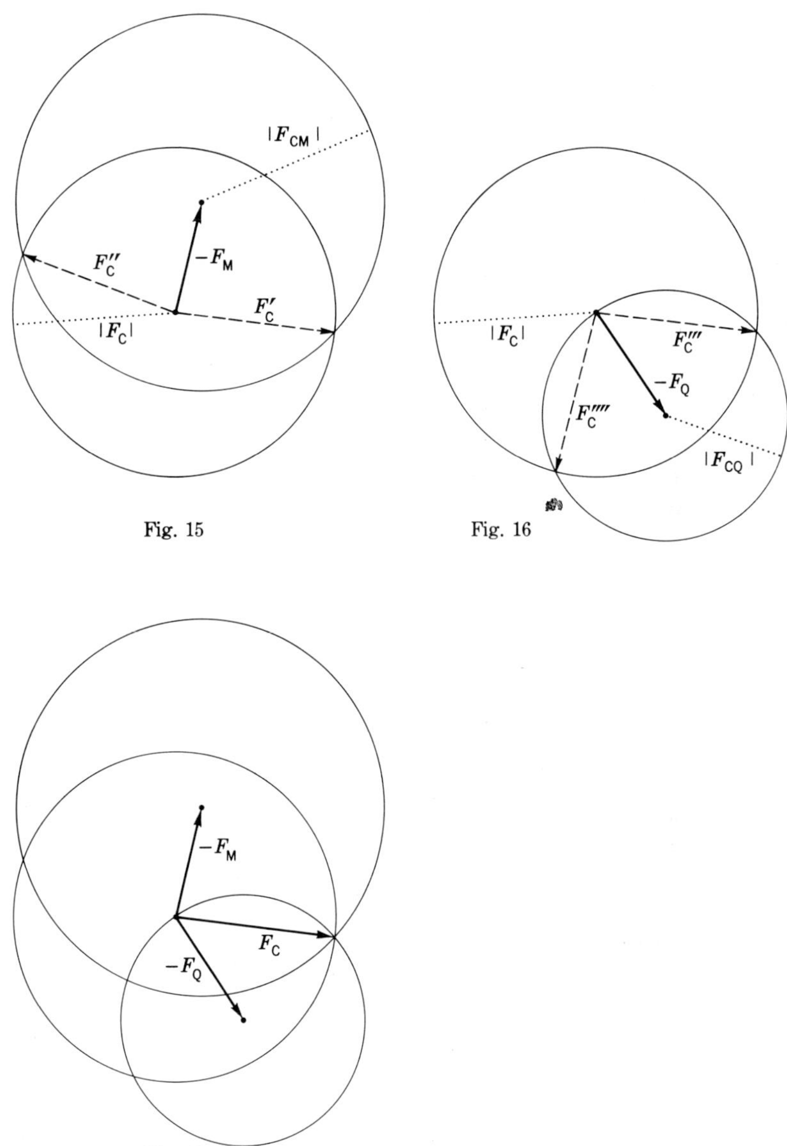

Fig. 15

Fig. 16

Fig. 17

mon solution is seen to be the common intersection of the three circles: one with radius $|F_{\mathrm{C}}|$ centered at the origin, one with radius $|F_{\mathrm{CM}}|$ centered at the end of vector $-F_{\mathrm{M}}$, and one with radius $|F_{\mathrm{CQ}}|$ centered at the end of vector $-F_{\mathrm{Q}}$. While this theory has been presented graphically by using the Argand diagram, it can also be expressed analytically. Accordingly, by making use of appropriate equations, the phases can be determined from the experimental intensities with the aid of an electronic computer.

This general method was used by Kendrew and collaborators to reveal the structure of the protein myoglobin, and by Perutz and collaborators to determine the structure of hemoglobin.

Myoglobin is one of the smaller proteins. Its type A crystals, which were studied by Kendrew et al., are monoclinic, with symmetry $P\,2_1$. The dimensions of the cell are

$$a = 64.6\ \text{Å} \qquad b = 31.3 \qquad c = 34.8 \qquad \beta = 105.5°.$$

This large cell, whose volume is 67,800 Å^3, contains 2 molecules of the protein, each having a molecular weight of 17,000 and consisting of some 1260 atoms (not counting hydrogen). The molecule has 153 amino acids. To permit determining the phases, three kinds of heavy-atom derivatives were attached to the molecule. In all, the intensities of the reflections from six kinds of crystals were measured:

1. Myoglobin + PCMBS (*p*-chloro-mercuri-benzenesulfonate)
2. Myoglobin + $HgAm_2$ (mercury diamine)
3. Myoglobin + Au (aurichloride)
4. Myoglobin + PCMBS + $HgAm_2$
5. Myoglobin + PCMBS + Au
6. Myoglobin

Crystals 4 and 5 are double derivatives. The heavy atoms were located by a modification of the Patterson-type Fourier synthesis (discussed in Chapter 13) but using as Fourier coefficients $\Delta F^2 = (|F_{M_1M_2}| - |F|)^2$, where M_1 and M_2 refer to heavy atoms in the double derivatives like 4 and 5 above. Some of the solutions for the phases of the unsubstituted myoglobin are shown in Fig. 18. In favorable cases, all circles were found to intersect in a small angular range, but in many cases the angular range was fairly large.

In order to explore the structure of myoglobin with a resolution of 6 Å, 400 sets of reflections had to be measured. These consisted of 100 centric $h0l$'s and 300 acentric hkl's. For the 6 Å investigation the intensity data were plotted and phases determined graphically, as in Fig. 18. To improve the resolution to 2 Å, some 9000 acentric reflections per crystal were later processed with the aid of an electronic computer.

Hemoglobin is a considerably larger protein. The material used by Perutz et al. was horse oxyhemoglobin, or methemoglobin. This is

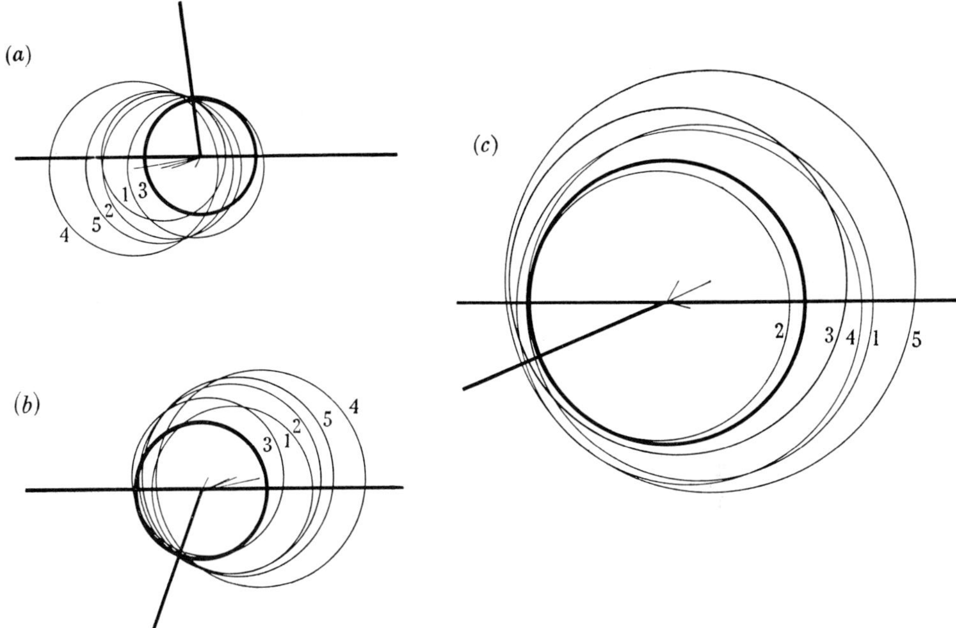

Fig. 18. Examples of phase determination for crystals of myoglobin and its derivatives. The heavy circle represents the amplitude of the reflection from the unsubstituted protein, and the light circles those from the following derivatives: 1, PCMBS; 2, $HgAm_2$; 3, Au; 4, PCMBS + $HgAm_2$; 5, PCMBS + Au. The short lines from the centers are the heavy-atom vectors; the heavy line indicates the phase angle eventually selected. The reflections illustrated are (*a*) 411, (*b*) $91\bar{1}$, (*c*) 212. [*From G. Bodo, H. M. Dintzis, J. C. Kendrew, and H. W. Wyckoff: Proc. Roy. Soc.* (*A*) **253** (1959) 91.]

monoclinic, symmetry $C\ 2$, with a cell whose dimensions are

$$a = 109.2\ \text{Å}$$
$$b = 63.2 \qquad \beta = 110.7°.$$
$$c = 64.7$$

This large cell has a volume of 352,000 Å^3. It contains 2 molecules, each with a molecular weight of 67,000, composed of some 10,000 atoms. The structure of this was solved by determining the phases by the replaceable-atom method, using the pure protein and its mercury and silver derivatives. The intensities of some 1200 reflections had to be measured to supply a resolution of 5.5 Å.

Direct methods

The heavy-atom method and replaceable-atom method provide strategies for determining crystal structures in spite of the phase problem.

There also exist other strategies which can be used for the same purpose under certain circumstances. But all these schemes provide indirect solutions of the phase problem, and cannot be depended upon to lead to the arrangement of atoms in an arbitrary crystal.

While there exists no general solution of the phase problem, there have been discovered several relations from which a set of unknown phases can be derived from the set of experimental diffraction magnitudes. Such methods have been used to determine the required phases for crystal structures having a limited number of atoms per unit cell. Two such general methods, which can be illustrated graphically, are noted below.

Harker-Kasper inequalities. Among the well-known inequalities of mathematics is Cauchy's inequality; this provides the relation between the square of the absolute value of a sum of products, and the products of the sums of the squared absolute values. The corresponding inequality for integrals (rather than sums) is known as Schwartz's inequality. David Harker and John Kasper made use of this in connection with the expression for the reflection amplitude, which has the general form

$$F_{hkl} = V \int_0^1 \int_0^1 \int_0^1 \rho(xyz) e^{i2\pi(hx+ky+lz)}\, dx\, dy\, dz. \tag{47}$$

The right side of this can be rewritten as the product of integrals in several ways. When these are substituted in Schwartz's inequality, a number of useful results can be obtained, especially when the right-hand side of (47) is expressed in forms which take account of the symmetry elements of the crystal. These results are made more useful if the F in (47) is divided by Z, the number of electrons in the cell. The quotient is called a *unitary structure factor*,

$$U_{hkl} = \frac{F_{hkl}}{Z}. \tag{48}$$

This quantity has the same phase as F, but its magnitude is such that its maximum value is $F_{000}/Z = 1$, since all electrons scatter in phase for F_{000}. The magnitude of the unitary structure factor is thus a measure of the efficiency of scattering for the reflection hkl.

A list of the simplest inequalities which result from these considerations is given in Table 2. A crystal whose space group contains several of the symmetry elements listed in the left column of Table 2 has its phases restricted by each of the limitations in the right column, plus their combinations. Many such relations exist for space groups with even a few symmetry elements, but no complete tabulations of these relations for the many space groups have yet been made.

Table 2
Harker-Kasper inequalities

Symmetry element $\parallel c$	Limitation
1	$\lvert U_{hkl}\rvert^2 \leq 1$
$\bar{1}$	$U^2_{hkl} \leq \frac{1}{2} + \frac{1}{2}U_{2h\,2k\,2l}$
2	$\lvert U_{hkl}\rvert^2 \leq \frac{1}{2} + \frac{1}{2}U_{2h\,2k\,0}$
2_1	$\lvert U_{hkl}\rvert^2 \leq \frac{1}{2} + \frac{1}{2}(-1)^l\, U_{2h\,2k\,0}$
$\bar{2} = m$	$\lvert U_{hkl}\rvert^2 \leq \frac{1}{2} + \frac{1}{2}U_{0\,2k\,0}$
a	$\lvert U_{hkl}\rvert^2 \leq \frac{1}{2} + \frac{1}{2}(-1)^h\, U_{0\,0\,2l}$
3	$\lvert U_{hkl}\rvert^2 \leq \frac{1}{3} + \frac{2}{3}\lvert U_{(h-k)(h+2k)\,0}\rvert \cos 2\pi\phi_{(h-k)(h+2k)0}$
3_1, 3_2	$\lvert U_{hkl}\rvert^2 \leq \frac{1}{3} + \frac{2}{3}\lvert U_{(h-k)(h+2k)0}\rvert \cos 2\pi(\phi_{(h-k)(h+2k)0} + \frac{1}{3}l)$
$\bar{3} = 3 + \bar{1}$	$U^2_{hkl} \leq \frac{1}{6} + \frac{1}{6}U_{2h\,2k\,2l} + \frac{1}{3}U_{h\,k\,2l} + \frac{1}{3}U_{(h-k)(h+2k)0}$
4	$\lvert U_{hkl}\rvert^2 \leq \frac{1}{4} + \frac{1}{4}U_{2h\,2k\,0} + \frac{1}{2}U_{(h-k)(h+k)0}$
4_1, 4_3	$\lvert U_{hkl}\rvert^2 \leq \frac{1}{4} + \frac{1}{4}(-1)^l\, U_{2h\,2k\,0} + \frac{1}{2}(\cos 2\pi\frac{1}{4}l)\,U_{(h-k)(h+k)0}$
4_2	$\lvert U_{hkl}\rvert^2 \leq \frac{1}{4} + \frac{1}{4}U_{2h\,2k\,0} + \frac{1}{2}(-1)^l\, U_{(h-k)(h+k)0}$
$\bar{4}$	$\lvert U_{hkl}\rvert^2 \leq \frac{1}{4} + \frac{1}{4}U_{2h\,2k\,0} + \frac{1}{2}\lvert U_{(h-k)(h+k)2l}\rvert \cos 2\pi\phi_{(h-k)(h+k)2l}$
6	$\lvert U_{hkl}\rvert^2 \leq \frac{1}{6} + \frac{1}{6}U_{2h\,2k\,0} + \frac{1}{3}U_{(h-k)(h+2k)0} + \frac{1}{3}U_{hk0}$
6_1, 6_5	$\lvert U_{hkl}\rvert^2 \leq \frac{1}{6} + \frac{1}{6}(-1)^l\, U_{2h\,2k\,0} + \frac{1}{3}(\cos 2\pi\frac{1}{3}l)\,U_{(h-k)(h+2k)0} + \frac{1}{3}(\cos 2\pi\frac{1}{6}l)\,U_{hk0}$
6_2, 6_4	$\lvert U_{hkl}\rvert^2 \leq \frac{1}{6} + \frac{1}{6}U_{2h\,2k\,0} + \frac{1}{3}(\cos 2\pi\frac{1}{3}l)\,U_{(h-k)(h+2k)}0 + \frac{1}{3}(\cos 2\pi\frac{1}{3}l)\,U_{hk0}$
6_3	$\lvert U_{hkl}\rvert^2 \leq \frac{1}{6} + \frac{1}{6}(-1)^l\, U_{2h\,2k\,0} + \frac{1}{3}U_{(h-k)(h+2k)0} + \frac{1}{3}(-1)^l\, U_{hk0}$
$\bar{6} = \dfrac{3}{m}$	$\lvert U_{hkl}\rvert^2 \leq \frac{1}{6} + \frac{1}{6}U_{0\,0\,2l} + \frac{1}{3}\lvert U_{(h-k)(h+2k)0}\rvert \cos 2\pi\phi_{(h-k)(h+2k)0} + \frac{1}{3}\lvert U_{(h-k)(h+2k)2l}\rvert \cos 2\pi\phi_{(h-k)(h+2k)2l}$

The use of these inequalities in phase determination can be illustrated by the relation in Table 2 for symmetry $\bar{1}$:

$$U^2_{hkl} \leq \tfrac{1}{2} + \tfrac{1}{2}U_{2h\,2k\,2l}. \tag{49}$$

The quantity U^2_{hkl} is available as the result of measuring F^2_{hkl} and transforming it by (48). Under certain circumstances the sign of $U_{2h\,2k\,2l}$ can be determined with the aid of (49), specifically, if U^2_{hkl} is $\frac{1}{2}$ or greater. In this event, (49) can be satisfied only if $U_{2h\,2k\,2l}$ is positive. But if U^2_{hkl} is less than $\frac{1}{2}$, no conclusion can be drawn.

A graphical justification of this general relation is shown in Fig. 19. Since the symmetry is $\bar{1}$, the crystal is centrosymmetrical, and all Fourier density waves, whose amplitudes are F/V, must have a cosine form and so must be either + or − at the origin. Now, if F_{hkl} is large, this implies that a considerable electron density (suggested by the black dots in Fig. 19) must occur near the crests of the density wave represented by F_{hkl}.

Whether F_{hkl} is positive (upper part of Fig. 19) or negative (lower part of Fig. 19), the amplitude of $F_{2h\,2k\,2l}$ must be positive beneath the black dot. Figure 19 shows that this requires $F_{2h\,2k\,2l}$ to be positive at the origin. Thus, whether F_{hkl} is positive or negative, if it is sufficiently large, $F_{2h\,2k\,2l}$ is positive. The more exact condition is provided by (49).

Unfortunately, the Harker-Kasper inequalities, even when improved by the device of transforming the F's (and corresponding U's) to the values they would have if the atoms were points, are not powerful enough to solve very complicated crystal structures. This is basically due to the fact that, as pointed out in a foregoing section of this chapter, if the number of atoms in a cell is large enough and they are uniformly distributed throughout the cell, the individual atoms tend to scatter wavelets which annul one another, thus producing a small magnitude for the resultant F_{hkl}. Hughes has shown that, when the number of atoms in the asymmetric unit is more than about 5, the Harker-Kasper inequalities do not ordinarily yield enough signs to solve the structure.

Relations between phases of reflections. A relation which has come into common use in determining the signs of reflections of centrosymmetrical crystals from their intensities was first discovered by Sayre and later confirmed through careful study by other investigators. Sayre considered a crystal composed of equal, resolved atoms, as suggested by the graphical representation of their electron density in the upper part of Fig. 20. He reasoned that the phases of the reflections were controlled chiefly by the positions of the atoms in the cell, and not by the shapes of

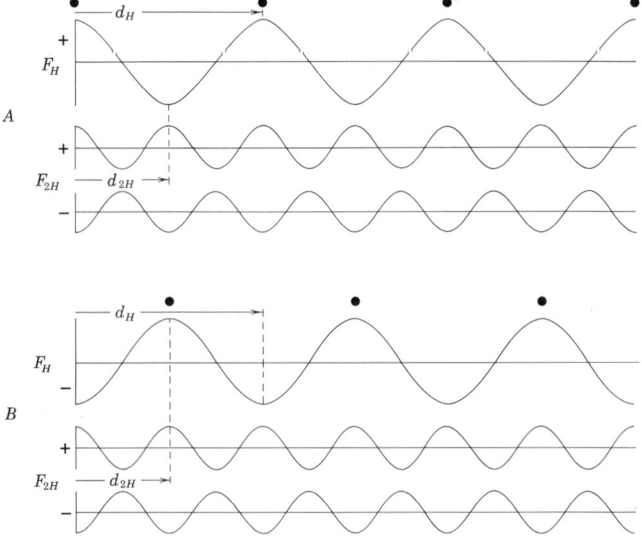

Fig. 19. [*From Martin J. Buerger: Crystal-structure analysis.* (*Wiley, New York,* 1960) 561.]

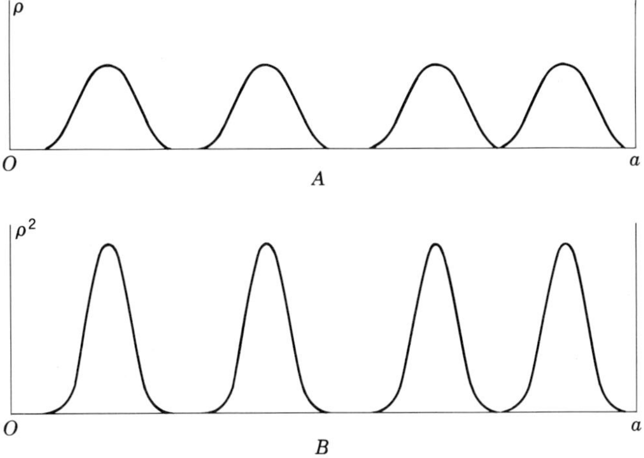

Fig. 20. Comparison of the functions ρ and ρ^2 for an equal-atom structure. [*From M. M. Woolfson: Direct methods in crystallography.* (*Clarendon Press, Oxford,* 1961) 43.]

the atoms. Accordingly, the electron density and its square ought to give reflections with the same sign. If hkl is abbreviated as H, these two electron densities are given by

$$\rho(x) = \frac{1}{V}\sum_H F_H e^{-i2\pi Hx} \tag{50}$$

and

$$\rho^2(x) = \frac{1}{V}\sum_H F_H^{sq} e^{i-2\pi Hx}. \tag{51}$$

When (50) is squared to produce (51), it can be reduced as follows:†

$$\rho^2(x) = \rho(x)\rho(\bar{x}) = \left[\frac{1}{V}\sum_H F_H e^{-i2\pi Hx}\right]\left[\frac{1}{V}\sum_H F_H e^{-i2\pi H\bar{x}}\right] \tag{52}$$

$$= \frac{1}{V^2}\sum_{H_1}\sum_{H_2} F_{H_1}F_{H_2}e^{-i2\pi(H_1-H_2)x}$$

$$= \frac{1}{V}\sum_{H_1}\left[\frac{1}{V}\sum_{H_2} F_{H_1}F_{H_2}\right]e^{-i2\pi(H_1-H_2)x}. \tag{53}$$

† There are some simple mathematical tricks in this sequence of lines. In the first line, advantage is taken of the fact that for a centrosymmetrical crystal the electron densities at x and $\bar{x}$ are identical. Therefore

$$\rho^2(x) = \rho(x)\rho(x) = \rho(x)\rho(\bar{x}).$$

Then when the two bracketed terms on the right of (52) are multiplied together, each is a sum of many terms like $F_H e^{-i2\pi Hx}$. Each term of one sum must be multiplied by each term in the other sum. The general term in the product has different H's; these are designated H_1 and H_2 in the second line.

If H is substituted for $H_1 - H_2$, and H' for H_2, (53) can be expressed as

$$\rho^2(x) = \frac{1}{V}\sum_H \left[\frac{1}{V}\sum_{H'} F_{H'}F_{H+H'}\right] e^{-i2\pi Hx}. \tag{54}$$

When this is compared with (51), it is seen that

$$F_H^{sq} = \frac{1}{V}\sum_{H'} F_{H'}F_{H+H'}. \tag{55}$$

Now the crystal and the artificial crystal with squared density have different Fourier coefficients. Nevertheless, if the original electron density is composed of resolved, equal atoms, the squared density is determined by coefficients whose magnitudes are each related by a factor g_H to those of the crystal, so that (55) can be simplified to

$$F_H = \frac{1}{gV}\sum_{H'} F_{H'}F_{H+H'}. \tag{56}$$

This relation must hold if the signs of F_H are correct. For example, when the left side of (56) is F_9, the right involves the sums of products like F_1F_{9+1}, F_2F_{9+2}, and F_3F_{9+3}.

Sayre investigated some crystal structures which had been solved by other methods and found that, in the sum of products which occurs on the right of (56), usually one term is so large that it dominates the sum, and this sum controls the sign of the right-hand side and therefore of F_H. This conclusion can be stated as

$$S_H = S_{H'}S_{H+H'}, \tag{57}$$

in which S_H signifies "the sign of reflection H", etc. This shows that many signs of reflections are related. A more useful form of (57) is

$$S_{H+H'} = S_H S_{H'}. \tag{58}$$

This rule has been found to hold well when the magnitudes of F_H, $F_{H'}$, and $F_{H+H'}$ are all large. The significance of (58) can be demonstrated graphically, as shown in Fig. 21. The upper diagram of Fig. 21 shows the vectors to the locations of the positions of indices H, H', and $H + H'$ in reciprocal space. The planes H, H', and $H + H'$ in the crystal are normal to the corresponding vectors in reciprocal space, and the spacings of these planes are proportional to the reciprocals of the lengths of the vectors. The lower four parts of Fig. 21 illustrate the four combinations of signs for H and H'. The crests of the waves are shown in full lines, the troughs in broken lines. When F_H and $F_{H'}$ have large magnitudes, they make large density contributions at the intersections of the crests of these waves. These density contributions are suggested by

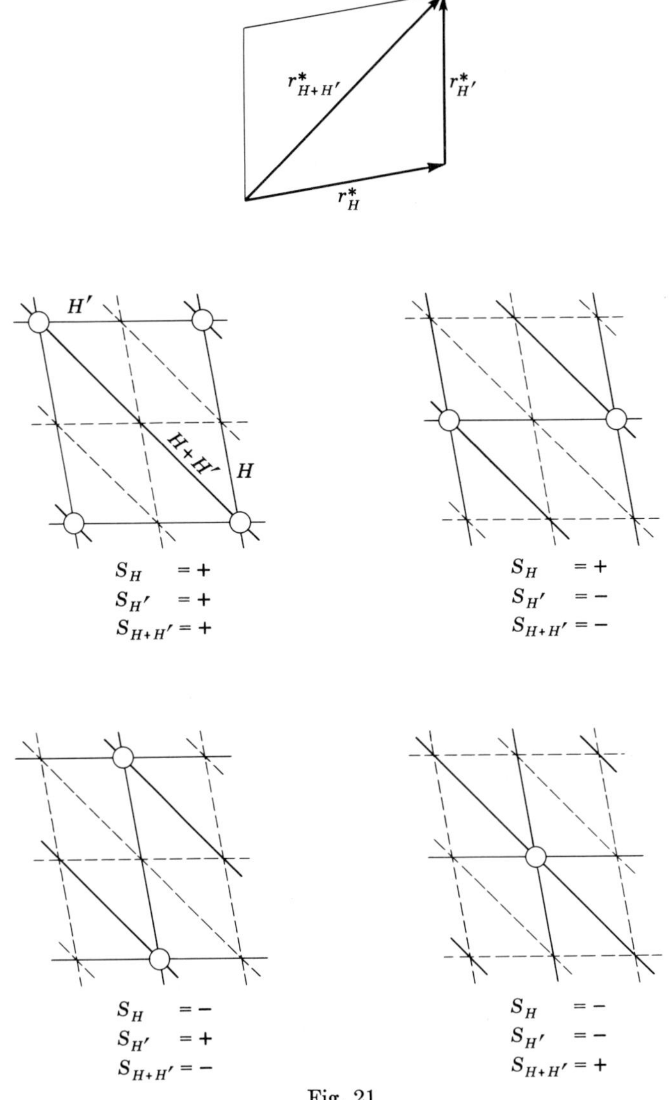

Fig. 21

the open circles. A crest of the density wave corresponding to $F_{H+H'}$ is accordingly called for at these circles. The signs of $F_{H+H'}$ which are required by these combinations are shown below each graphical representation. It is evident that the product of the signs of F_H and $F_{H'}$ is the sign of $F_{H+H'}$, as required by (58).

Zachariasen derived a similar relation, which was placed on a statistical basis. Zachariasen's result may be expressed as

$$S_{H+H'} = \overline{S(S_H S_{H'})}^{H+H'}. \tag{59}$$

This says that the sign of $H + H'$ is found by averaging the result of $S(S_H S_{H'})$ over all reflections whose indices add to $H + H'$. This takes account of the fact that a point in reciprocal space $H + H'$ can be reached by a number of routes corresponding to the sums of different vectors $\mathbf{r}_H + \mathbf{r}_{H'}$. These do not always give consistent results for $S_{H+H'}$. Zachariasen's method averages the result for various routes.

Karle and Hauptman derived an equivalent of Sayre's relation incidental to a general statistical investigation of how signs are limited by magnitudes for centrosymmetrical crystals. Their results have been extensively used by J. Karle and I. Karle to derive the structures of organic crystals, in which the C, O, and N atoms have substantially equal scattering powers. In this method it is customary to convert the F's to what are called *normalized structure factors*, designated as E's. This is done by dividing F_H by the average value of the magnitude of the F's in the set.

The average magnitude of F can be found by a method due to A. J. C. Wilson: Since F is a complex number, its squared magnitude is found by multiplying it by its complex conjugate:

$$|F_{hkl}|^2 = F_{hkl}\tilde{F}_{hkl} \tag{60}$$

$$= \Big(\sum_j f_j e^{i2\pi(hx_j+ky_j+lz_j)}\Big)\Big(\sum_{j'} f_{j'} e^{-i2\pi(hx_{j'}+ky_{j'}+lz_{j'})}\Big)$$

$$= \sum_j \sum_{j'} f_j f_{j'} e^{i2\pi(h[x_j-x_{j'}]+k[y_j-y_{j'}]+l[z_j-z_{j'}])}. \tag{61}$$

It is convenient to separate this into two parts, accordingly, as $j = j'$ and $j \neq j'$:

$$|F_{hkl}|^2 = \sum_j f_j^2 + \sum_j \sum_{\substack{j' \\ j\neq j'}} f_j f_{j'} e^{i2\pi(h[x_j-x_{j'}]+k[y_j-y_{j'}]+l[z_j-z_{j'}])}. \tag{62}$$

If the average is taken over all hkl, there are as many negative as positive components, and so these cancel in the double summation, and there remains

$$\langle |F_{hkl}|^2\rangle_{hkl} = \sum_j f_{j,hkl}^2 \tag{63}$$

so that

$$\langle |F_{hkl}|^2\rangle_{hkl}^{\frac{1}{2}} = \Big(\sum_j f_{j,hkl}^2\Big)^{\frac{1}{2}}. \tag{64}$$

Thus the normalized structure factor is found by transforming the F's in the following way:

$$E_H = \frac{F_H}{\langle |F|^2 \rangle^{\frac{1}{2}}} \tag{65}$$

$$= \frac{F_H}{\epsilon \left(\sum_j f^2\right)^{\frac{1}{2}}}. \tag{66}$$

The coefficient ϵ (which is 1 for space group $P\,\bar{1}$) is needed when F occupies a symmetry location in reciprocal space. This system has the property that the average value of $|E|^2$ is 1 for all crystals.

Karle and Hauptman derived, among others, the two following relations, which are in common use for centrosymmetrical crystals, especially for symmetry $P\,\bar{1}$:

$$\Sigma_1: \quad SE_{2H} \approx S(E_H^2 - 1), \tag{67}$$

$$\Sigma_2: \quad SE_H \approx S \sum_K E_K E_{H+K}. \tag{68}$$

The first of these is a generalization of the Harker-Kasper relation for symmetry $\bar{1}$; the second is a generalization of the Sayre relation.

In using sign relations for centrosymmetrical crystals, it is important to recognize that, since the structure is centrosymmetrical, it has several distinct sets of symmetry centers, any one of which could be used as an origin. A change of origin from one symmetry center to another changes the signs of some F's but leaves others, called *structure invariants*, unchanged. This implies that a change of signs of some F's, namely, those which are not structure invariants, merely has the effect of changing the origin of the Fourier synthesis of the electron density. Thus, the signs of a certain number of F's can be set arbitrarily, since this simply refers an unknown structure to a particular one of its permissible symmetry centers as origin. For space group $P\,\bar{1}$ there are eight symmetry centers, located at the cell corners: 000; at the centers of cell edges: $00\frac{1}{2}$, $0\frac{1}{2}0$, $\frac{1}{2}00$; at the centers of cell faces: $0\frac{1}{2}\frac{1}{2}$, $\frac{1}{2}0\frac{1}{2}$, $\frac{1}{2}\frac{1}{2}0$; and at the cell center: $\frac{1}{2}\frac{1}{2}\frac{1}{2}$. If a particular amplitude F_{hkl} has a certain sign when referred to 000 as an origin, the sign it has when referred to one of the other permissible origins depends on the even or odd nature of the indices h, k, and l. The specific sign for another origin is found by multiplying the sign for F_{hkl} referred to 000 by the sign given in Table 3.

It is seen that the sign is unchanged for reflections having h even, k even, and l even. Such reflections are said to be *structure-invariant* (which really means origin-invariant, that is, invariant under transformation of origin). For triclinic, monoclinic, and orthorhombic space groups with primitive cells, the signs of any three F_{hkl}'s which are not structure-invariant can be set arbitrarily provided that the hkl's are

Table 3
Sign changes involved in referring F_{hkl} to various origins
(e = even; o = odd)

Origin at	Parity-group number								
	1	2	3	4	5	6	7	8	
	e	e	e	o	e	o	o	o	h
	e	e	o	e	o	e	o	o	k
	e	o	e	e	o	o	e	o	l
000	+	+	+	+	+	+	+	+	
$00\frac{1}{2}$	+	−	+	+	−	−	+	−	
$0\frac{1}{2}0$	+	+	−	+	−	+	−	−	
$\frac{1}{2}00$	+	+	+	−	+	−	−	−	
$0\frac{1}{2}\frac{1}{2}$	+	−	−	+	+	−	−	+	
$\frac{1}{2}0\frac{1}{2}$	+	−	+	−	−	+	−	+	
$\frac{1}{2}\frac{1}{2}0$	+	+	−	−	−	−	+	+	
$\frac{1}{2}\frac{1}{2}\frac{1}{2}$	+	−	−	−	+	+	+	−	

"linearly independent mod 2." This means that any reflection $h_1k_1l_1$ from a column other than column 1 of Table 3 can be assigned a phase at pleasure. When this is done, four of the eight possible origins listed on the left column are discarded. Another reflection $h_2k_2l_2$ can be selected from another column and assigned a phase provided that the sums h_1+h_2, k_1+k_2, and l_1+l_2 are not all even, that is, not structure-invariant. This choice eliminates half the four origins which were not eliminated by the choice of the first reflection. Finally, a third reflection $h_3k_3l_3$ can be chosen from another column and assigned a phase provided that the sums $h_1+h_2+h_3$, $k_1+k_2+k_3$, and $l_1+l_2+l_3$ are not all even. This assignment of signs to these reflections fixes the origin of the structure to be synthesized by Fourier summation.

In the practical application of sign relations, it is convenient to divide all observed reflections into the eight parity groups of Table 3. For each group, the reflections are listed in order of decreasing $|E_{hkl}|$. From these columns are first chosen three linearly independent origin-determining reflections which have the largest values and are otherwise suitable in that they can be used in many sign-determining combinations. In addition to assuming known phases for these three reflections, some other reflections usually have to be assigned phases which are unknown, but which can be represented symbolically by the letters $a, b, c, d, \ldots$, each of which will eventually turn out to be either $+$ or $-$.

The determination of signs based upon these reflections can now proceed by making use of

(68): $$SE_H \approx S \sum_{H'} S_{H'}S_{H-H'},$$

where H is shorthand for *hkl*. This sign determination is preferably started by limiting consideration initially to reflections for which $|E| \geq 1.5$. While the following relation cannot be derived here, it can be shown that the probability that the sign of the reflection is positive is given by

$$P_+(H) = \tfrac{1}{2} + \tfrac{1}{2} \tanh \left[\frac{\sigma_3}{\sigma_2^{\frac{3}{2}}} |E_H| \sum_{H'} E_{H'}E_{H-H'}\right], \tag{69}$$

where $$\sigma_m = \sum_{j=1}^{N} f^m. \tag{70}$$

The coefficient in brackets is equal to $N^{-\frac{1}{2}}$ for structures with N equal atoms per cell. In general, a sign determination is not accepted unless this probability exceeds 0.97.

After the E's with large magnitudes are exhausted, those with smaller magnitudes are gradually used. The phase assignment ends when most E's with larger magnitudes have their signs determined as $+$, $-$, a, b, c, d, ... Sometimes one or more of the unknowns can be determined as $+$ or $-$ by means of

(67): $$SE_{2H} = S(E_H^2 - 1).$$

There may remain several possible combinations of signs for the entire set of E's involving some unknown signs a, b, ... Some of these may be eliminated by external evidence. For example, if most phases are $+$, a heavy atom at the origin will be built up by the Fourier summation. This may be known to be unacceptable on chemical grounds because the crystal is an organic one without heavy atoms. In the end, however, several Fourier syntheses may have to be computed, one with each combination of the Fourier coefficients whose signs have not been fixed. Ordinarily the correct structure can be easily recognized by its obvious correspondence with the known chemical composition of the crystal.

This general method of sign determination is commonly called the *symbolic addition method.* It has been much used by J. Karle and I. L. Karle for the solution of the structures of organic crystals, some of which have had up to 35 atoms per asymmetric unit. An example of a structure solved by them is that of the alkaloid jamine, $C_{21}H_{35}N_3$. This forms triclinic crystals, space group $P\,\bar{1}$, with cell dimensions

$$\begin{aligned} a &= 6.79 \text{ Å} & \alpha &= 95°00' \\ b &= 10.61 & \beta &= 97°20' \\ c &= 13.41 & \gamma &= 103°55'. \end{aligned}$$

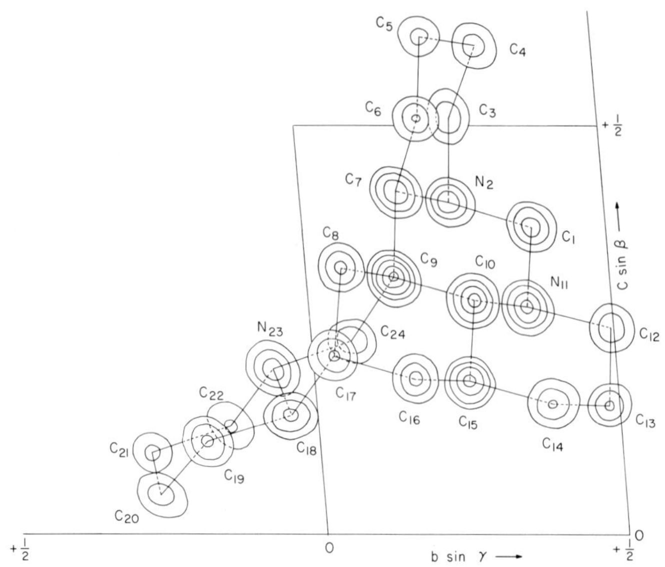

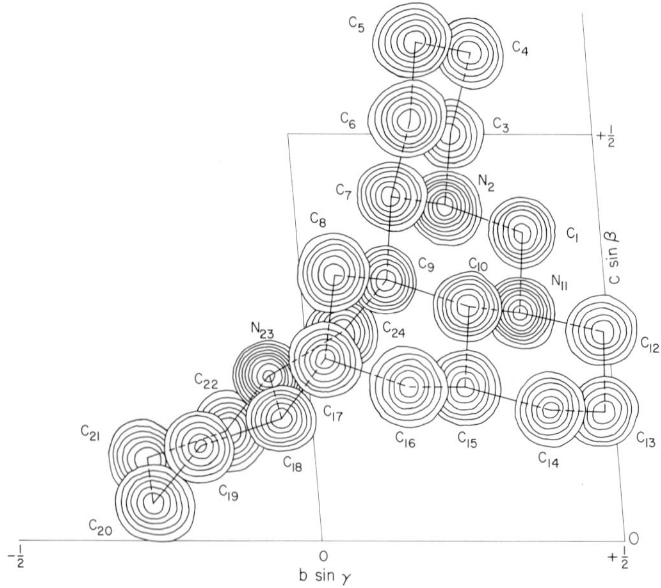

Fig. 22. Comparison of the *E* map (*above*) with the electron density (*below*) of the alkaloid jamine $C_{21}H_{35}N_3$. [*From Isabella L. Karle and J. Karle: Acta Cryst.* **17** (1964) 1358, 1359.]

There are 2 molecules of $C_{21}H_{35}N_3$ in this cell. The sign-determining procedure left four sets of unknown signs, labeled a, b, d, and g. This permitted $2^4 = 16$ possible structures. These sign combinations were used with the experimental $|E_{hkl}|$'s to produce three-dimensional Fourier syntheses called E maps. One of these, shown in Fig. 22, which was based upon 286 signs with $|E|$'s > 1.35, was judged to correspond to the chemistry in an appropriate way and was accepted as correct. This revealed the locations of all atoms; from these the signs of all F_{hkl}'s could be computed. After refinement of the atom parameters by the method of least squares (discussed in Chapter 14), the electron density was computed by using all 1831 F_{hkl}'s, with the result shown in Fig. 22.

This method of phase determination has been extended to noncentrosymmetrical structures. It was used successfully to solve the structure of the alkaloid $C_{20}H_{33}N_3 \cdot 2ClO_4$ by Isabella L. Karle and J. Karle. This has the noncentric monoclinic symmetry $P\,2_1$ with 29 nonhydrogen atoms per cell.

Summary

The general problem of finding the arrangement of atoms in a crystal from the diffraction by the crystal can be neatly formulated in terms of a mathematical function called a *Fourier transform.* An ordinary lens produces the physical analog of a Fourier transformation by focusing at its back focal plane the parallel rays which reach it from the object. This is the diffraction image of the object. If the rays are allowed to continue beyond the diffraction image, they eventually converge again, but in a different combination, to produce an image of the inverse object. The original object and its diffraction image are physical analogs of the mathematical relationship known as *Fourier mates.* Each contains the same information, but this is assembled in two different ways. If everything is known about either one, the other can be formed by a Fourier transformation. This mathematical transformation has the same characteristics as the converging of rays by the lens.

For most purposes in crystallography, a Fourier transformation takes the form of summing a series of terms. Each term is the product of an amplitude and an exponential (or trigonometric) term which acts as an operator that shifts the phase of the amplitude term. The series, which builds up the image of the electron density of the crystal structure from the diffraction amplitudes F_{hkl}, has terms of the form $(F_{hkl}/V)e^{-i2\pi(hx+ky+lz)}$, where V is the volume of the cell, and x, y, and z are the fractional coordinates of the points in the cell where the electron density is required.

Unfortunately, F_{hkl} is not just a positive number but, in general, is a complex number having both magnitude and phase. The magnitude can be measured (in the form of the intensity of the reflection), but there is no experimental way of observing the phase. Thus, in attempting to build up the electron density of the crystal by summing the terms of a Fourier series, it becomes obvious that the experimental data do not supply the Fourier coefficients fully. This deficiency constitutes the *phase problem* of x-ray crystallography.

In spite of the phase problem, crystal structures can be determined from data derived by x-ray diffraction. The methods for accomplishing this fall into two classes. One kind of method, discussed in the next chapter, treats the subject in crystal space; the other, discussed in this chapter, treats it in diffraction (Fourier) space. Among the latter there are many varieties; some are useful only for certain special classes of crystals; others, called direct methods, are applicable to any crystal.

When the chemical composition of a crystal is composed of a heavy atom plus a number of lighter ones, it is ordinarily possible to find the location of the heavy atom by several means, in spite of the phase problem. This is basically because it is not much more difficult to find the locations of a set of equivalent heavy atoms immersed in a matrix or light atoms than to find the locations of a set of equivalent atoms in an otherwise empty cell. This problem can be solved easily by several means from diffraction intensities alone. Once the heavy atom has been located, its contribution to the wave scattered by the entire cell can be computed in both magnitude and phase. This wavelet scattered by the heavy atom is the chief contribution to the wave scattered by the entire cell, and dominates it. Accordingly, the F_{hkl} of the crystal tends to have the same phase as that scattered by the heavy atom alone. This permits assigning tentative phases to all (or most) F_{hkl}'s, and so a Fourier series can be computed to reveal the electron density. This always shows the heavy atom upon which the phases were based, as well as some or all of the lighter ones, perhaps in a crude way. A second Fourier series, based upon phases computed from the locations of all the atoms which can now be distinguished, reveals these atoms in an improved way and possibly others as well. If no phase changes occur after several repetitions of the cycle of computing phases followed by Fourier summation, the process of revealing the electron density by successive Fourier syntheses is complete. The heavy-atom method is indirect, in that it can be used only to solve a crystal structure which is specialized in having one set of equivalent heavy atoms (or at most a few such sets) in the cell.

If two crystals of the same structure are available whose compositions are MABCD and NABCD, the atom M is said to be a replaceable atom; the crystal structures of such a pair can usually be determined by another indirect method known as the *replaceable-atom method.* The phase of a reflection can be determined by comparison of the magnitudes of the same reflection from the two crystals. This is a fairly simple procedure for centrosymmetrical crystals. If it is assumed that atom M is heavier than atom N, then the substitution of M for N increases the scattering power of the replaceable site. If this change produces an increase in the reflection intensity of the crystal, then the phase of the wave scattered by the entire cell is the same as the phase of the wavelet contributed by the atom in the replaceable site.† If the replacement produces a decrease, the phases of scattering by the cell and the replaceable atom are probably opposite. For a noncentrosymmetrical crystal the same general relations hold, but the phases can no longer be manipulated as positive and negative, but must be treated as vectors in the complex plane. Unfortunately, this generally gives two complementary solutions, only one of which is correct. The incorrect one can be eliminated by making use of a series of crystals having several sets of replaceable atoms in different sites. The replaceable-atom method has been extensively used to solve structures with large numbers of atoms in the cell, and was the method used to solve the first protein structures.

The direct methods are those which can be used to determine the phases of reflections of an arbitrary crystal. The simplest of the direct methods makes use of the Harker-Kasper inequalities. If the crystal has any symmetry, an appropriate inequality can be used to determine, or to limit the range of, the phase of a reflection when the intensity of a certain related reflection is large enough. The simplest inequality is often useful in determining the sign of a reflection $2h\ 2k\ 2l$ as positive if the intensity of hkl is sufficiently large. Unfortunately, conditions for the successful use of these inequalities are unlikely to occur except for crystals with a small number of atoms per cell.

A much more powerful method of determining phases is the relation between the probable signs of reflections discovered by David Sayre. This can be expressed as

$$S_{h_1+h_2,\,k_1+k_2,\,l_1+l_2} \approx S_{h_1k_1l_1}S_{h_2k_2l_2}.$$

This relation was put on a firmer statistical footing by Zachariasen, and later by Karle and Hauptman. It has become an essential feature of what is now known as the *symbolic addition method* of sign determination. This method permits determining the phases of crystals with up to 50 or 100 atoms per cell.

Notes on history

In mathematics and physics it has long been customary to represent a function which is repeated periodically in one dimension by a Fourier

† Except for reflections having small magnitudes; for such reflections the magnitude may have been reduced through zero to the opposite sign by the replacement.

series. That such a series could be used to represent the electron density of a crystal was suggested by W. H. Bragg in a lecture of 1915. In 1925 Duane outlined the theory of the representation of the electron density of a crystal by a three-dimensional Fourier series in essentially its modern form. He indicated a number of technical details in the use of x-ray diffraction for this purpose. At the same time Havighurst applied this to determining the electron densities of NaCl, KI, NH_4I, and NH_4Cl, as projected along a cell axis. The first application to a two-dimensional electron-density projection was made by Zachariasen in 1929 in his determination of the structure of $KClO_3$. In these earlier days, Fourier syntheses had to be calculated by adding the required terms, aided, at best, by desk computers, so that the extensive labor of computing a three-dimensional synthesis was rarely attempted. Since about 1955, however, high-speed electronic computers have come into use for Fourier summations, and with present technology, three-dimensional summations can be computed at reasonably spaced sample points in the cell of a typical inorganic crystal in 5 minutes or less.

The heavy-atom method has been in use since the very earliest days of the application of Fourier synthesis to crystal-structure analysis. The first explicit use of the method appears to have been West's determination of the structure of KH_2PO_4 in 1930. It quickly came into extensive use, especially by organic chemists who wished to determine the arrangement of atoms in organic molecules. In order to use the method for this purpose it was merely necessary to prepare the molecule with a heavy-atom substituent, crystallize the substance, and then solve its crystal structure. By such use, the heavy-atom method probably became the routine which accounted for more solutions of crystal structures than any other method.

The replaceable-atom method was first tried in 1927 by Cork, who compared the reflection amplitudes of the Tl, Rb, K, and NH_4 alums. He attempted to use the information from the *hhh* reflections to find the location of the sulfur atoms along the 3-fold axis by a one-dimensional Fourier synthesis of the electron density projected on this axis. The results were not very encouraging, because electron densities of other atoms in the structure, projected on the same 3-fold axis, swamped the desired electron density of the sulfur. The replaceable-atom method was successfully applied to the alums some 8 years later by Lipson and Beevers, and this initiated a more general use of the method for crystals which had centrosymmetrical projections. Its successful general use for noncentrosymmetrical crystals awaited the publication by Harker in 1956 of a simple graphical description of the problem and its solution. The first application of Harker's solution to revealing a structure as complicated as that of the proteins was published by Kendrew et al. in

1958. Since this successful use, the replaceable-atom method has become the standard routine for the solution of protein structures.

The stimulus for seeking relations which would permit determining sets of phases of reflections from sets of the reflection intensities came from the development of implication theory. In 1945 Buerger showed how the Harker sections of the Patterson function (discussed in Chapter 13) could be interpreted as maps of crystal structures projected along a symmetry axis, with ambiguities. For certain space groups there was no ambiguity, so that a projection of the crystal structure was furnished by a section of the Patterson function, which is a Fourier series using magnitudes $|F_{hkl}|^2$ alone. The surprising conclusion from this was that for certain symmetrical crystals, at least, structures could be solved from intensity data alone, so that, in some way, phase information was carried by sets of intensities.

Harker and Kasper deliberately searched mathematical compilations for relations between numbers and their squares. In 1946 they showed that, when Schwartz's inequality is applied to the equation for the reflection amplitude, there result useful inequalities involving F's and $|F|^2$'s. The first fruit of this development was the solution of the structure of decaborane by Kasper, Lucht, and Harker.

Thus began an era of investigation in which many crystallographers sought other relations between the phases of reflections. The earliest such investigations supplemented the results of Harker and Kasper by finding other more complicated inequalities relating F's to $|F|^2$'s, although some linear inequalities were also discovered. The first real departure came in 1952, when Sayre discovered the sign relation between amplitudes of reflections with related indices. Sayre's results stimulated others, notably Zachariasen, Hughes, Woolfson, Cochran, and Karle and Hauptman, to investigate statistically the relations between the signs of related reflections. Most of the results turned out to be variations, with statistical improvements, on the Sayre relation, which has now become the main theme in the direct determination of crystal structures.

Additional reading

A. D. Booth. *Fourier technique in x-ray organic structure analysis.* (Cambridge University, Cambridge, England, 1948) 106 pages.

Werner Nowacki. *Fouriersynthese von Kristallen.* (Birkhäuser, Basel, 1952) 237 pages.

H. Lipson and W. Cochran. *The determination of crystal structures.* (G. Bell, London, 1953) 76–109, 199–276.

M. J. Buerger. *Crystal-structure analysis.* (Wiley, New York, 1960) 352–584.

M. M. Woolfson. *Direct methods in crystallography.* (Oxford, 1961) 141 pages.

Significant literature

Early papers

W. H. Bragg. *X-rays and crystal structure.* Phil. Trans. Roy. Soc. **215** (1915) 253–274.

William Duane. *The calculation of the x-ray diffracting power at points in a crystal.* Proc. Nat. Acad. Sci. U.S. **11** (1925) 489–493.

R. J. Havighurst. *The distribution of diffracting power in sodium chloride.* Proc. Nat. Acad. Sci. U.S. **11** (1925) 502–507.

R. J. Havighurst. *The distribution of diffracting power in certain crystals.* Proc. Nat. Acad. Sci. U.S. **11** (1925) 507–512.

W. Lawrence Bragg. *The determination of parameters in crystal structures by means of Fourier series.* Proc. Roy. Soc. (London) (A) **123** (1929) 537–559.

W. H. Zachariasen. *The crystal structure of potassium chlorate.* Z. Kristallogr. **71** (1929) 501–516.

W. L. Bragg and J. West. *A note on the representation of crystal structure by Fourier series.* Phil. Mag. **10** (1930) 823–841.

W. H. Zachariasen. *The crystal lattice of oxalic acid dihydrate,* $H_2C_2O_2 \cdot 2H_2O$ *and the structure of the oxalate radical.* Z. Kristallogr. **89** (1934) 442–447.

J. Monteath Robertson. *X-ray analysis of the structure of dibenzyl. II. Fourier analysis.* Proc. Roy. Soc. (London) (A) **150** (1935) 348–362.

S. B. Hendricks and M. E. Jefferson. *Electron distribution in* $(NH_4)_2C_2O_4 \cdot H_2O$ *and the structure of the oxalate group.* J. Chem. Phys. **4** (1936) 102–107.

The heavy-atom method

J. West. *A quantitative x-ray analysis of the structure of potassium dihydrogen phosphate* (KH_2PO_4). Z. Kristallogr. **74** (1930) 306–332.

W. A. Wooster. *On the crystal structure of gypsum,* $CaSO_4 \cdot 2H_2O$. Z. Kristallogr. **94** (1936) 375–396.

J. Monteath Robertson and Ida Woodward. *An x-ray study of the phthalocyanines. Part IV. Direct quantitative analysis of the platinum compound.* J. Chem. Soc. (1940) 36–48.

The replaceable-atom method

J. M. Cork. *The crystal structure of some of the alums.* Phil. Mag. (7) **4** (1927) 688–698.

H. Lipson and C. A. Beevers. *The crystal structure of the alums.* Proc. Roy. Soc. (London) (A) **148** (1935) 664–680.

J. Monteath Robertson. *An x-ray study of the phthalocyanines. Part II. Quantitative structure determination of the metal-free compound.* J. Chem. Soc. (1936) 1195–1209.

David Harker. *The determination of the phases of the structure factors of non-centrosymmetric crystals by the method of double isomorphous replacement.* Acta Cryst. **9** (1956) 1–9.

J. C. Kendrew, G. Bodo, H. M. Dintzis, R. G. Parrish, H. Wyckoff, and D. C. Philips. *A three-dimensional model of the myoglobin molecule obtained by x-ray analysis.* Nature **181** (1958) 662–666.

G. Bodo, H. M. Dintzis, J. C. Kendrew, and H. W. Wyckoff. *The crystal structure of myoglobin. V. A low-resolution three-dimensional Fourier synthesis of sperm-whale myoglobin crystals.* Proc. Roy. Soc. (A) **253** (1959) 70–102.

M. F. Perutz, M. G. Rossmann, Ann F. Cullis, Hilary Muirhead, Georg Will, and A. C. T. North. *Structure of haemoglobin; a three-dimensional Fourier synthesis at 5.5-Å resolution, obtained by x-ray analysis.* Nature **185** (1960) 416–442.

J. C. Kendrew, R. E. Dickerson, B. E. Strandberg, R. C. Hart, and D. R. Davies, *Structure of myoglobin; a three-dimensional Fourier synthesis at 2Å resolution.* Nature **185** (1960) 422–427.

Direct methods

David Harker and J. S. Kasper. *Phases of Fourier coefficients directly from crystal diffraction data.* J. Chem. Phys. **15** (1947) 882–884.

D. Harker and J. S. Kasper. *Phases of Fourier coefficients directly from crystal diffraction data.* Acta Cryst. **1** (1948) 70–75.

E. W. Hughes. *Limitations on the determination of phases by means of inequalities.* Acta Cryst. **2** (1949) 34–37.

D. Sayre. *The squaring method: a new method for phase determination.* Acta Cryst. **5** (1952) 60–65.

W. H. Zachariasen. *A new analytical method for solving complex crystal structures.* Acta Cryst. **5** (1952) 68–73.

Herbert Hauptman and Jerome Karle. *The solution of the phase problem. I. The centrosymmetric crystal.* American Crystallographic Association Monograph 3 (Edwards Brothers, Ann Arbor, Mich., 1953) 87 pages.

Edward W. Hughes. *The signs of products of structure factors.* Acta Cryst. **6** (1953) 871.

R. K. Bullough and D. W. J. Cruickshank. *Some relations between structure factors.* Acta Cryst. **8** (1955) 29–31.

H. Hauptman and J. Karle. *A unified algebraic approach to the phase problem. I. Space group* $P\bar{1}$. Acta Cryst. **10** (1957) 267–270.

J. Karle and I. L. Karle. *The symbolic addition procedure for phase determination for centrosymmetric and noncentrosymmetric crystals.* Acta Cryst. **21** (1966) 849–859.

Isabella L. Karle and J. Karle. *An application of the symbolic addition procedure to space group* $P2_1$ *and the structure of the alkaloid panamine,* $C_{20}H_{33}N_3$. Acta Cryst. **21** (1966) 860–868.

13

The Patterson function and image theory

Background

The use of Fourier synthesis in obtaining a picture of the electron density in a crystal for the purpose of determining the locations of atoms in the structure began in 1929. As Fourier synthesis came into routine use, it became clear that the determination of crystal structures was restricted by a lack of knowledge of the phases of the Fourier coefficients F_{hkl}. The crystal structures which could be solved at the time were those in which certain special conditions obtained, especially the presence of a heavy atom in the chemical composition. Whenever the cell contents required the heavy atom to be located in a special position of the space group, that atom made a contribution to the F_{hkl}'s which could be computed. In favorable cases of centrosymmetrical crystals or centrosymmetrical projections, this contribution, which was large because the atom was heavy, dominated the magnitudes of the F_{hkl}'s and so determined their signs. For example, if the heavy atom must be at the origin, its contribution is large and always positive, so that the phases of all F_{hkl}'s may be assumed positive for the purposes of a first Fourier synthesis. Many crystal structures were solved by application of this condition.

But the possibility of solving crystal structures in such special cases merely emphasized the impossibility of solving an arbitrary crystal structure from a set of measured intensities. From a point of view of Fourier synthesis, the complete crystal structure is determined only if both the magnitudes and the phases of the F_{hkl}'s are known, since a complex number involves both of these parameters, specifically,

$$F_{hkl}' = |F_{hkl}|e^{i\phi(hkl)}. \tag{1}$$

It occurred to A. L. Patterson to raise the question as to what features of a structure *are* determined by the magnitudes alone. He was able to give a specific and useful answer to this question which marked a turning point in the theory and practice of crystal-structure analysis.

In view of the fact that Patterson first derived the nature and properties of the Fourier series whose coefficients are $|F_{hkl}|^2$, it is customary to call this series the *Patterson function*, while the graphical representation of it for a particular crystal is sometimes informally called "the Patterson" of that crystal. Patterson's derivation of this function is given after an alternative derivation, which is somewhat simpler.

The Fourier series whose coefficients are the $|F_{hkl}|^2$'s

The squared magnitude of a complex number is readily found by multiplying the number, F_{hkl} in this case, by its complex conjugate $\tilde{F}_{hkl}$. This is easily demonstrated with the aid of the Argand diagram (Fig. 1), in which F_{hkl} and its complex conjugate are resolved into their real component A_{hkl} and imaginary component iB_{hkl}. In terms of these components, the number and its complex conjugate are seen to be given by

$$F_{hkl} = A_{hkl} + iB_{hkl}, \tag{2}$$

$$\begin{aligned} \tilde{F}_{hkl} &= F_{\bar{h}\bar{k}\bar{l}} \\ &= A_{hkl} - iB_{hkl}. \end{aligned} \tag{3}$$

If these are multiplied together, the result is

$$\begin{aligned} F_{hkl}\tilde{F}_{hkl} &= (A_{hkl} + iB_{hkl})(A_{hkl} - iB_{hkl}) \\ &= A_{hkl}^2 - iA_{hkl}B_{hkl} + iA_{hkl}B_{hkl} + B_{hkl}^2 \\ &= A_{hkl}^2 + B_{hkl}^2. \end{aligned} \tag{4}$$

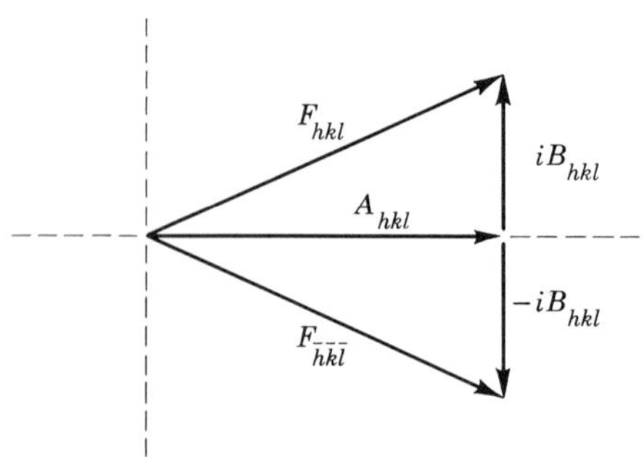

Fig. 1

The right of (4) is numerically equal to the square of the magnitude of the hypotenuse of a right triangle whose orthogonal sides have magnitudes A and B. Accordingly, (4) is equivalent to

$$F_{hkl}\tilde{F}_{hkl} = |F_{hkl}|^2. \tag{5}$$

To understand the properties of a Fourier series whose coefficients are the $|F_{hkl}|^2$'s, let the left side of (5), namely, $F_{hkl}\tilde{F}_{hkl}$, be expressed in terms of the scattering by the individual atoms of the structure. In (22) of Chapter 3 it was seen that the complex amplitude F of a scattered wave, indexed as hkl, is given by

$$F_{hkl} = \sum_j f_j e^{i2\pi(hx_j+ky_j+lz_j)}, \tag{6}$$

where the summation is over the j atoms of the structure. The complex conjugate of the wave is

$$\tilde{F}_{hkl} = \sum_j f_j e^{-i2\pi(hx_j+ky_j+lz_j)}. \tag{7}$$

The squared magnitude of the wave amplitude, according to (5), is the product of these summations, namely,

$$F_{hkl}\tilde{F}_{hkl} = \left(\sum_j f_j e^{i2\pi(hx_j+ky_j+lz_j)}\right)\left(\sum_j f_j e^{-i2\pi(hx_j+ky_j+lz_j)}\right). \tag{8}$$

In the general term of this product, the j's are not necessarily the same; that is, they refer to different atoms. Suppose these are represented by subscripts m and q. Then the product in the right of (8) can be written

$$F_{hkl}\tilde{F}_{hkl} = \sum_m \sum_q f_m f_q e^{i2\pi(h[x_m-x_q]+k[y_m-y_q]+l[z_m-z_q])}. \tag{9}$$

To make the meaning of this summation more definite, suppose that the structure contains only two atoms, designated 1 and 2. Then equations of the sequence (6) to (9) take the specific forms

(6′): $$F_{hkl} = f_1 e^{i2\pi(hx_1+ky_1+lz_1)} + f_2 e^{i2\pi(hx_2+ky_2+lz_2)}$$

(7′): $$\tilde{F}_{hkl} = f_1 e^{-i2\pi(hx_1+ky_1+lz_1)} + f_2 e^{-i2\pi(hx_2+ky_2+lz_2)}$$

(8′): $$F_{hkl}\tilde{F}_{hkl} = (f_1 e^{i2\pi(hx_1+ky_1+lz_1)} + f_2 e^{i2\pi(hx_2+ky_2+lz_2)}) \times (f_1 e^{-i2\pi(hx_1+ky_1+lz_1)} + f_2 e^{-i2\pi(hx_2+ky_2+lz_2)})$$

(9′): $$F_{hkl}\tilde{F}_{hkl} = f_1^2 + f_1 f_2 e^{i2\pi(h[x_2-x_1]+k[y_2-y_1]+l[z_2-z_1])} + f_1 f_2 e^{i2\pi(h[x_1-x_2]+k[y_1-y_2]+l[z_1-z_2])} + f_2{}^2.$$

If the summations in (6′) and (9′) are examined, they are seen to have comparable forms in the following sense: In (6′) a composite wave is represented in which the components have amplitudes f and phases determined by the coordinates xyz of the scattering atoms. The right-hand side of (9′) is also a sum which can be interpreted as a composite wave in which the components have amplitudes that are all products of the pairs of f's of (6′), with phases determined by the differences in coordinates of the corresponding pairs of atoms. The original wave, (6′), is scattered by 2 atoms, while the wave of (9′) appears to be scattered by 4 atoms, of which 2 are at the origin and the other 2 are centrosymmetrically related.

The more general form of the product in (9) can be interpreted along similar lines. If there are n atoms in the structure, suggested by the summation over j in (6), the products in (9) can be arranged in a square array, having the following form:

$$\begin{array}{ccc} f_1f_1e^{i2\pi(h[x_1-x_1]+k[y_1-y_1]+l[z_1-z_1])} & f_1f_2e^{i2\pi(h[x_1-x_2]+k[y_1-y_2]+l[z_1-z_2])} & \cdots \\ f_2f_1e^{i2\pi(h[x_2-x_1]+k[y_2-y_1]+l[z_2-z_1])} & f_2f_2e^{i2\pi(h[x_2-x_2]+k[y_2-y_2]+l[z_2-z_2])} & \cdots \\ \vdots & \vdots & \end{array} \qquad (10)$$

The various terms can be interpreted as the contributions to a composite wave $F_{hkl}\tilde{F}_{hkl}$. The individual contribution has an amplitude which is the product of a pair of f's in (6), and a phase which is determined by the difference of the coordinates of the corresponding pair in (6). The terms on the main diagonal have amplitudes of the form f_j^2 and phases appropriate to coordinates 000, that is, zero phase angle.

When all the F_{hkl}'s of (6) are used as the coefficients of a Fourier series, an electron density is built up in which there appear atoms each of which has one of the sets of coordinates $x_jy_jz_j$ of (6) and an electron count Z_j which corresponds to the scattering power f_j of (6). In exactly the same way, if all the $F_{hkl}\tilde{F}_{hkl} = |F_{hkl}|^2$'s of (9) are used as coefficients of a Fourier series, a density is built up as if it were composed of atoms which are related to pairs of atoms in (9). For example, a particular atom (which might be labeled mq) has an electron count Z_mZ_q (corresponding to scattering power f_mf_q) and is located at coordinates $x_m - x_q$, $y_m - y_q$, $z_m - z_q$.

All these conclusions are simple consequences of the algebra of finding the squared magnitude of a complex number F by forming the product $F\tilde{F}$.

Some properties of the Patterson function

Peak locations. The significance of the differences in coordinates which appear in (9) and (10) can be appreciated by specifying the locations of the atoms in the crystal structure with the aid of vectors. For two dimensions the relations of interest are illustrated in Fig. 2. Here two atoms, labeled 1 and 2, are located at the ends of vectors $\mathbf{r}_1$ and $\mathbf{r}_2$, whose components on the (not necessarily orthogonal) axes are x_1y_1 and

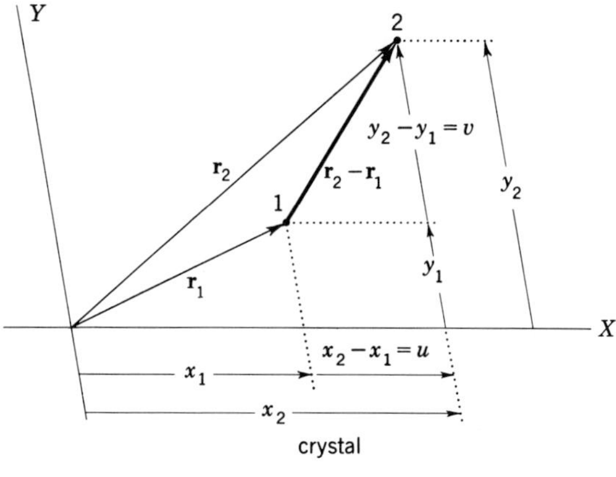

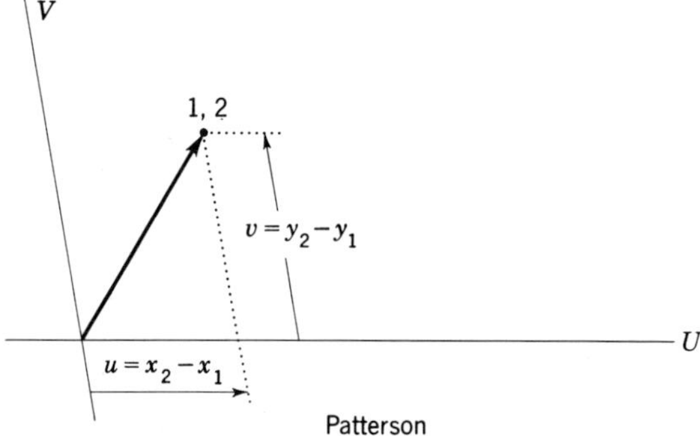

Fig. 2. The relation between the location of a peak in the Patterson function and the locations of the two atoms 1, 2 of a pair in the crystal.

x_2y_2. The difference between vectors $\mathbf{r}_2$ and $\mathbf{r}_1$ is $\mathbf{r}_2 - \mathbf{r}_1$. This difference is another vector whose components are the differences in the components of the first two vectors. Such differences in components are the differences which appear in the Patterson function. In three dimensions these differences are customarily labeled u, v, and w, as shown in Table 1.

The identification of the differences in coordinates in (10) as the components of vectors not only permits a simplification of notation but also permits a terse description of the locations of high density in the Patterson function. The simplification of nomenclature amounts to substituting the uvw coordinates in the right column of Table 1 for the differences in (9), which gives

(9):
$$F_{hkl}\tilde{F}_{hkl} = \sum_m \sum_q f_m f_q e^{i2\pi(hu_{mq}+kv_{mq}+lw_{mq})}. \tag{11}$$

Table 1
Relations between high density in the Patterson function and atom locations in the crystal

Atom locations in the crystal structure		Locations of high density in the Patterson function	
Vector location	Vector components	Vector location	Vector components
$\mathbf{r}_2$ $\mathbf{r}_1$	$\left.\begin{matrix} x_2y_2z_2 \\ x_1y_1z_1 \end{matrix}\right\}$	$\mathbf{r}_2 - \mathbf{r}_1$	$x_2-x_1,\quad y_2-y_1,\quad z_2-z_1 = u_{12}\,v_{12}\,w_{12}$

The terse description of locations of high density in the Patterson function is illustrated for two dimensions in Fig. 3. Suppose the crystal contains three atoms, labeled 1, 2, and 3. The vectors which locate these atoms do not appear explicitly in the Patterson function, but the differences between these locating vectors determine the location of high value of the Patterson function. Every interatomic vector in the crystal is a vector which locates a high value, with respect to the origin location, in the Patterson function. These high-value locations can be predicted by transferring every interatomic vector in the crystal to the origin of the Patterson function, as shown in Fig. 3.

Cell. If the F of (6) and the $|F|^2$ of (9) are compared,

$$(6):\qquad F_{hkl} = \sum_j f_j\, e^{i2\pi(hx \quad +ky \quad +lz \quad)},$$

$$(9):\qquad |F_{hkl}|^2 = \sum\sum f_m f_q e^{i2\pi(h[x_m-x_q]+k[y_m-y_q]+l[z_m-z_q])},$$

it is clear that both are functions of the same integral reciprocal-lattice coordinates h, k, and l, and both are functions of the same fractional cell-edge coordinates x, y, and z. This requires the two functions to have the same periodicities.

Thus the Patterson function has the same periodicity as the crystal, and both are referred to the same cell. For example, a vector from an atom in one cell of the crystal to an atom in another cell, such as $\overrightarrow{1\,2'}$ (Fig. 4), is the same as the sum of the vectors $\overrightarrow{1\,2} + \vec{t}$. In the Patterson function Fig. 4, the point associated with vector $\overrightarrow{1\,2'}$ falls at the same point that is associated with vector $\overrightarrow{1'\,2'}$, whose origin is displaced by vector $\vec{t}$ from the first origin. That is, the parallelogram 1,2,2′,1′, shown in the crystal, has periodicity t in both the crystal and the Patterson function.

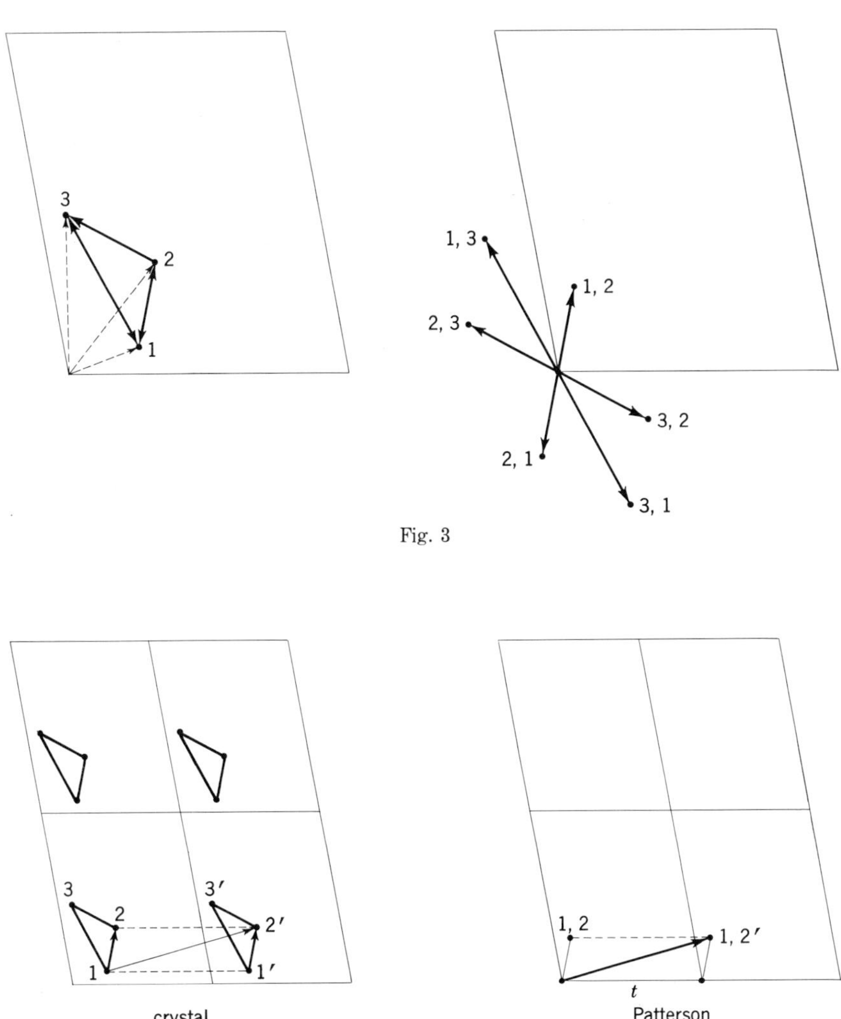

Fig. 3

Fig. 4

Peak weights. In Chapter 12 it was seen that a Fourier series whose coefficients are the F_{hkl}'s of a crystal builds up a picture of the structure of that crystal in terms of the distribution of electron density in it. Wherever an atom occurs in the structure, the Fourier series shows this as a concentration of electron density as it appears in the atom. The atoms are therefore marked out as peaks on a featureless background whose density is substantially zero. If the crystal cell has n atoms, the Fourier series shows n peaks which are at the ends of vectors from the origin and have the following electron counts:

Vectors from the origin: $\mathbf{r}_1,\ \mathbf{r}_2,\ \mathbf{r}_3,\ \ldots,\ \mathbf{r}_n;$ (12)
Electron counts: $Z_1,\ Z_2,\ Z_3,\ \ldots,\ Z_n.$ (13)

In a similar manner the Fourier series whose coefficients are the $|F_{hkl}|^2$'s of a crystal builds up a picture of the Patterson function of that crystal. This has high values of Patterson density at coordinates which are differences of pairs of vectors in (12). The locations of these high values of the Patterson are therefore given by the ordered array in (14) and the electron-product count associated with each location at a corresponding place in the ordered array in (15).

Vectors from origin to peaks:

$$\begin{array}{ccccc} \mathbf{r}_1 - \mathbf{r}_1 & \mathbf{r}_2 - \mathbf{r}_1 & \mathbf{r}_3 - \mathbf{r}_1 & \cdots & \mathbf{r}_n - \mathbf{r}_1 \\ \mathbf{r}_1 - \mathbf{r}_2 & \mathbf{r}_2 - \mathbf{r}_2 & \mathbf{r}_3 - \mathbf{r}_2 & \cdots & \mathbf{r}_n - \mathbf{r}_2 \\ \mathbf{r}_1 - \mathbf{r}_3 & \mathbf{r}_2 - \mathbf{r}_3 & \mathbf{r}_3 - \mathbf{r}_3 & \cdots & \mathbf{r}_n - \mathbf{r}_3 \\ \cdot & \cdot & \cdot & & \cdot \\ \cdot & \cdot & \cdot & & \cdot \\ \cdot & \cdot & \cdot & & \cdot \\ \mathbf{r}_1 - \mathbf{r}_n & \mathbf{r}_2 - \mathbf{r}_n & \mathbf{r}_3 - \mathbf{r}_n & \cdots & \mathbf{r}_n - \mathbf{r}_n \end{array} \tag{14}$$

Electron counts:

$$\begin{array}{ccccc} Z_1Z_1 & Z_1Z_2 & Z_1Z_3 & \cdots & Z_1Z_n \\ Z_2Z_1 & Z_2Z_2 & Z_2Z_3 & \cdots & Z_2Z_n \\ Z_3Z_1 & Z_3Z_2 & Z_3Z_3 & \cdots & Z_3Z_n \\ \cdot & \cdot & \cdot & & \cdot \\ \cdot & \cdot & \cdot & & \cdot \\ \cdot & \cdot & \cdot & & \cdot \\ Z_nZ_1 & Z_nZ_2 & Z_nZ_3 & \cdots & Z_nZ_n. \end{array} \tag{15}$$

It is convenient to call each location specified by a location in (14) a peak to which is attributed an electron count equal to the corresponding product in (15). But it is important to note that the word "peak" used in this sense does not imply "discrete peak"; that is, it does not imply that the peak stands out separately from other peaks as does a peak which represents an atom in the electron-density function. Indeed, peaks in the Patterson function characteristically coalesce with other peaks to form composite, high-topographic features.

If the crystal has n atoms per cell, as in (12) and (13), the square arrays in (14) and (15) show that the Patterson function has n^2 peaks per cell. Because of the physical nature of the atoms in the crystal cell, the n peaks corresponding to the n atoms are discrete. But it cannot be expected that the n^2 peaks of the Patterson function are, in general, resolved as discrete maxima. An actual example of how these maxima ordinarily run together is given later.

Origin peak. Of the n^2 peaks, those which appear along the main diagonals of (14) and (15) are located at the ends of vectors of the form $\mathbf{r}_j - \mathbf{r}_j$, which are vectors of zero length, and so all these peaks coincide identically at the origin of the Patterson cell. Their electron-product counts are of the form Z_j^2, so that the composite peak at the origin has a total count of $\Sigma_j Z_j^2$. This great composite peak at the origin is one of the most striking features of the Patterson function. The n^2 Patterson peaks are accordingly separable into two categories, the n coincident origin peaks on one hand, and $n^2 - n$ nonorigin peaks on the other.

Centrosymmetry. An important characteristic of every Patterson function can be deduced from the square arrays (14) and (15). For every peak location at the end of a vector in (14), such as $\mathbf{r}_2 - \mathbf{r}_1$, there corresponds a peak at the end of a vector symmetrical with it with respect to the main diagonal of the array; this is a reverse vector such as $\mathbf{r}_1 - \mathbf{r}_2 = -(\mathbf{r}_2 - \mathbf{r}_1)$. These peaks are at equal and opposite locations with respect to the origin, and they have the same electron-product counts. This is true of every peak, so that the Patterson function must be centrosymmetrical in the origin.

Alternative derivation of the Patterson function†

The discussion originally given by Patterson of the nature of a Fourier series whose coefficients are the $|F_{hkl}|^2$'s followed a different course from the one just given. It is somewhat more complicated, but it does bring out other useful features of the function.

Electron-density products. To devise a Fourier series whose coefficients involve only absolute magnitudes, these coefficients must be

† This section may be omitted by the reader without interfering with the general argument of the chapter.

products of the form $F_{hkl}\tilde{F}_{hkl} = F_{hkl}F_{\bar{h}\bar{k}\bar{l}}$. Patterson was able to arrange this by multiplying together two Fourier series representing the same electron density at different locations. For simplicity the derivation is outlined here for the one-dimensional case, but it can be readily generalized to two and three dimensions. It is also convenient to use absolute coordinates in this derivation. The fractional coordinate x and absolute coordinate X are related by

$$x = \frac{X}{a}. \tag{16}$$

The electron densities of a one-dimensional crystal at two points, specifically, X and $X + U$, are given by

$$\rho(X) = \frac{1}{a}\sum_h F_h e^{-i2\pi hX/a}, \tag{17}$$

$$\rho(X+U) = \frac{1}{a}\sum_h F_h e^{-i2\pi h(X+U)/a}. \tag{18}$$

If these two series are multiplied together to give $\rho(X)\rho(X+U)$, the result is the product of two electron densities within the same crystal separated by a chosen interval U. If this product is integrated with respect to X over the period of X, namely, from $X = 0$ to $X = a$, it provides a measure of the amount of electron density which is separated from other electron density by the chosen interval U. This integration is

$$P(U) = \int_0^a \rho(X)\rho(X+U)\,dX \tag{19}$$

$$= \int_0^a \left(\frac{1}{a}\sum_h F_h e^{-i2\pi hX/a}\right)\left(\frac{1}{a}\sum_h F_h e^{-i2\pi h(X+U)/a}\right) dX. \tag{20}$$

The separate terms in this product, in general, have different indices h. If these are designated m and q, and the variable part of the exponential separated, (20) can be rewritten

$$P(U) = \int_0^a \left(\frac{1}{a}\sum_m F_m e^{-i2\pi mX/a}\right)\left(\frac{1}{a}\sum_q F_q e^{-i2\pi qX/a} e^{-i2\pi qU/a}\right) dX \tag{21}$$

$$= \frac{1}{a^2}\sum_m\sum_q F_m F_q e^{-i2\pi qU/a}\int_0^a e^{-i2\pi(m+q)X/a}\,dX. \tag{22}$$

When $q = -m$, the exponential reduces to zero, and the value of the integral is a. For all other values of q the exponential represents a fraction of a cycle, due to the fractional nature of X/a, and in all these other cases the value of the integral is zero. Since the only nonvanishing terms are those for which $q = -m$, there remain only the terms

$$P(U) = \frac{1}{a^2} \sum_h F_h F_{\bar{h}} e^{-i2\pi(-h)U/a} a$$
$$= \frac{1}{a} \sum_h F_h^2 e^{i2\pi hU/a}. \tag{23}$$

Expressed in fractional coordinates $u = U/a$, this becomes

$$P(u) = \frac{1}{a} \sum_h F_h^2 e^{i2\pi hu}. \tag{24}$$

This result can be further simplified by expanding the exponential with the aid of the Eulerian relation

$$e^{i\phi} = \cos\phi + i\sin\phi. \tag{25}$$

The expansion changes (24) to

$$P(u) = \frac{1}{a} \sum_h F_h^2 (\cos 2\pi hu + i \sin 2\pi hu). \tag{26}$$

Because of the symmetry of the sine, and because $F_h^2 = F_{-h}^2$, the summation over h gives rise to paired terms in $+h$ and $-h$:

$$F_h^2 i \sin 2\pi hu$$

and

$$F_{-h}^2 i \sin 2\pi \bar{h} u = -F_h^2 i \sin 2\pi hu. \tag{27}$$

These terms cancel each other in the summation, leaving

$$P(u) = \frac{1}{a} \sum_h F_h^2 \cos 2\pi hu. \tag{28}$$

This function has the same value for $+u$ and $-u$, and so it is symmetrical in the origin. This means that all Patterson functions have a center of symmetry in the origin.

The two- and three-dimensional forms of (28) are, respectively,

$$P(uv) = \frac{1}{A} \sum_h \sum_k F_{hk}^2 \cos 2\pi(hu + kv) \tag{29}$$

and

$$P(uvw) = \frac{1}{V} \sum_h \sum_k \sum_l F_{hkl}^2 \cos 2\pi(hu + kv + lw), \tag{30}$$

where A is the area of the two-dimensional cell, and V is the volume of the three-dimensional cell.

General significance. It was pointed out in the beginning of this derivation of the Patterson function that (19) has a high value whenever

U (and therefore u) is the separation of two large values of the electron density in the crystal. Since high values of the electron density occur at the centers of atoms, it is evident that the Patterson function has a high value whenever u coincides with the separation between a pair of atoms. Conversely, whenever the Patterson function displays a high value at the parameter u, it is evident that somewhere within the crystal there are one or more pairs of atoms separated by this value of u. Figure 5*A* shows two periods of an arbitrary one-dimensional electron-density sequence. If a general separation, such as u_g, is chosen and allowed to range over the sequence, it is found not to span any peaks (atom centers) in the electron-density pattern (Fig. 5*A*). But if a separation which

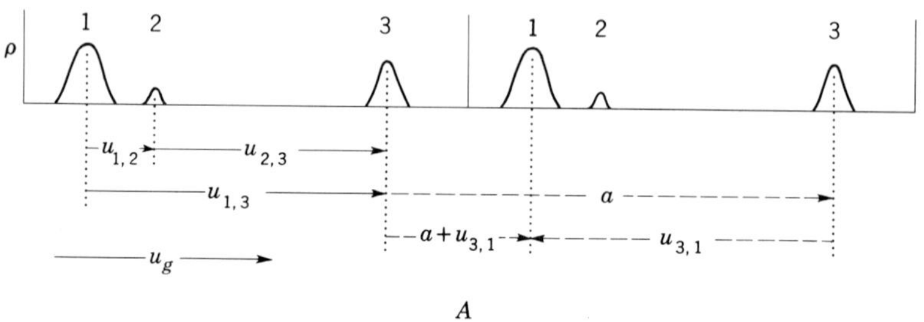

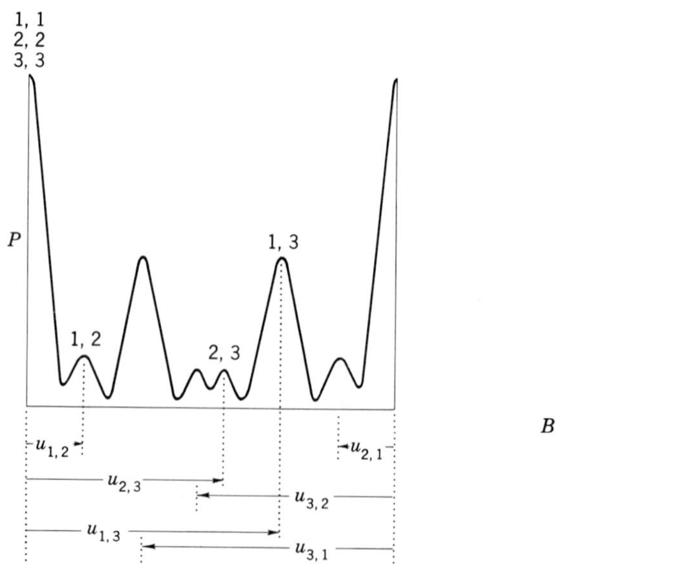

Fig. 5. [*Modified from Martin J. Buerger: Vector space and its application in crystal-structure investigation* (*Wiley, New York,* 1959) 14.]

spans peaks in Fig. 5*A* is chosen, such as u_{12}, u_{13}, or u_{23}, then when x is at one of the peaks, $x + u$ is at some other peak, so for that value of u the product $\rho(x)\rho(x + u)$ has a high value and, consequently, the integral over all x's has a high value. For such values of u the Patterson function has high values, as shown in Fig. 5*B*. Conversely, each high value in Fig. 5*B* indicates that a pair of atoms in Fig. 5*A* is separated by the value of u of that high value.

In the one-dimension illustration of Fig. 5, the separation u is simply a positive or a negative number. In two and three dimensions the separation must be expressed by a vector. The Patterson function for these can be written in terms of the vectors themselves or in terms of their components along the crystal axes. The component form, in which the separation vector is expressed as its components written as fractions u, v, and w of the cell axes a, b, and c, was given in (29) and (30).

Origin peak. A characteristic feature of every Patterson function is that when the separation vector has zero length, its two ends are on high values of the electron density for every atom in the crystal. This gives rise to a very high peak at the end of a vector of zero length, that is, at the Patterson origin. This peak has the maximum possible magnitude because the modulating function $\cos 2\pi(hx + ky + lz)$ has its maximum value of unity, so that (30) becomes $P(000) = \Sigma F_{hkl}^2/V$.

General nature of the information provided by the Patterson function

A Fourier synthesis whose coefficients are the F_{hkl}'s provides an essentially true picture of the arrangement of atoms in a crystal structure. On the other hand, a Fourier synthesis whose coefficients are merely the magnitudes of the F_{hkl}'s yields only the geometrical relations between the pairs of atoms in the structure. An analogy can be drawn between these two conditions and the two states of knowledge which can be set down about a steel structure composed of girders, beams, struts, and braces. A detailed working drawing, showing the specific locations of these members and the pins which join them, corresponds to the map of the arrangements of atoms in the crystal, that is, to the map provided by a Fourier synthesis whose coefficients are the F_{hkl}'s. On the other hand, a mere drawing of the individual linear steel members of the structure arranged, say, in order of increasing length, but with no information as to how they are to be joined together, corresponds to a Patterson synthesis, that is, to a Fourier synthesis whose coefficients are the magnitudes $|F_{hkl}|^2$. The relation between pairs of pins is known, but the specific locations of the pins are unknown.

Symmetry information provided by the Patterson function

Restrictions on the symmetries of the Patterson functions. It was demonstrated that every Patterson function contains a center of symmetry at the origin. Of the 230 space groups, only 92 are centrosymmetrical; therefore the Patterson symmetries must be restricted to these few symmetries.

But there is another limitation, illustrated in Fig. 6, which further restricts the permissible space groups. In the crystal structure all features, including the atom locations as well as the interatomic vectors connecting them, must conform to the space-group symmetry. In the transfer of an interatomic vector from the crystal to the Patterson function, the relative positions of the tails of the vectors are lost since all are placed at the origin. If two vectors, as in Fig. 6*A*, are related in the crystal by a symmetry operation having a translation component, that component is lost when the vector is transferred to the Patterson function (Fig. 6*B*), although the directions of the vectors are necessarily preserved. It follows that symmetry operations which relate vectors in the crystal structure appear in the Patterson function only as their translation-free residues. This applies strictly to those screw axes and glide planes which are isogonal with the rotation axes and reflection planes of the point group. It does not apply to the screws and glides which relate a centered lattice point (if present) to the origin lattice point, because the Patterson function has the same translations as the crystal.

These additional restrictions, though they permit the Patterson symmetries to be those of the space groups based upon all lattice types, nevertheless limit the space groups to those not containing characteristic screw axes or glide planes. These are the space groups which are obtained by placing the 11 centrosymmetric point groups at the points of the permissible lattices. There are 24 space groups of this type; they are enumerated in Table 2.

Recovery of symmetry information. Since a crystal may have any one of the 230 space-group symmetries but its Patterson function may have only 24 space-group symmetries, it might be supposed that the symmetry of a crystal cannot be determined from its Patterson function since the translations of any characterizing screw axes and glide planes are lost. Fortunately, however, this is not true; indeed, except in unfavorable cases the Patterson function is able to reveal the complete symmetry of the crystal.

The basis of determining symmetry from the Patterson function is illustrated in Figs. 7 and 8. In Fig. 7*A* the atoms in a set related by a glide plane in the crystal structure are shown, together with their inter-

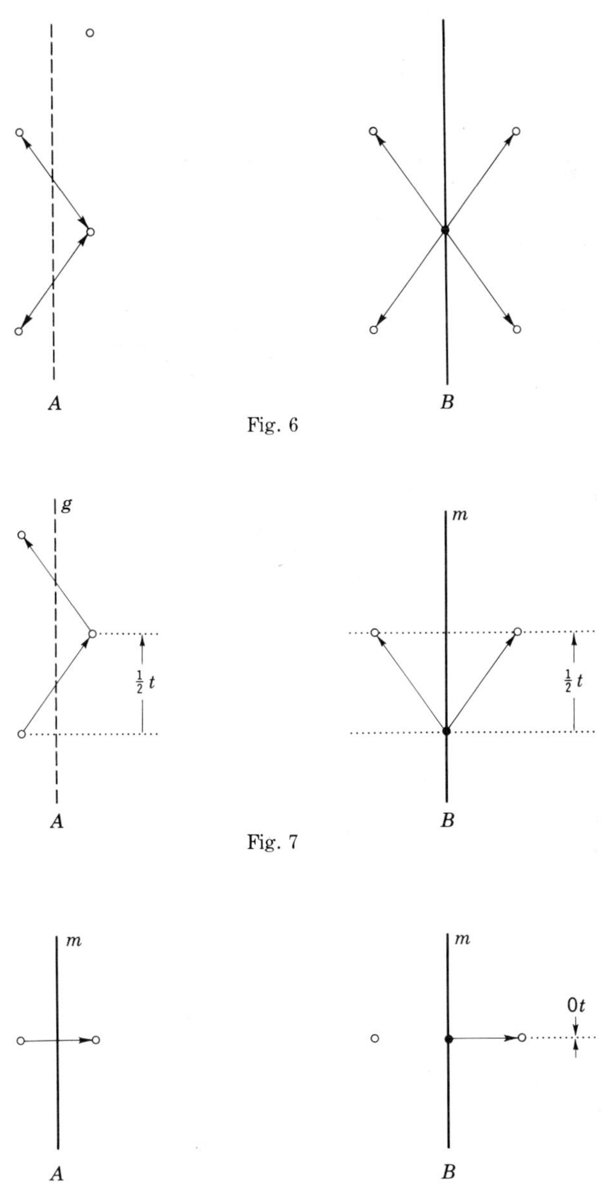

Fig. 6

Fig. 7

Fig. 8

atomic vectors. When the vectors are transferred to the Patterson function, they appear as in Fig. 7*B*. There they have lost the translation component of the glide operation which relates them in Fig. 7*A*, but every such vector has retained the common component of $\frac{1}{2}t$ (where t is a lattice translation) between head and tail. As a consequence, there is a

Table 2
Symmetries possible for Patterson function

Crystal system	Centrosymmetrical crystal class	Patterson space group
Triclinic	1	$P\bar{1}$
Monoclinic	$\frac{2}{m}$	$P\frac{2}{m}$, $A\frac{2}{m}$
Orthorhombic	$\frac{2}{m}\frac{2}{m}\frac{2}{m}$	$P\frac{2}{m}\frac{2}{m}\frac{2}{m}$, $C\frac{2}{m}\frac{2}{m}\frac{2}{m}$, $I\frac{2}{m}\frac{2}{m}\frac{2}{m}$, $F\frac{2}{m}\frac{2}{m}\frac{2}{m}$
Tetragonal	$\frac{4}{m}$	$P\frac{4}{m}$, $I\frac{4}{m}$
	$\frac{4}{m}\frac{2}{m}\frac{2}{m}$	$P\frac{4}{m}\frac{2}{m}\frac{2}{m}$, $I\frac{4}{m}\frac{2}{m}\frac{2}{m}$
Hexagonal	$\bar{3}$	$P\bar{3}$, $R\bar{3}$
	$\bar{3}\frac{2}{m}$	$P\bar{3}\frac{2}{m}1$, $P\bar{3}1\frac{2}{m}$, $R\bar{3}\frac{2}{m}$
	$\frac{6}{m}$	$P\frac{6}{m}$
	$\frac{6}{m}\frac{2}{m}\frac{2}{m}$	$P\frac{6}{m}\frac{2}{m}\frac{2}{m}$
Isometric	$\frac{2}{m}\bar{3}$	$P\frac{2}{m}\bar{3}$, $I\frac{2}{m}\bar{3}$, $F\frac{2}{m}\bar{3}$
	$\frac{4}{m}\bar{3}\frac{2}{m}$	$P\frac{4}{m}\bar{3}\frac{2}{m}$, $I\frac{4}{m}\bar{3}\frac{2}{m}$, $F\frac{4}{m}\bar{3}\frac{2}{m}$

strong concentration of peaks along a line whose direction is normal to the glide plane, but which intersects the glide plane at a point removed from the origin by a distance equal to the translation component of the glide, namely, $\tau = \frac{1}{2}t$.

In a similar way, Fig. 8*A* shows a pair of atoms related by a pure reflection plane in the crystal structure. For a pure reflection plane, of course, $\tau = 0$; therefore, when the vectors connecting such pairs are transferred to the Patterson function (Fig. 8*B*), it is seen that they are localized in a line normal to the reflection plane and through the origin.

It can be concluded that a symmetry plane in the crystal structure, whether a pure reflection or a glide reflection, leaves a record of its presence in the Patterson function in the form of a concentration of peaks along a line normal to the reflecting plane. This line is displaced from the origin by a vector equal to the translation component of the glide. Of course, this displacement is zero when the translation component is zero, that is, when the symmetry operation is a pure reflection.

In a similar way, it can be readily demonstrated that vectors between atoms related by a rotation are concentrated in a plane at right angles to the rotation axis. If the operation is a pure rotation, this plane of concentration contains the origin; if the operation contains a translation component, the plane of concentration is displaced from the origin by a vector equal to the translation component of the screw.

By examining the Patterson function for concentrations of peaks in planes normal to possible (or expected) symmetry axes, and in lines normal to possible (or expected) symmetry planes, the presence of characteristic rotation axes and reflection planes can ordinarily[†] be established. This feature of a Patterson function, which has been derived from the magnitudes $|F_{hkl}|$ of the x-ray reflections alone, ordinarily permits determination of the space group even though the use of extinctions alone allows only 122 diffraction symmetries to be recognized, only 58 of which uniquely characterize a space group, as shown in Chapter 5. The general reason that the Patterson function is superior as a determiner of symmetry is that the extinction method makes use only of a qualitative feature of the $|F_{hkl}|$'s, namely, their Friedel symmetry and a list of those whose magnitudes are identically zero, whereas the Patterson-concentration method makes use of the magnitudes of all reflections (and, of course, the Friedel-symmetry relations between them). In this way the Patterson function provides a unique characterization of pure rotation axes and pure reflection planes, and also distinguishes between 4 and $\bar{4}$, for example.

To aid space-group determination by the use of concentrations in the Patterson function, tables of such concentrations are available. Merely by noting the locations of planar and linear concentrations in the Patterson function, it is ordinarily possible to distinguish all space groups except between the members of the 11 enantiomorphic pairs (like $P\,3_1$ and $P\,3_2$, for example) and 8 other pairs.[‡] The members of these other

[†] Complications may occur if the crystal structure contains a substructure. In this event some, but not all, of the atoms have specialized positions, and these atoms may give rise to concentrations that suggest symmetries which do not, however, relate all atoms.

[‡] These pairs are:

$$\begin{Bmatrix} P\,1 & P\,3 & R\,3 & P\,4 & I\,4 \\ P\,\bar{1} & P\,\bar{3} & R\,\bar{3} & P\,\bar{4} & I\,\bar{4} \end{Bmatrix}$$

$$\begin{Bmatrix} P\,\bar{6} \\ P\,6/m \end{Bmatrix}$$

$$\begin{Bmatrix} I\,2\,2\,2 & I\,2\,3 \\ I\,2_1\,2_1\,2_1 & I\,2_1\,3 \end{Bmatrix}$$

8 pairs may be distinguished from one another by certain relations between their concentrations.

Harker lines and sections. That the peaks due to pairs of atoms which are related by reflections and rotations are localized in certain lines and planes, respectively, as just discussed, was first recognized by David Harker. They are accordingly called *Harker lines* and *Harker sections.*

While the general volume within the cell of the Patterson function contains peaks due to arbitrary pairs of atoms, the Harker lines and sections ideally contain only peaks due to pairs of atoms related by symmetry. Such peaks are easier to interpret than the peaks in the general body of the cell, and accordingly are much used in crystal-structure analysis. An especially important use is finding the location of a set of heavy atoms whose coordinates are required for the solution of the crystal structure by the heavy-atom method. Harker sections, in which the peaks are better resolved than in Harker lines, are especially easy to interpret. There also exists a complete theory, known as *implication theory,* with the aid of which the exact nature of the structure information that can be gained by an examination of any Harker section can be formulated. In an ideal case a Harker section can provide the locations of atoms in a projection of a crystal structure, subject to certain ambiguities dependent on the value of n of the n-fold axis to which the Harker section corresponds.

Unfortunately, Harker sections contain, in addition to the peaks which are present due to pairs of atoms related by symmetry, other peaks due to other pairs of atoms whose separations parallel to the translation component of the symmetry operation are nearly identical to the translation component. Such peaks, called the *non-Harker peaks,* give rise to a background which causes confusion and severely limits the usefulness of Harker sections.

Peaks due to pairs of symmetry-related atoms. It has just been seen that peaks due to pairs of atoms related by certain symmetry operations occur in specialized loci called Harker lines and Harker sections. If the symmetry of the crystal contains more than one symmetry operation, the coordinates of all these peaks are related.

The origin of this relation is illustrated in Fig. 9, where the symmetry elements in the neighborhood of an inversion center of space group $P\,2/m$ and a point having arbitrary coordinates $\bar{x}\bar{y}\bar{z}$ are shown. The several symmetry operations within the cell are listed in the first column of Table 3. These operations, illustrated by the symmetry elements they determine, are shown in Fig. 9*A*. Each operation generates a symmetrically related point from the initial point at $\bar{x}\bar{y}\bar{z}$. The new points and the vectors from the old point to the new ones are shown in Fig. 9*A* and

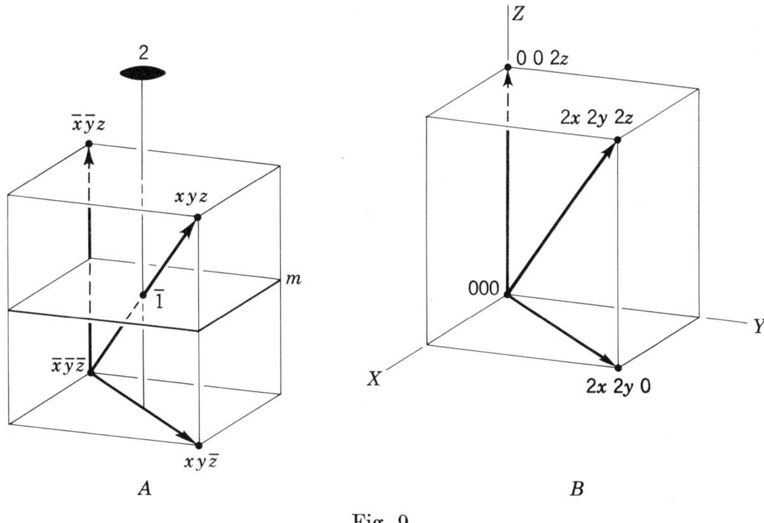

Fig. 9

listed in the second and third columns of Table 3. A set of such vectors issuing from a specific point to other points is called a *sheaf* of vectors issuing from that point. When the sheaf connects a specific point with the set of points symmetrical with it, the sheaf is called a *characteristic sheaf,* since it is characteristic of the action of the symmetry operations on a point in the crystal. If each vector in the crystal is translated to the Patterson function, shown in Fig. 9*B*, it is evident that the whole characteristic sheaf is translated as a unit. It is also apparent that the vectors of the sheaf are geometrically related, which implies that the coordinates of the peaks at the ends of the vectors are related.

It is convenient to label the vectors of a characteristic sheaf, and the

Table 3
Points produced by operations within a cell of symmetry $P\,2/m$ from an initial point whose coordinates are $\bar{x}\bar{y}\bar{z}$

Crystal		Patterson function	
Operation	Point produced from point at $\bar{x}\bar{y}\bar{z}$	Coordinates of peak	Name
1	$\bar{x}\bar{y}\bar{z}$	$\bar{x}-\bar{x},\quad \bar{y}-\bar{y},\quad \bar{z}-\bar{z} = 0\ \ 0\ \ 0$	identity
m	$\bar{x}\bar{y}z$	$\bar{x}-\bar{x},\quad \bar{y}-\bar{y},\quad z-\bar{z} = 0\ \ 0\ \ 2z$	reflection
2	$xy\bar{z}$	$x-\bar{x},\quad y-\bar{y},\quad \bar{z}-\bar{z} = 2x\ 2y\ 0$	rotation
$\bar{1}$	xyz	$x-\bar{x},\quad y-\bar{y},\quad z-\bar{z} = 2x\ 2y\ 2z$	inversion

peaks at the ends of these vectors, by the symmetry operation which relates the pair of atoms between which the vector is drawn, as in Fig. 9. Such peaks are called *symmetrical peaks* generally, and are specifically called *rotation, reflection, screw, glide, rotoinversion,* and *inversion peaks,* according to whatever symmetry operation relates the pair. The three nontrivial operations of $P\ 2/m$ in Fig. 9*A*, for example, give rise to reflection, rotation, and inversion peaks, as listed in the last column of Table 3.

The vectors of the characteristic sheaf, and the peaks at their ends, are related in an obvious way in this simple example. They are also characteristically related for every symmetry. The basic reason for this is that in every symmetry the operations form a group. This requires that the combination of any two operations in the group be equivalent to another operation in the group. Accordingly, the vector sum of the symmetry vectors for any two operations is the same as the symmetry vector for some other operation.

It is convenient to regard the symmetrical peaks with the most general coordinates as fundamental, but those with coordinates which are specializations of these as *satellites,* in the sense that the coordinates of the satellites are dependent upon the coordinates of the more general peaks. Commonly, the most general symmetrical peaks are inversion peaks, so that if the operation of inversion occurs in a space group, other symmetrical peaks are satellitic to the inversion peaks. Thus, in Fig. 9 and the second-to-last column of Table 3, the coordinates of the reflections and rotation peaks are specializations of the coordinates of the inversion peaks, and are satellites of it. But note that neither the rotation nor the reflection peak is satellitic to the other in this case. Other symmetries exist, however, in which reflection peaks are satellitic to rotation peaks.

If there are p atoms in a symmetrical set, there are p vectors from each atom to the atoms of the set. Such a characteristic sheaf of p vectors issues from each of the p atoms of the set, so that the Patterson function contains p^2 such vectors for every symmetrical set of p atoms in the crystal cell. To obtain this full set of p^2 vectors, the characteristic sheaf of p vectors may be placed at the Patterson origin and multiplied by the Patterson symmetry operations which intersect in the origin. In this way the characteristic sheaf may be used as a motif to be inserted in the Patterson symmetry in the process of finding all peaks in the Patterson function corresponding to all pairs of atoms in the symmetrical set. In this process, however, a symmetrical peak may come to coincide identically with another one of the same sheaf after the sheaf has been transformed by the Patterson symmetry to another position. In this case the resulting peak has a double weight (or, more generally, a multiple weight).

In centrosymmetrical crystals any centrosymmetrical pair of atoms is self-inverse, whereas all other pairs, both symmetrical and unsymmetrical, are duplicated by the inversion center. In the Patterson function of centrosymmetrical crystals, therefore, all inversion vectors are of single weight but all other vectors are of double (or possibly even-fold) weight.

Image description of the Patterson function

The geometry of the Patterson function can be described in a manner which lends itself to methods of extracting the crystal structure from the function. It has been seen that, if every interatomic vector in the crystal is translated so that its tail is at the origin, and an appropriate peak is placed at the head, the resulting set of peaks constitutes the Patterson function. Consider some particular atom A in the crystal structure. If vectors are drawn from that atom to all the atoms of the structure, and then all these vectors are translated so as to radiate from the origin, the geometrical result is the same as translating the whole structure so that atom A occupies the origin. This much of the Patterson function is the same as the way the crystal structure would appear as viewed from that particular atom A. It can be neatly described as the *image* of the whole crystal structure as seen from atom A. According to an earlier section, every peak in this image has an electron count which is the same as it has in the crystal structure, multiplied by the electron count of the atom A. Suppose the electron count of the atom A is designated by Z_A, and those of the other various atoms in the structure by Z_B, Z_C, Z_D, ... The images of these atoms in the Patterson function have weights

$$Z_AZ_A, \quad Z_AZ_B, \quad Z_AZ_C, \ . \ . \ . \ = Z_A(Z_A, \quad Z_B, \quad Z_C, \ . \ . \ .).$$

This means that the whole image as seen from A is weighted by the electron count Z_A of the viewing atom A.

In a similar way, the vectors from some other atom B to the various atoms of the crystal structure contribute another image of the same crystal structure, but this time as seen from the particular atom B. The peaks at the ends of these vectors in the Patterson function have electron counts

$$Z_BZ_A, \quad Z_BZ_B, \quad Z_BZ_C, \ . \ . \ . \ = Z_B(Z_A, \quad Z_B, \quad Z_C, \ . \ . \ .).$$

This process can be applied to every atom in the crystal structure. As a consequence, the geometry of the peak locations of the Patterson function can be described as the collection of the several images of the crystal structure as seen from each atom of the structure. Each image is weighted by the electron count of the atoms from which the image is seen. If there are n atoms in the crystal cell, the Patterson function can be described as n displaced images of this crystal structure, in each of which one of the n atoms is at the origin, that entire image being weighted by the electron count of the atom placed at the origin. This is illustrated for a simple example in Fig. 10.

This viewpoint has the advantage of bringing order to what at first had appeared to be a hodgepodge of vectors in the Patterson function. The function is now seen to be organized neatly into n images of the crystal structure.

This description of the Patterson function makes it evident that the loss of phase information in measuring the x-ray diffraction effects results in splitting the desired three-dimensional picture of the crystal structure into n weighted ghosts of itself. If there were some way of singling out

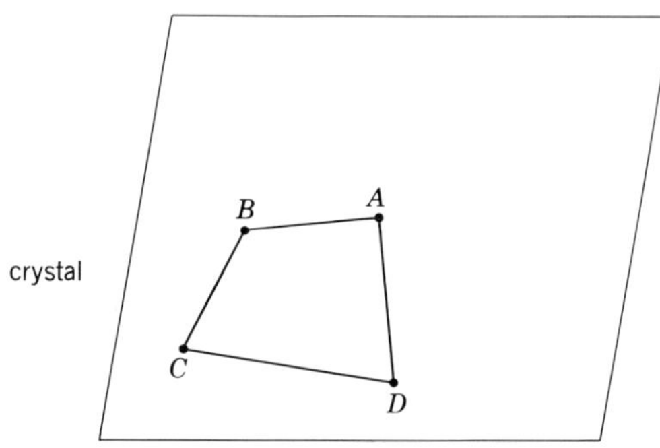

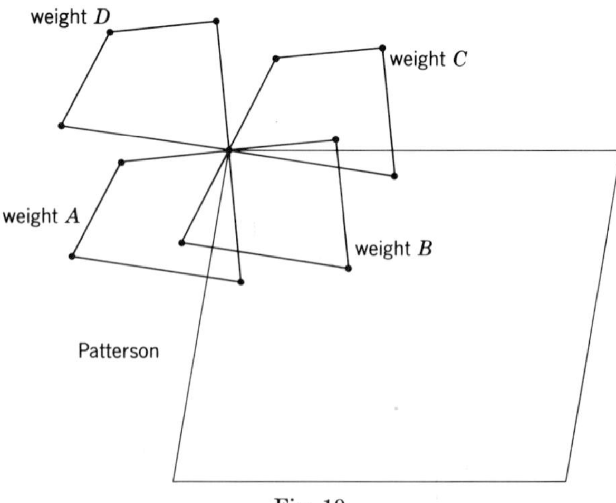

Fig. 10

one ghost, or enhancing the weight of one ghost, or making the several ghosts coalesce, the Patterson function would be solved for the crystal structure to which it corresponds.

Patterson function for sets of points

About 1950 it was discovered independently by several investigators that some relations exist between a set of points and its Patterson function which permit deducing the original set of points from its Patterson function. This discovery immediately raised the hope that a crystal structure could be solved by manipulation of its Patterson function. The methods based upon sets of points are noted briefly here to draw attention to the necessity for using more powerful methods when dealing with actual crystal structures.

Point structures. Instead of an actual crystal structure, consider an artificial one in which there is substituted for each atom in the structure a point whose weight is equal to the number of electrons in the atom. This is equivalent to placing all the scattering matter of an atom, namely, its Z electrons, at the position occupied by the atom center. This structure of points has an electron-density map composed of discrete points, to each of which is attributed a weight equal to the number of electrons in the atom.

The Patterson function of such a structure consists of peaks which are also weighted points. Since points are discrete unless they coincide identically, the Patterson function of a point structure consists of discrete peaks. The origin peak, of course, is a multiple peak, and in the case of centrosymmetrical crystals, all peaks except inversion peaks are double peaks.

Extraction of the point structure by coincidences. It was discovered in 1950 that the original set of points could be found in the Patterson function of the set by a simple prodecure which could be applied easily in two dimensions. In this procedure the points of the Patterson function are first marked on two sheets of paper. One of these is then translated until a nonorigin point of one superposes on the origin point of another (Fig. 11). In image language, this always causes an image of the structure in one map to coincide with another image in the other. The points which are thus found to coincide determine the arrangements of the original set of points, provided that the original set was centrosymmetrical and provided that the original nonorigin point was nonmultiple. If these conditions are not met, the coincidences mark out a pair of related structures, one of which can be eliminated by repeating the procedure with a second nonorigin point.

Other methods. The method just described has the advantage of

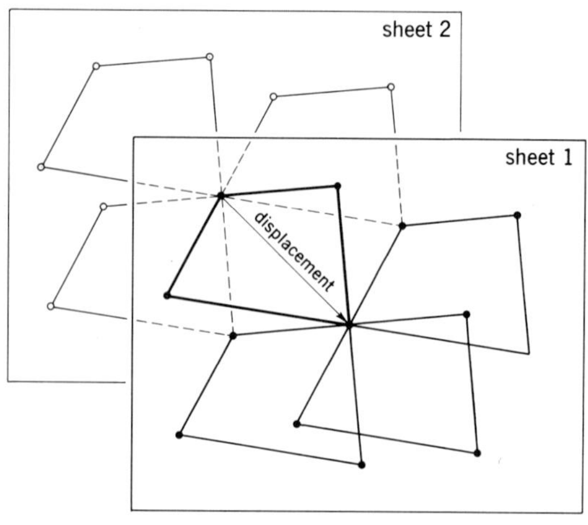

Fig. 11

simplicity. There are, however, at least two other methods which are more elegant in the sense that each utilizes, and therefore exhausts, all the Patterson points. These methods are superior for use with point structures because they prove that the solution found is the correct one, and not an accidental set of coincidences. Such methods are not detailed here because, as will be seen below, all methods based upon point structures are unrealistic when applied to the solution of actual crystal structures.

Characteristics of actual Patterson functions. Methods which succeed with point structures are not, in general, suitable for use with actual crystals unless some special condition is satisfied. Such special conditions include the case of a structure consisting of one equivalent set of heavy atoms and additional light atoms, as in some organic crystals. In such cases the methods just described may be successful because the n^2 images of the structure in the Patterson function are dominated by the few intense images of the structure as seen from the few heavy atoms. But the Patterson function of a more general crystal, even when n is comparatively small, is too complicated to treat by such methods, as seen in the following example.

The structure of the mineral berthierite, $FeSb_2S_4$, has only 7 atoms in the asymmetric unit, which occupies a quarter cell. The full cell accordingly contains $n = 4 \times 7 = 28$ atoms. The Patterson cell contains $n^2 = 28^2 = 784$ peaks, of which $n = 28$ coincide in the origin, leaving $784 - 28 = 756$ in the body of the cell. Since the crystal is centrosymmetrical, each of its 28 atoms has an inversion peak, while the remaining 728 peaks coincide identically in $728 \div 2 = 364$ double peaks. Thus

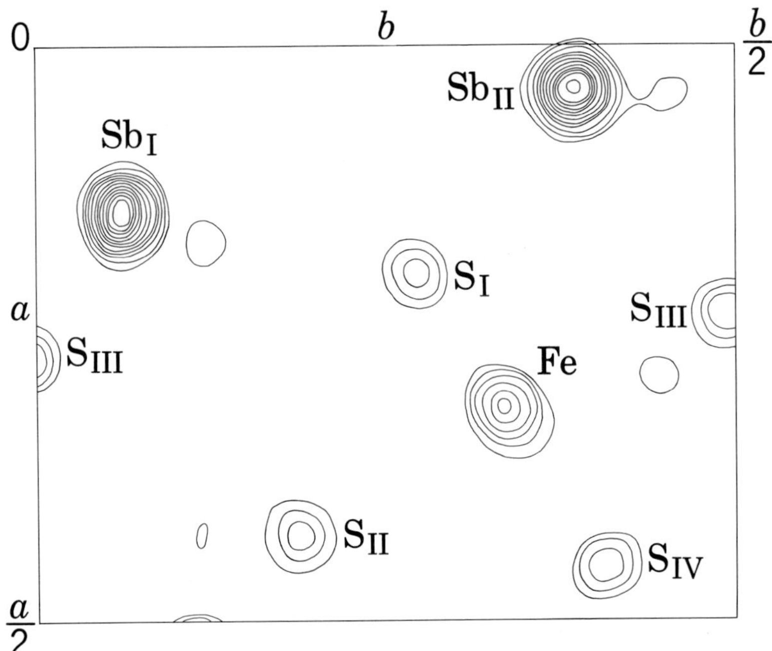

Fig. 12. Electron density $\rho(xy)$ of berthierite, $FeSb_2S_4$.

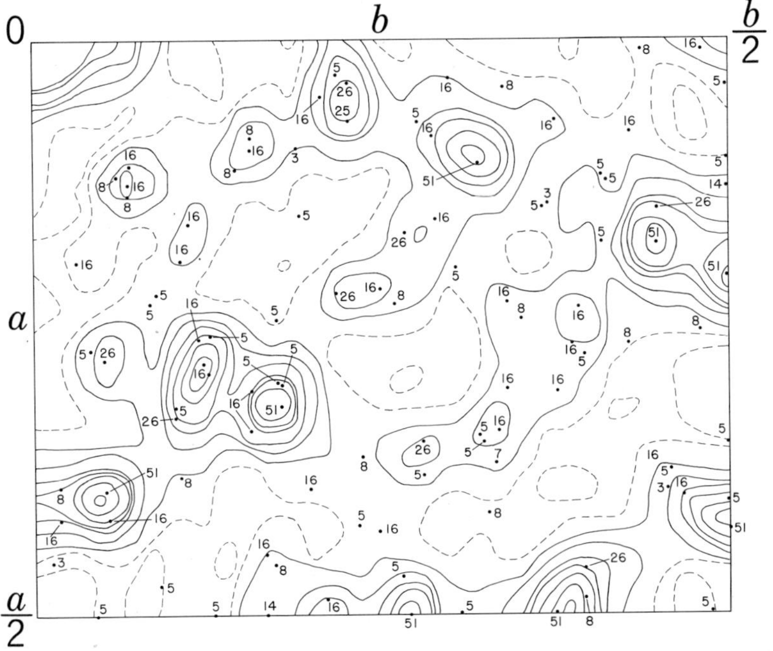

Fig. 13. Patterson projection $P(xy)$ of berthierite. Each dot is a calculated peak location and the accompanying number is the electron-product count of the peak. [*Figures* 12 *and* 13 *from M. J. Buerger and Theodor Hahn: Am. Mineral.* 40 (1955) 228, 229.]

there are 364 double and 28 single peaks in the body of the cell, making 392 in all, so that the asymmetric unit in the quarter cell contains 392 ÷ 4 = 98 peaks.

In the crystal structure the 7 atoms in the quarter cell are neatly resolved as discrete peaks. The Patterson quarter cell contains over twelve times as many peaks, plus an origin peak, each of which has a radius approximately double the radius of a peak in the crystal structure. Every maximum in the Patterson function, therefore, must be a coalescence of many Patterson peaks. This contrast between the discrete peaks of the crystal structure and the smear of the peaks in the Patterson function is brought out in Figs. 12 and 13. To attempt to solve even such a simple structure as berthierite by point-structure methods is obviously unrealistic.

Attempts to transform a real structure to a point structure. Since point structures can be readily extracted from their Patterson functions, attempts have been made to transform the Patterson function of a real crystal to the Patterson function of the corresponding point structure. There are many ways of partly accomplishing this, which will not be detailed here except to note that they attempt to allow for the thermal motions of the atoms and for the decline of the scattering curves with sin θ. If these could be perfectly allowed for, then a triply infinite Fourier series would build up a Patterson function consisting of a set of discrete points.

But this cannot be done, for many reasons. The chief reason is that coefficients $|F_{hkl}|^2$ for the triply infinite set of indices hkl are not available, nor can they ever be, for only reflections with small indices can be observed. Secondary reasons are that the thermal motions and the decline of the scattering curve can be only roughly approximated. The theoretical impossibility of computing a Patterson function whose peaks are sharpened enough for application of point-structure methods is confirmed by the failure of attempts to produce acceptable Patterson maps which actually show discrete points.

Image-seeking methods

The simple methods which can be used to extract a set of points from their Patterson function fail because, in general, the high regions in the Patterson function do not correspond to individual peaks, but to the coalescence of many peaks. The desired peak locations and heights are always unresolved in a background which interferes with finding them. Nevertheless, the crystal-structure information is embedded in the Patterson function and ought to be accessible in some way. To this end there has been devised another category of methods which depend on seeking and finding images even when they are unresolved in the background. These methods, which make use of the image-seeking functions described in a later section, can extract the desired crystal-structure information provided the number of atoms per cell is not too large.

Image-locations theorem. The several images of the crystal structure are located in the Patterson function in positions which bear a simple relation to the geometry of the image itself. It is this relation which makes possible the use of image-seeking methods.

Consider a simple crystal structure whose atoms A, B, C, and D are located at the ends of vectors OA, OB, OC, and OD which radiate from the origin O, as in Fig. 14. To produce the image of the structure $ABCD$ as seen from A, the quadrilateral $ABCD$ must be translated so that A comes to O, that is, translated by vector AO. To produce the image of $ABCD$ as seen from B, the quadrilateral in Fig. 14 must be translated so that B comes to O, that is, translated by vector BO. If this process is repeated for all images in the Patterson, it is seen that the structure must be translated from its original position in Fig. 14 by vectors AO, BO, CO, and DO. The vectors of this set are each exact opposites of ones which locate the atoms in the structure, specifically, OA, OB, OC, and OD. The set of vectors which locates the several images, shown in Fig. 15, therefore determines a set of positions which is the inverse of the set of positions of the atoms of the structure (Fig. 14). If a weight is attributed to an image location equal to the weight of the image, this is

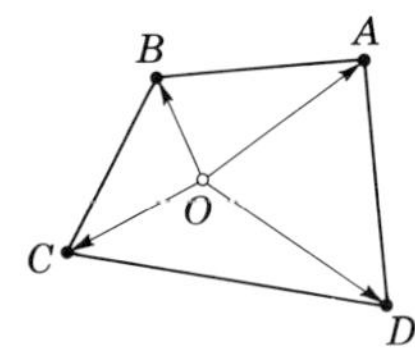

Fig. 14

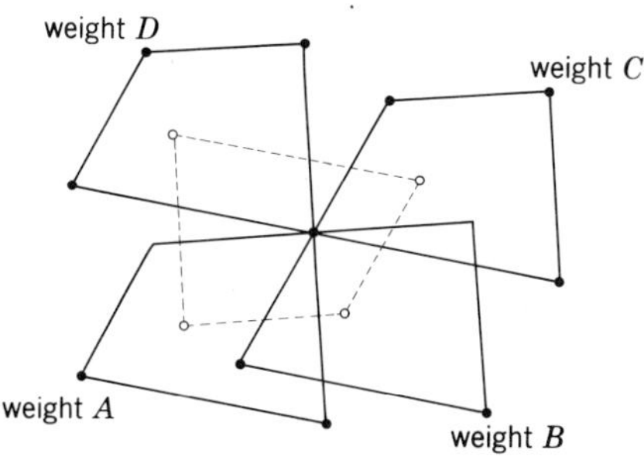

Fig. 15

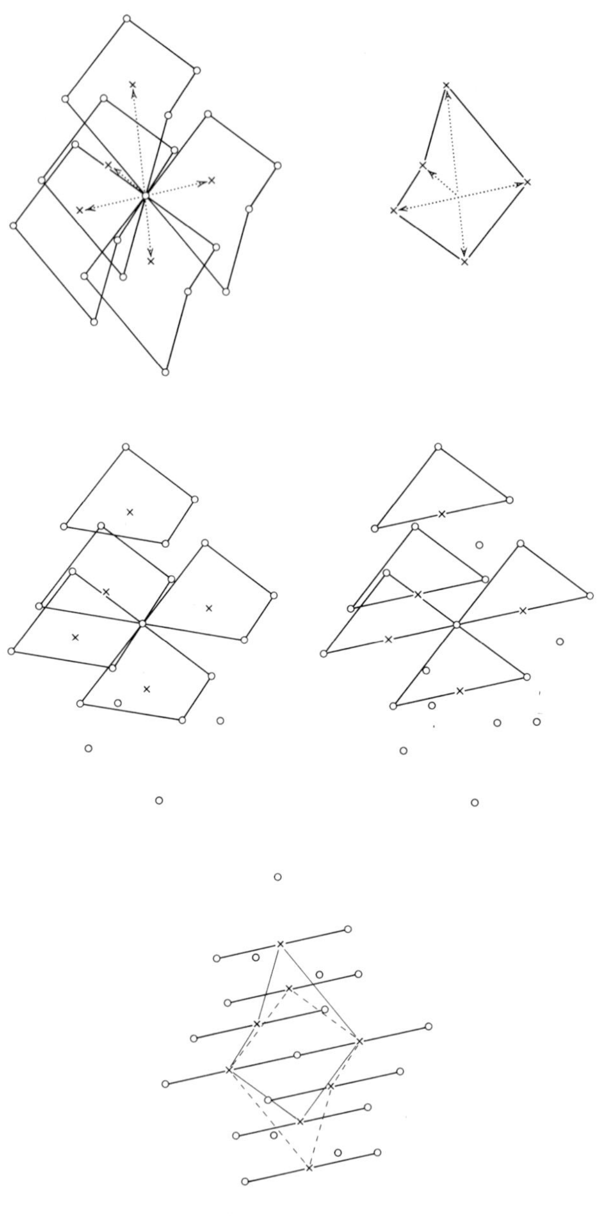

Fig. 16. The upper left diagram shows the Patterson function resolved into 5 images of the structure. The upper right diagram indicates the locations of these images by small crosses; this set of locations is a map of the inverse of the points in the structure. The middle two diagrams show that the locations of partial images of the structure define the same locations as the full images. [*From M. J. Buerger: Acta Cryst.* 4 (1951) 533, 534.]

also the weight of the atoms corresponding to that location in the crystal structure.

Theorem: The locations of a complete set of images, each location weighted by the weight of the image, constitute the arrangement and weighting of atoms in the inverse structure.

Suppose that the crystal structure is known. If this arrangement of atoms is systematically translated over the Patterson function, it will register with the images in the Patterson function, one by one. If a point is placed at each registry location and assigned a weight equal to the weight of the image found there, the set of points marks out the inverse of the original crystal structure.

Locations of fragments of images. This result is interesting but useless as it stands, because the crystal structure is unknown. Fortunately, if even an appropriate minimum fragment of the structure is known, that fragment will register with the part of the image which corresponds to itself. The collection of the relative locations of such image fragments is the same as the collection of locations of the full images, as shown in Fig. 16.

The minimum fragment is generally three noncollinear points, but if the crystal is centrosymmetrical, it may be merely two points, one of which may be the Patterson origin. (It will be seen later, in connection with the discussion of Fig. 18, that the other point is one which has unit weight. If such a point can be identified, the interpretation of a Patterson function can be started from information contained in the Patterson function itself.) Thus the knowledge of as little about the structure as the location of one pair of atoms may lead to the solution of the crystal structure with the aid of the Patterson function.

Alternatively, since the chemical composition of the crystal being studied is usually known, the geometrical configuration of a part of a molecule or other small coordinated group of atoms is usually easy to predict. If the orientation of the group is fixed by symmetry, it can be used as a fragment for image seeking. If its orientation is not known, it can often be found by various means which cannot be discussed here.

Image-seeking functions

Image-seeking methods can be placed on a useful basis by specifying the coordinates of the points of the image fragment in the Patterson function. Suppose that the relative locations of two atoms A and B in the crystal structure are known. The coordinates of these are $x_Ay_Az_A$ and $x_By_Bz_B$. In the Patterson function the vector from atom A to atom B appears with its tail at 000 and its head at $u = x_B - x_A$, $v = y_B - y_A$,

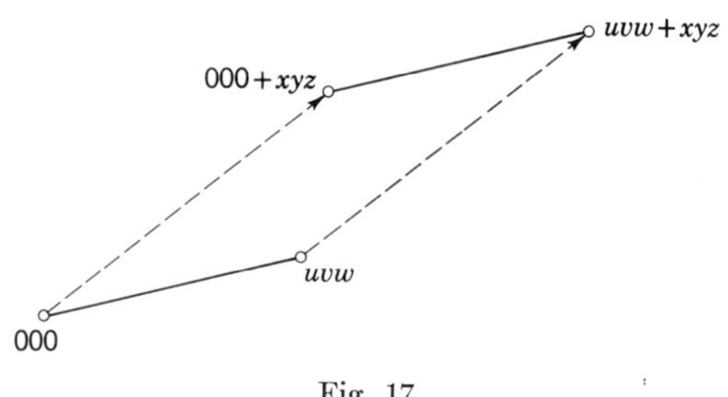

Fig. 17

and $w = z_B - z_A$. In image language, this can be described by saying that the images of atom A and atom B, as seen from atom A, have Patterson coordinates of 000 and uvw. This is a minimum image fragment. Let it be regarded as a pattern of the fragment of the image which is to be translated over the Patterson function. It can be translated over the Patterson function by displacing it to the end of a vector whose components are xyz (Fig. 17). After such a trial displacement its two points, which were originally located at

$$\begin{array}{c} 000 \\ \text{and } uvw, \end{array} \tag{31}$$

then have new coordinates

$$\begin{array}{lll} 0 + x, & 0 + y, & 0 + z, \\ u + x, & v + y, & w + z, \end{array} \quad \text{or, more briefly:} \quad \begin{array}{l} 000 + xyz, \\ uvw + xyz. \end{array} \tag{32}$$

An *image-seeking function* is now defined as some function of the Patterson densities at these new coordinates which reveals that the translation xyz displaces the original image to a position where it makes registry with duplicate image in the Patterson function.

To see how this should work out, consider Fig. 18A, which shows the location of 6 atoms in the cell of a crystal structure. The Patterson function of this structure has peaks at locations shown in Fig. 18B. Suppose that the locations of the 2 atoms connected by the vector in Fig. 18A are known. This vector appears in the Patterson function, as seen in Fig. 18B. The coordinates of its two ends are 000 and uvw. If these two points are translated by variable coordinates xyz, the new values of the coordinates of the two points are those given in (32), and whenever these coordinates cause these points to register with an image of themselves, they have attained the positions marked out by lines in Fig. 18C. If a cross is placed at a convenient place on this line, such as its center, it is seen that this set of crosses marks out the same positions as in the original structure (Fig. 18A).

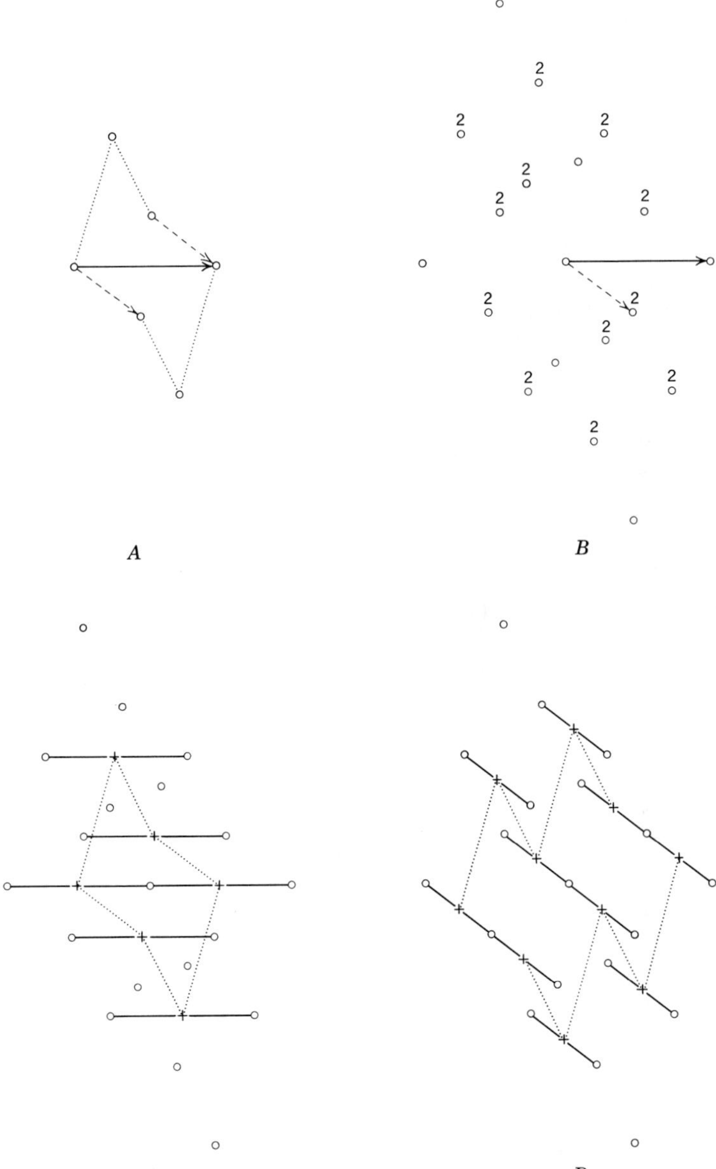

Fig. 18. [*From M. J. Buerger: Acta Cryst.* **4** (1951) 536.]

An image-seeking function is a function of the Patterson densities at (32), which reveals that xyz has translated the line in Fig. 18B to one of the locations of a line in Fig. 18C. Since the Patterson density at the points of Fig. 18C has high values, it is easy to devise functions of the coordinates of these points which also have high values. Three simple functions of this sort have been proposed and used, namely, the *sum function*, the *product function*, and the *minimum function*.

The sum function. If the sum of the Patterson densities at the end of the roving image is continuously laid down at the locations of the roving pattern of the image location, it maps out the sum function $S(xyz)$. It can be readily shown that this function bears an inequality relation to the electron density, as follows:

$$S(xyz) \geq K_1\rho(xyz), \tag{33}$$

where K_1 is a constant. The sum function is a poorly behaved image-seeking function because the desired peaks of the electron-density map which occur in it are surrounded by halos of undesired false peaks.

The product function. If the product of the Patterson densities at the ends of the roving image is continuously laid down at the image-pattern locations, it maps out the product function $\Pi(xyz)$. It can be readily shown that this function also bears an inequality relation to the electron density, as follows:

$$\sqrt[p]{\Pi_p(xyz)} \geq K_2\rho(xyz), \tag{34}$$

where p is the number of points in the image fragment, and K_2 is a constant. This function provides a more faithful representation of the electron density than the sum function.

The minimum function. Both the sum function and the product function are adversely affected by the unwanted background, and behave comparatively poorly because of the multiple-peak nature of Patterson maxima. A better behavior in these respects is obtained if the minimum value of the Patterson density at the points of the roving image is continuously laid down at the locations of the roving pattern of the image. This function, known as the minimum function $M(xyz)$, also bears an inequality relation to the electron density:

$$M(xyz) \geq K_3\rho(xyz), \tag{35}$$

where K_3 is a constant.

It commonly occurs that the atoms at the points of the image pattern have different electron densities. In this case the values of the Patterson function found at the points of the roving-image pattern must be weighted before their minimum value is recorded, in order to give each point an equal chance in determining an image-registry location.

Image-function rank. In the foregoing discussion it has been assumed that a pattern consisting of a minimum fragment of an image has been used to search the Patterson function for image locations. There are precautions which must be observed in the choice of this initial image pattern. For example, if a multiple peak is a part of the image, as in the dashed lines in Fig. 18*A* and *B*, the result is a pair of related solutions, as in Fig. 18*D*. But if there is a proper selection, the result is a crude representation of the electron density, the crudeness resulting from using only a small fragment of the full image, which may find coincidences that are purely accidental. If the crude approximation to the electron density can be at least partly interpreted, however, so that one or more new atom locations in the crystal structure are found, the number of points in the image-fragment pattern can be increased, and the registry of the new enlarged pattern tried again. Ordinarily this second try finds fewer accidental coincidences and therefore improves the approximation to the electron density.

The number of points in an image fragment is indicated by a subscript. For example, $M_2(xyz)$ is a minimum function for an image fragment consisting of two points. By the process of improvement just described, the *rank* of the function can be increased to $M_3(xyz)$ or $M_4(xyz)$, for example, and even considerably further. In this connection any symmetry other than a symmetry center can be put to use in a simple, systematic way.

Computation. In the earliest days of the use of image-seeking functions they were laboriously computed by desk computers from Patterson-function values which had been computed for the contouring of the Patterson function. Later, methods of graphically sketching the contours of the minimum function came into common use, and many crystal structures were solved by such graphical methods when two-dimensional projections were sufficient to determine a structure.

Graphical methods were also used with three-dimensional Patterson functions by treating them in sections, but this application was extremely tedious. At present, three-dimensional Patterson functions are routinely searched for images with the aid of high-speed digital computers. By such means the three-dimensional Patterson function can be readily searched by minimum functions of any rank, and the rank can be increased as the approximation to the electron density reveals increasing detail in the structure.

Examples of solutions by minimum functions. Some examples of minimum functions and the crystal structures corresponding to them, as finally represented by Fourier synthesis of their electron densities, are shown in Figs. 19 to 21.

Limitations of image-seeking methods. The Patterson function, and its unraveling by image-seeking functions, casts light on the limita-

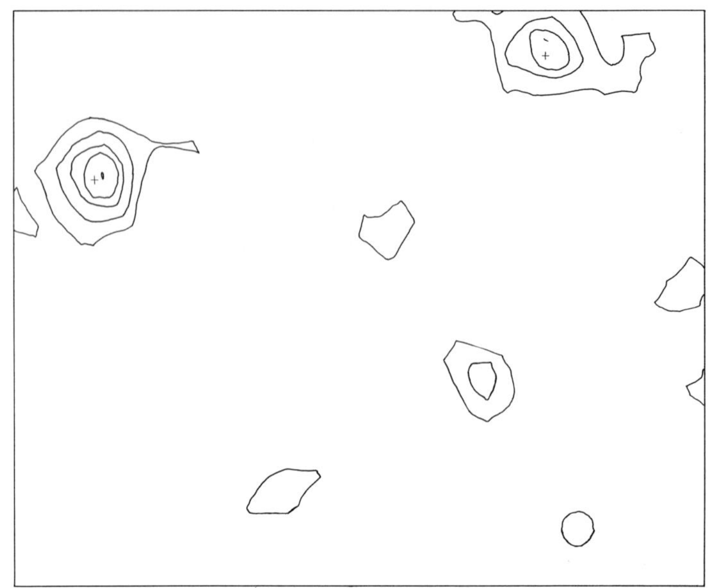

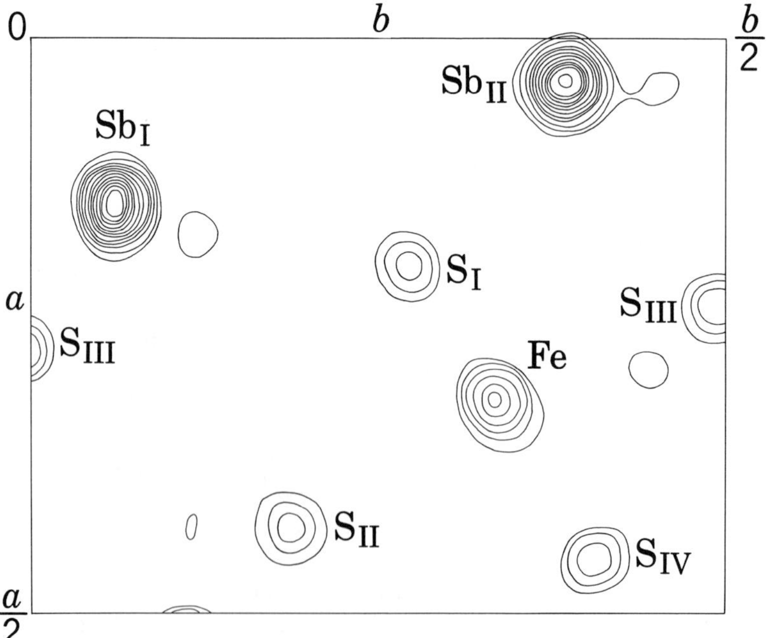

Fig. 19. Comparison of the minimum function $M_8(xy)$ (*above*) with the electron-density projection $\rho(xy)$ for berthierite, $FeSb_2S_4$. [*From M. J. Buerger: Acta Cryst.* **4** (1951) 543.]

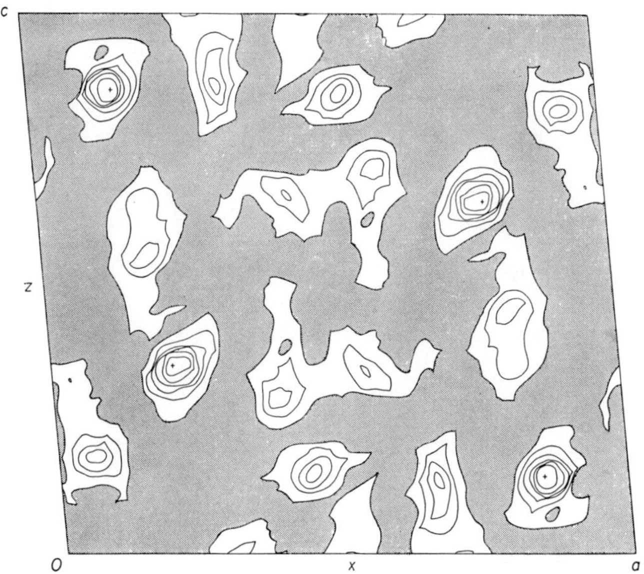

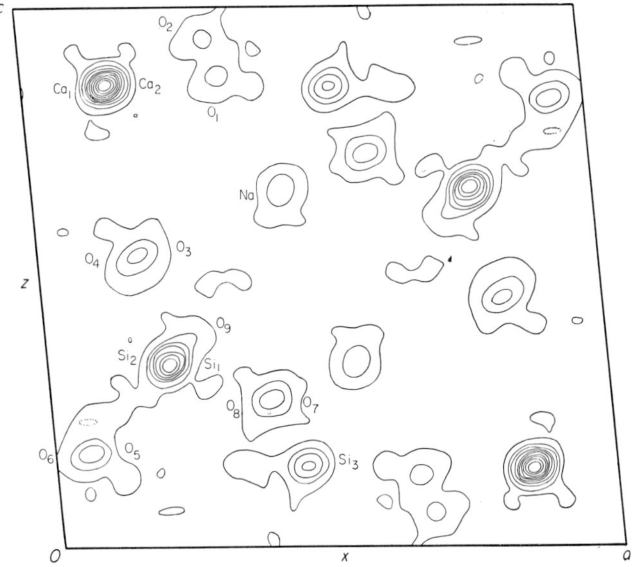

Fig. 20. Comparison of the minimum function $M_4(xz)$ (*above*) with the electron-density projection $\rho(xz)$ for pectolite, $NaHCa_2Si_3O_9$. [*From M. J. Buerger: Z. Kristallogr.* **108** (1956) 254, 255.]

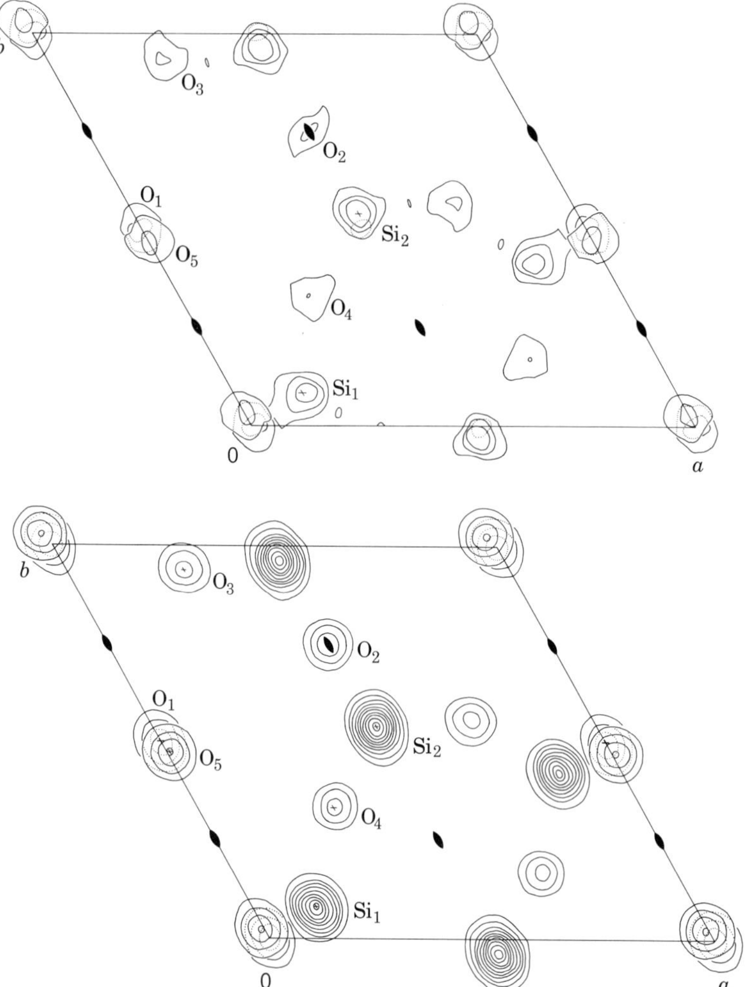

Fig. 21. Comparison of the projections on (001) of the largest cross sections of the peaks found on various levels of the minimum function $M_8(xyz)$ and those found on the electron-density function $\rho(xyz)$ for coesite, SiO_2. [*From Tibor Zoltai and M. J. Buerger: Z. Kristallogr.* **111** (1959) 136, 138.]

tions of solving crystal structures without phase information. Image-seeking methods are useful only if image peaks stand out from their unwanted background. This is analogous to the problem of detecting a signal in the presence of unwanted noise. Solutions of crystal structures having up to some 30 atoms per cell have been successfully achieved by image-seeking methods, which call for finding images in a smear of more than 900 Patterson peaks per cell. Such solutions require knowing some minimum feature of the structure. This can often be determined from features of the Patterson function itself by various strategies which cannot be detailed here. A solution can also be initiated if the structure is known to contain a specific image fragment, for example, an AlO_6 octahedron or a benzene ring. Several computer programs have been devised to find and utilize such information.

Summary

It is not possible to find the electron density of a crystal by using the experimental amplitude data in a Fourier synthesis, because only the magnitudes of the amplitudes are measurable. Patterson showed, however, that a Fourier series whose coefficients are the squares of these magnitudes yields useful information about the crystal structure. This series, known as the Patterson function, has the property of showing a peak for every pair of atoms in the crystal. The weight attributed to a peak (that is, the value of the function integrated over the volume of the peak) is equal to the product of the atomic numbers of the atoms of the pair. The peak is located at the end of a vector from the origin which is the interatomic vector of the pair of atoms. Unfortunately, if there are n atoms in the crystal cell, there are n^2 pairs of atoms, so that there are n^2 peaks in the Patterson function; n of these coincide identically at the origin. The cell has so many peaks that many must coalesce unless n is a very small integer. Every Patterson function is inherently centrosymmetrical.

Although crystals can have any one of the 230 space-group symmetries, Patterson functions can have only 24 of these. Yet symmetry information is not lost in the Patterson function, but merely appears in a curious way: Peaks due to atom pairs which are symmetrical with respect to a mirror or glide plane appear along a line that is perpendicular to the plane and whose distance from the origin is the translation component of the glide; peaks due to pairs of atoms which are symmetrical with respect to an axial symmetry element occur on a plane that is perpendicular to the axis and whose location is characteristic of the translation component of the screw. By taking advantage of such information, the space-group symmetry of a crystal can be determined from the features of its Patterson function.

The interpretation of Patterson functions is facilitated by the use of image theory. In these terms, the function can be described as the image of the crystal structure as seen from one atom, superimposed upon the images as seen from all the other atoms. If one image could be isolated, it would constitute the solution of the crystal structure. Provided that there are not too many atoms in the cell, this isolation can be effected with the aid of image-seeking functions. Such a function can be based upon a single peak of the Patterson function. In not unfavorable cases, this simple image-seeking function reveals the positions of a few atoms in the crystal, and a more powerful image-seeking function can be set up with this new information. Iteration of this sequence ends with a revelation of most of the atoms in the cell.

Notes on history

The Patterson function was presented by A. L. Patterson in 1934. This was a vector generalization of a method which had been used as early as 1927 by Zernike and Prins to investigate the local arrangement of atoms in fluids, and which was extensively used later to find the arrangement of atom neighbors in glasses. Only two years after Patterson's first paper, David Harker pointed out that, for symmetrical crystals, there existed certain sections and lines in the Patterson function in which, ideally, only peaks due to symmetrically related pairs of atoms should occur. These became known as Harker sections and Harker lines. The peaks in these loci were especially easy to interpret with respect to locations of atoms in the structure. In 1946 Buerger formulated the exact structural interpretation of Harker sections in *implication theory* and pointed out how space groups could be determined, in spite of Friedel's law, by using Harker lines and sections.

Dorothy Wrinch brought to Patterson theory the viewpoint of images. This was developed by Buerger, who was able to show that by making use of the nature of images, any set of points could be recovered from its Patterson function by several means. This led to the image-locations theorem and image-seeking functions, which eventually solved many crystal structures directly from $|F|^2$ data.

Additional reading

Martin J. Buerger. *Vector space and its application in crystal-structure investigation.* (Wiley, New York, 1959) 347 pages.

M. J. Buerger. *Image methods in crystal-structure analysis.* In *Advanced methods of crystallography*, edited by G. N. Ramachandran. (Academic, London, 1964) 1–24.

Significant literature

A. L. Patterson. *A Fourier series method for the determination of the components of interatomic distances in crystals.* Phys. Rev. **46** (1934) 372–376.

A. L. Patterson. *A direct method for the determination of components of interatomic distances in crystals.* Z. Kristallogr. (A) **90** (1935) 517–542.

David Harker. *The application of the three-dimensional Patterson method and the crystal structure of proustite,* Ag_3AsS_3, *and pyr argyrite,* Ag_3SbS_3. J. Chem. Phys. 4 (1936) 381–390.

D. M. Wrinch. *The geometry of discrete vector maps.* Phil. Mag. **27** (1939) 98–122.

M. J. Buerger. *The interpretation of Harker syntheses.* J. Appl. Phys. **7** (1946) 579–595.

M. J. Buerger. *Vector sets.* Acta Cryst. **3** (1950) 87–97, 243.

M. J. Buerger. *A new approach to crystal-structure analysis.* Acta Cryst. **4** (1951) 531–544.

14

Refinement

Introduction

In determining how the atoms are arranged in a crystal, the major difficulty which must be overcome is that of arriving, somehow, at a rough but essentially correct set of locations for the atoms. Once this has been accomplished, the chief obstacle has been left behind, and it is a routine matter to refine this roughly known structure to any precision which the data permit. There are many ways of accomplishing this refinement, but only a few in common use will be discussed here. Some remarkable information, which one would probably not anticipate, can be the incidental fruit of refinement, as will be seen at the end of this chapter.

The discrepancy index

The progress in improving the precision of the rough model to its final refined state is commonly followed by noting the value of the discrepancy index

$$R = \frac{\Sigma \big| |F_{\text{obs}}| - |F_{\text{cal}}| \big|}{\Sigma |F_{\text{obs}}|},$$

which was introduced in (31) of Chapter 11 and discussed in that chapter. If each F_{obs} has been measured with complete accuracy and if the model is in perfect agreement with the actual crystal structure, then the numerator vanishes, and the value of R is zero. The value of R, therefore, reflects the errors that still exist in the model and in the intensity data. The errors in the observed F's can be appraised by comparing the observed F's of reflections which are equivalent by symmetry. For intensities carefully measured with a diffractometer and appropriately corrected for Lorentz, polarization, absorption, and extinction effects, the differences in observed F's may be as small as 2 percent. The difference represents errors in measurement which set a lower limit for R. The differences in observed F's for symmetrically equivalent reflections can be neatly noted by measuring the intensities from a specimen ground to a spherical shape and rotating it about the axis of maximum symmetry during the data collection. With this arrangement the correction factors are equal for all reflections which are equivalent by this symmetry, so that any differences in intensity are immediately obvious.

The rough structure

The preliminary model which is to be refined may originate in any of a number of ways. It may have been derived from a minimum-function study based upon a Patterson map; it may have been derived from application of one of the direct methods of phase determination or from one of the indirect methods, such as the replaceable-atom method. The preliminary model may even have arisen as a result of a guess based upon the way in which the atoms of the crystal are known (from other crystals) to surround each other as coordination groups, combined with discovering a scheme for connecting these coordination groups to conform to the space-group symmetry and to the cell dimensions.

Whatever the origin of the model, the value of its R is likely to be high. For a completely wrong structure the expected value of R is 82.8 percent if the crystal is centrosymmetrical, or 58.6 percent if it is noncentrosymmetrical. When the value of R for a model proves to be 45 percent or lower, the structure may well be correct, its poor R value being due to moderate departures of the coordinates and other parameters (such as the temperature coefficients) from their correct values. Refinement will readily improve such a structure so that it will come to have a small R value. On the other hand, the fairly high initial R value may be due to the structure's being only partly correct. Ordinarily such a structure cannot be refined purely by the least-squares method, which is described later, but the incorrect aspect can be discovered and corrected by cautious application of various Fourier methods.

Adjustment of parameters

The earliest method of refining a structure was to study a list of observed and computed intensities, along with the computations, to see whether some obvious adjustment in a parameter, such as a coordinate of one of the atoms, could be made which would produce an improvement in agreement of intensities. This is a straightforward method when the number of parameters is very small, and especially where the intensities of one set of reflections are a function of the parameters of one atom only. At present this method is rarely used, for two reasons. In the first place, in structures currently being solved, the number of atoms and their variable parameters is so large that a reflection intensity is affected by the parameters of many atoms. Secondly, computations are almost never made by hand now because of the availability of electronic computers; these do not ordinarily display the results of intermediate steps in the computations, so that the effects on a set of reflections of varying the parameters of one atom are not normally revealed.

Successive Fourier syntheses

In the discussion of the heavy-atom method in Chapter 12 it was seen that a set of heavy atoms makes a contribution to a reflection such that, under favorable circumstances, the phase of the heavy-atom contribution dominates the phase of the reflection. For centrosymmetrical crystals the set of heavy atoms controls the phases of most reflections if the atoms are heavy enough. As a consequence, these phases can be used to compute a Fourier synthesis which will reveal the locations of some or all of the rest of the atoms. In this case, successive Fourier syntheses improve a very limited initial knowledge of the structure until the locations of all atoms are known.

For somewhat similar reasons, if the locations of many, but not all, of the atoms of a structure are known, a Fourier synthesis based upon the phases of these known atoms has the important property of not only displaying peaks at the positions of the known atoms but also showing less accentuated peaks at the positions of the unknown ones. In this case also, successive Fourier syntheses provide a program for improving a limited initial knowledge of a structure. There is an important rule for using this method of finding additional atoms: The phases determined by a set of atoms return those atoms in the Fourier synthesis at approximately the places where they are assumed to be, whether they are actually there or not. Thus, a Fourier synthesis perpetuates any error in a proposed structure, so that great care must be taken not to use in the phase calculation an atom whose location is in any way doubtful. On the other hand, if such a doubtful atom is actually where surmised, a Fourier synthesis based upon phases computed without that atom will reveal it.

If the positions of all atoms in a structure are approximately known, then a program of successive Fourier syntheses will improve the knowledge of their locations. This action is easiest to understand in the case of a centrosymmetrical crystal. For such a crystal the phase of every reflection must be either 0 or π, which corresponds to assigning a positive or negative sign to the F of every reflection.

The process of refinement can be illustrated by an artificial example.† A centrosymmetrical crystal has a simple structure consisting of a pair of atoms located on the a axis with an undetermined coordinate x. Initially, there is reason to believe that x is approximately 0.18. From this trial coordinate a set of signs can be computed for the $F(hk0)$'s. When these signs are attributed to the correct (corresponding to experimental) magnitudes $|F(hk0)|$, the several Fourier coefficients A_n required for a Fourier synthesis of the electron density along the a axis can be computed. These are given in the line of Table 1 indicated as "Stage I." The resulting synthesis is shown by the dotted curve in Fig. 1. Although

† This is based upon an imaginary centrosymmetrical crystal having an orthogonal cell with $a = b = c = 10$ Å. The structure contains only a centrosymmetrical pair of alkali-ion-like atoms whose maximum scattering power is normalized to $Z = 1$. It is assumed that $hk0$ reflections are available for which the range of h and k is from -10 to $+10$.

Table 1
Sequence of steps in the example of refinement by successive Fourier syntheses

Stage	x	Signs of Fourier coefficients											Fourier peak found at
		A_0	A_1	A_2	A_3	A_4	A_5	A_6	A_7	A_8	A_9	A_{10}	
I	0.18 (assumed)	+	+	−	−	−	+	+	−	−	−	+	0.19
II	0.19	+	+	−	−	+	+	+	−	−	−	+	0.195
III	0.195	+	+	−	−	+	+	+	−	−	+	+	0.20
IV	0.20	+	+	−	−	+	+	+	−	−	+	+	0.20

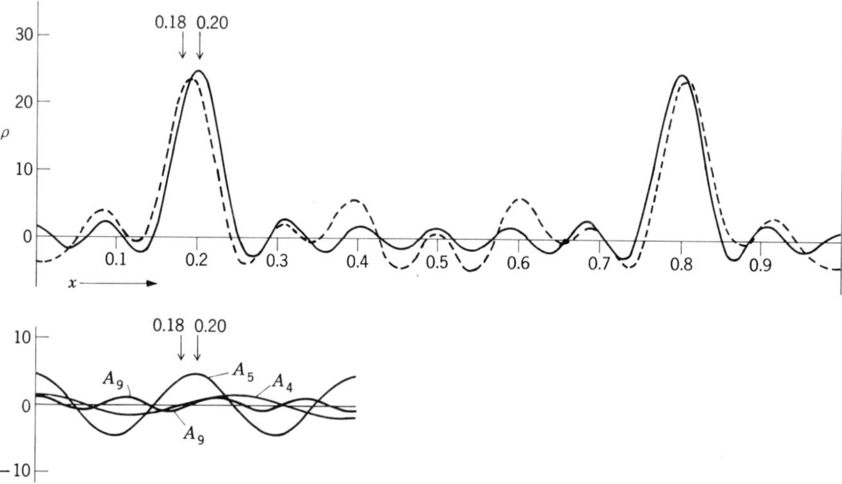

Fig. 1. An example of the refinement of a simple structure by successive Fourier syntheses. An atom was initially assumed to be located at $x = 0.18$. After three cycles of refinement by successive Fourier syntheses, the coordinate attained a final correct value of $x = 0.20$. In this process, outlined in Table 1, two Fourier coefficients, A_4 and A_9, changed signs. As seen in the lower part of the figure, these coefficients had small amplitudes near the correct coordinate. Other Fourier coefficients, represented by A_5, had large amplitudes which varied little in this region.

the incorrect coordinate $x = 0.18$ was used, the peak of the electron density appears in the Fourier synthesis at $x = 0.19$. This value is used in computing signs to be used in Stage II of Table 1. When the signs of the $F(hk0)$'s for this new x are computed, it is found that the Fourier coefficient A_4 for use in the synthesis of the projected electron density has changed sign. A second Fourier synthesis with the correct magnitudes and with this corrected set of signs now shows a peak at about $x = 0.195$. For this value of x another computation shows that an additional sign change occurs in the Fourier coefficient A_9 for the synthesis of the projected electron density. A third Fourier synthesis with this third set of signs has the peak at 0.20 (Fig. 1). For this value of x, a computation of signs shows that all Fourier coefficients have the same sign as in the last computation, so that if a new Fourier synthesis were made it would be the same as the last synthesis. The coordinate x of the peak has consequently converged to $x = 0.20$.

The nature of the action in this process depends on the magnitudes of the observed F's, which are correct for the correct location of the atom, and on the signs attributed to the F's, of which most are correct but one or a few are initially incorrect. This partially correct Fourier synthesis can be thought of as decomposed into two syntheses: one based upon the set of F's whose signs and magnitudes are both correct, plus another based upon the few F's whose signs are incorrect but whose magnitudes are correct. So long as the number of F's in the latter is few, the correct part of the Fourier synthesis dominates, and the full Fourier synthesis shows peaks which have drifted from the assumed positions toward the correct positions. This is due to the fact that if an atom is in nearly the correct position and if it is separated from the correct position by a sign change in one or a few F's, then these F's must be small in order that they can pass through zero and change sign because of a small shift in coordinates. This is illustrated in the foregoing example by A_4 and A_9, which changed signs during the refinement. It is basically because of this situation (specifically, that F's which are incorrectly phased are small when the atom is *near* its correct location) that successive Fourier syntheses converge to the correct final location.

Successive Fourier syntheses of a noncentrosymmetrical structure also lead to refinement of the structure. For such crystals the convergence to final atom location is slower because the phases are continuously variable, not merely 0 or π. There are schemes for hastening the convergence, however, such as assuming a new location for an atom at twice the distance it appears to have drifted, as judged by the Fourier synthesis.

Unfortunately, after convergence of a succession of Fourier syntheses, the final peak locations are not the accurate positions of the atoms. This is because a perfect Fourier synthesis of the atoms would require the use of all Fourier coefficients F_{hkl}, with h, k, and l ranging from $-\infty$ through 0 to $+\infty$. Only a limited number of these F_{hkl}'s with small values of h, k, and l, however, are experimentally available, say about 2000 in a three-dimensional Fourier synthesis. The lack of observed F_{hkl}'s for larger values of h, k, and l produces what is known as a *series-termination effect,* one aspect of which is to produce small errors in the peak location of the atoms. This effect causes the method of successive

Fourier syntheses to be unsatisfactory for complete refinement of the atom locations.

Difference Fourier syntheses

After a series of successive Fourier syntheses has converged, the phases of all the F_{hkl}'s are fixed. With these phases, two Fourier syntheses can be prepared, one using observed F's, the other computed F's as follows:

$$\rho_{\text{obs}} = \frac{1}{V} \sum_h \sum_k \sum_l F_{\text{obs}} e^{-i2\pi(hx+ky+lz)}, \tag{1}$$

$$\rho_{\text{cal}} = \frac{1}{V} \sum_h \sum_k \sum_l F_{\text{cal}} e^{-i2\pi(hx+ky+lz)}. \tag{2}$$

If the model ρ_{cal} is the same as the true structure ρ_{obs}, then these should be identical. Actually this may not be so, for the magnitudes of F_{obs} and F_{cal} may differ slightly. If (2) is subtracted from (1), the difference, known as a difference Fourier synthesis, may result in a slight difference in electron density, namely,

$$\rho_{\text{obs}} - \rho_{\text{cal}} = \frac{1}{V} \sum_h \sum_k \sum_l (F_{\text{obs}} - F_{\text{cal}}) e^{-i2\pi(hx+ky+lz)}. \tag{3}$$

It is seen that this synthesis is easily computed by using as Fourier coefficients the differences in magnitudes of F_{obs} and F_{cal} and attributing to the resulting difference the phase of F_{cal}, which is the same as that of F_{obs}.

This difference Fourier synthesis has some remarkable properties. If the model matched the actual structure exactly, the difference density (3) should be zero everywhere, except that random errors in measuring $|F_{\text{obs}}|$ should produce a small random fluctuation in the function. Generally, it presents a map of the anomalies which exist between the true structure and the structure assumed in the final model. How can such differences exist? They could not exist if an infinite number of Fourier coefficients were available. But when only a finite number are available and used, a small but actual latitude exists in the position each atom can assume without changing the sign of an F, as well as a latitude in its thermal behavior.

An example of how the difference density can be used is shown in Fig. 2. Suppose that the positions of the peak in the actual structure ρ_{obs} differs from the peak in the model ρ_{cal} by a small displacement ϵ. Then the difference density shows a positive anomaly in the direction toward which the model's atom should be shifted, and a negative anomaly

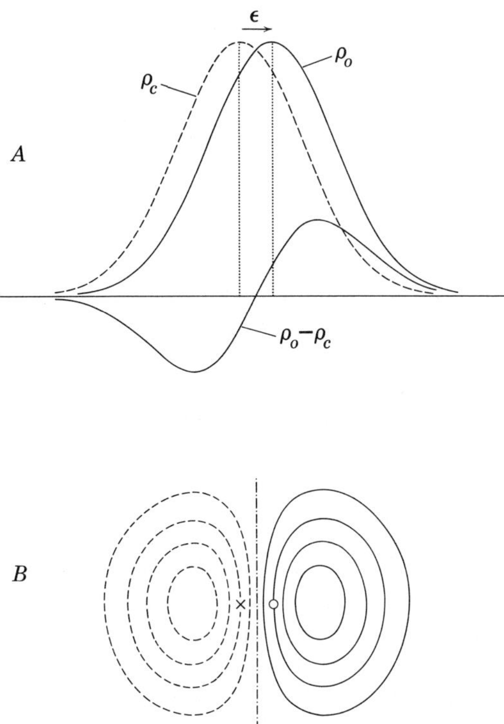

Fig. 2. Difference electron density due to an atom whose location should be corrected by a shift ε.

in the opposite direction. The magnitude of the anomaly is a measure of the amount of shift required to bring the model into conformity with the true structure. When the proper adjustment has been achieved, the location of an atom should be on a position of zero gradient on a new difference-density map.

Another simple error which the difference density reveals is the use of an incorrect temperature coefficient in the model. Suppose that a greater thermal motion occurs in the actual structure than is allowed for in the model. Then the electrons of an atom in the actual structure are distributed by the thermal motion over a larger volume, so that the peak density of the atom is smaller than assumed (Fig. 3). The difference density then shows a negative region in the immediate vicinity of the atoms's center, surrounded by a positive ring. Anisotropic thermal motion is ordinarily displayed by a kind of 4-fold clover leaf (Fig. 4) with alternating positive and negative sectors. (A somewhat similar, but cruder, effect results from improper correction for anisotropic absorption in the specimen; this reveals an error in $|F_{\text{obs}}|$.)

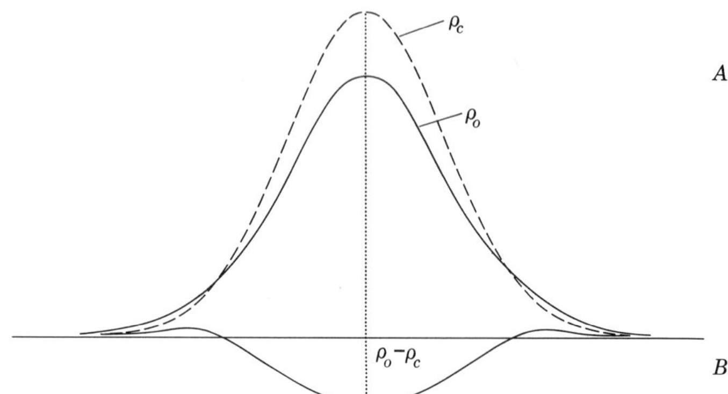

Fig. 3. Difference electron density resulting from overestimation of temperature motion.

Difference-density maps are nearly free from series-termination errors, because these errors are about the same for both ρ_{obs} and ρ_{cal} and therefore tend to cancel on subtraction. Because they are relatively unaffected by series-termination errors, difference-density maps are more useful for refinement than are electron-density maps.

Difference-density maps have another very useful application. When a crystal contains atoms having a wide range of atomic numbers, very light atoms are difficult or impossible to notice in an ordinary Fourier synthesis of the electron density, because the peaks of the light atoms

Fig. 4. Difference electron density resulting from anisotropic thermal motion with maximum displacement left and right. [*Figures 2 to 4 from Martin J. Buerger: Crystal-structure analysis.* (*Wiley, New York,* 1960) 605, 606.]

are about the same heights as the background fluctuations. But since the background fluctuations occurring in ρ_{obs} and ρ_{cal} are nearly the same, they tend to cancel in the difference density, leaving as a residue any real difference in the electron densities of the structure and model. Even hydrogen, which donates its lone electron to the hydrogen bond, can be readily located in difference maps in the presence of C, N, and O. Indeed, the difference-density technique has become the chief direct method of locating hydrogen atoms in organic crystals. In a similar way, Be and B may be located in the presence of heavier atoms, and even oxygen atoms ($Z = 8$) can be located in the presence of uranium ($Z = 92$).

Least squares

General principles. When a measurement is made of some quantity, there is always an error (hopefully small) associated with the measurement, so that the results of many measurements of the same quantity are always found to scatter about the correct value, which is, of course, unknown. Suppose it is known that a physical quantity q is linearly related to some variables $x, y, z, \ldots$ by the relation

$$q = ax + by + cz + \cdots, \tag{4}$$

where a, b, and c are different but known constants. If q is measured, there is always an error E associated with it, so that what is actually found is $q + E$. Then the left of (4) appears to be $q + E$. On making a specific measurement, designated by the numeral 1, the specific relation appears to be

$$q_1 + E_1 = a_1x + b_1y + c_1z + \cdots, \tag{5}$$

so that in this case the error is

$$E_1 = a_1x + b_1y + c_1z + \cdots - q_1. \tag{6}$$

Other observations can be made, each with a different value of a, b, c, . . . and a different observation q with which is associated a different error E, to give a set of equations like (6), namely,

$$\begin{aligned} E_1 &= a_1x + b_1y + c_1z + \cdots - q_1 \\ E_2 &= a_2x + b_2y + c_2z + \cdots - q_2 \\ E_3 &= a_3x + b_3y + c_3z + \cdots - q_3 \\ &\vdots \\ E_m &= a_mx + b_my + c_mz + \cdots - q_m. \end{aligned} \tag{7}$$

Such observations can be carried out until the number of observations

m exceeds the number of variables x, y, z, ... These variables are thus overdetermined. According to Legendre's principle of least squares, the most acceptable values of these variables x, y, z, . . . are those which make the sum of the squares of the errors a minimum, namely, the sum

$$\sum_{j=1}^{m} E_j^2 = E_1^2 + E_2^2 + E_3^2 + \cdots + E_m^2 \tag{8}$$

$$= \sum_{j=1}^{m} (a_jx + b_jy + c_jz + \cdots - q_j)^2. \tag{9}$$

This minimum can be readily found by differential calculus; this leads to the most acceptable values of the variables x, y, z, ...

Application to refinement of atom coordinates. In crystal-structure analysis, the measured quantity is F_{hkl}, and this is related to the sets of coordinates of the locations of the atoms by

$$F_{hkl} = \Sigma f_j e^{i2\pi(hx_j+by_j+lz_j)}. \tag{10}$$

Unfortunately, this is not the linear relation required by the least-squares method of finding the best values of the variables x_j, y_j, and z_j. If the atom locations are almost correct, however, only small changes in the coordinates are required, and for these shifts the value of F may be regarded as nearly linear in the xyz's. (Mathematically, this can be formalized by expanding the exponential in Taylor's series and retaining only the first two terms of the series.)

Computations concerned with finding the best values of the xyz's for many atoms with the aid of least squares are too tedious to be handled effectively by hand computation. The development of high-speed digital electronic computers, however, has made such computing relatively easy, and it is now customary to refine crystal structures with the aid of such computers.

Thermal motion

General features. In the absence of a disturbing force, each atom of a crystal structure is fixed at an equilibrium position in the structure. Thermal agitation, however, causes each atom to execute vibrations about its equilibrium position. If u indicates the displacement of the atom, then the magnitude of the thermal motion can be expressed by the mean-square displacement $\langle u^2 \rangle$ from the equilibrium position. If the restoring forces are the same in all directions of vibration, then the field of restoring forces is said to be *isotropic*. In this case the mean-square displacement is the same in all directions and marks out the surface of a sphere.

In general, however, the field of restoring forces varies with direction. As a result, the mean-square displacement vector also varies with direction and can be shown to mark out a quadric surface. Provided that the restoring forces are linear, this surface is an ellipsoid having three orthogonal axes OX, OY, and OZ. Unless restricted by the symmetry of the atom's location, these axes do not coincide, in general, with the axes of the cell of the crystal.

Correction for isotropic thermal motion. If the atoms of a crystal were motionless, then ideally each atom of a set of translation-equivalent atoms would be fixed relative to one another as if their centers occupied the points of a lattice. In considering a reflection from a plane (hkl), every atom would be located so that its center would be exactly in one plane of the stack (hkl). Under these circumstances every atom on a particular plane scatters exactly in phase with all other atoms on the plane, and exactly 2π out of phase with the atoms in the next neighboring planes. As a result, all atoms of the translation-equivalent set scatter in phase with an amplitude corresponding to f.

With isotropic thermal motion, however, the atoms do not remain strictly in their own planes of the stack (hkl). Those belonging to a particular plane actually are somewhat displaced from it, and this spoils the in-phase relations of their combined scattering. As a result, the amplitude they scatter is reduced from one corresponding to f to a lesser one. The reduction can be shown to have the form e^{-M}, called the *temperature factor*, as noted in Chapter 11.

The out-of-phase scattering by an atom is directly proportional to the component of its displacement from its plane, and inversely proportional to the spacing to its nearest neighboring plane. The exponent in the correction term thus involves $\langle u^2 \rangle$ and $1/d_{hkl}$; the latter is recognized as the distance from the origin to the reciprocal-lattice point hkl. The exact form of the temperature factor for isotropic thermal vibrations is

$$T_{hkl} = e^{-2\pi^2 \langle u^2 \rangle / d_{hkl}{}^2} \tag{11}$$

$$= e^{-2\pi^2 \langle u^2 \rangle r_{hkl}{}^{*2}} \tag{12}$$

$$= e^{-2\pi^2 \langle u^2 \rangle (4 \sin^2 \theta)/\lambda^2} \tag{13}$$

$$= e^{-B(\sin^2 \theta)/\lambda^2}, \tag{14}$$

where

$$B = 8\pi^2 \langle u^2 \rangle, \tag{15}$$

and

$$r^*_{hkl} = \frac{1}{d_{hkl}} = \frac{2 \sin \theta_{hkl}}{\lambda}. \tag{16}$$

These forms may be compared with (29) and (30) of Chapter 11.

Correction for anisotropic thermal motion. As pointed out above, if the restoring forces are linear but vary with direction, then the mean-square displacement marks out an ellipsoid. The lengths and directions of its principal axes describe the average thermal motion of the atom. To investigate this ellipsoid, the temperature coefficient as expressed in (12) must be generalized to take account of the variation of restoring force with direction. To do this the product $\langle u^2 \rangle r_{hkl}^{*2}$ must be expanded. The reciprocal-lattice vector $\mathbf{r}_{hkl}^*$ can be expressed in terms of the unit vectors of the reciprocal lattice, $\mathbf{a}^*$, $\mathbf{b}^*$, and $\mathbf{c}^*$, as follows:

$$\mathbf{r}_{hkl}^* = h\mathbf{a}^* + k\mathbf{b}^* + l\mathbf{c}^* \tag{17}$$

$$|r_{hkl}^*|^2 = r_{hkl}^* \cdot r_{hkl}^* \tag{18}$$

$$\begin{aligned} &= (ha^* + kb^* + lc^*) \cdot (ha^* + kb^* + lc^*) \\ &= h^2a^{*2} + k^2b^{*2} + l^2c^{*2} + 2hka^* \cdot b^* + 2klb^* \cdot c^* + 2lhc^* \cdot a^*. \end{aligned} \tag{19}$$

In generalizing the isotropic product $\langle u^2 \rangle r_{hkl}^{*2}$, each of the terms in (19) is a component of $|\mathbf{r}_{hkl}|^2$, and each must be multiplied by its characteristic $\langle u^2 \rangle$. The result of generalizing $\langle u^2 \rangle r_{hkl}^{*2}$ is then

$$\begin{aligned} u^2 r_{hkl}^{*2} \rightarrow\ & u_{11}h^2a^{*2} + u_{22}k^2b^{*2} + u_{33}l^2c^{*2} \\ & + 2u_{12} \cos (a^*b^*)hka^*b^* \\ & \qquad + 2u_{23} \cos (b^*c^*)klb^*c^* + 2u_{31} \cos (c^*a^*)lhc^*a^* \\ \rightarrow\ & U_{11}h^2a^{*2} + U_{22}k^2b^{*2} + U_{33}l^2c^{*2} \\ & + 2U_{12}hka^*b^* + 2U_{23}klb^*c^* + 2U_{31}lhc^*a^*, \end{aligned} \tag{20}$$

where

$$U_{12} = u_{12} \cos (a^*b^*), \text{ etc.}$$

The more general form of the anisotropic temperature coefficient is therefore

$$T_{hkl} = e^{-2\pi^2(U_{11}h^2a^{*2}+U_{22}k^2b^{*2}+U_{33}l^2c^{*2}+2U_{12}hka^*b^*+2U_{23}klb^*c^*+2U_{31}lhc^*a^*)}. \tag{21}$$

There are two other common forms of this anisotropic temperature correction:

$$T_{hkl} = e^{-\frac{1}{4}(B_{11}h^2a^{*2}+B_{22}k^2b^{*2}+B_{33}l^2c^{*2}+B_{12}2hka^*b^*+B_{23}2klb^*c^*+B_{31}2lha^*c^*)}, \tag{22}$$

and

$$T_{hkl} = e^{-(\beta_{11}h^2+\beta_{22}k^2+\beta_{33}l^2+\beta_{12}hk+\beta_{23}kl+\beta_{31}lh)}. \tag{23}$$

The form given in (22) has coefficients B_{ij} which are comparable to the isotropic temperature factor B of (14). The form given in (23) is ordinarily used in computing. The several forms are related by

$$\begin{aligned} \beta_{11} &= \tfrac{1}{4}B_{11}a^{*2} &= 2\pi^2U_{11}a^{*2}, \text{ etc.} \\ \beta_{12} &= \tfrac{1}{2}B_{12}a^*b^* &= 4\pi^2U_{12}a^*b^*, \text{ etc.} \end{aligned} \tag{24}$$

The exponents of (21) to (23) each have the form of the expression representing the surface of an ellipsoid referred to an arbitrary coordinate system. The equation of such a surface is usually written

$$Ax^2 + By^2 + Cz^2 + 2Dxy + 2Eyz + 2Fzx = 1. \tag{25}$$

The term Ax^2 in (25) is represented in (21) by the term $2\pi^2U_{11}h^2a^{*2}$, in which the constant A is $2\pi^2U_{11}$, and the variable is h^2a^{*2}. In the next section it will be pointed out that such constants, which are characteristics of a given atom in a crystal, can be found from the diffraction data by least-squares refinement.

The equation of an ellipsoid given in (25) is referred to a coordinate system supplied by the reciprocal lattice of the crystal. This arbitrary coordinate system bears no simple relation, in general, to the principal axes of the ellipsoid. If the principal axes of the ellipsoid are found, however, and if the ellipsoid is referred to these natural axes, the coefficients D, E, and F in (25) vanish, and the equation becomes simply

$$Ax^2 + By^2 + Cz^2 = 1. \tag{26}$$

When the following substitution is made:

$$A \equiv \frac{1}{a^2} \qquad B \equiv \frac{1}{b^2} \qquad C \equiv \frac{1}{c^2}, \tag{27}$$

(26) takes the well-known form

$$\frac{x^2}{a^2} + \frac{y^2}{b^2} + \frac{z^2}{c^2} = 1, \tag{28}$$

in which a, b, and c are the lengths of the three orthogonal principal axes of the ellipsoid.

The transformation from the general equation of an ellipsoid given in (25) to the *canonical form*, in which the ellipsoid is referred to its orthogonal principal axes, is most neatly represented with the aid of the algebra of matrices but is often accomplished by successive approximations. When this has been done for the thermal ellipsoid whose coefficients are the $2\pi^2U_{ij}$'s of (21), the cross terms with $i \neq j$ vanish, leaving only three terms, $2\pi^2U'_{11}$, $2\pi^2U'_{22}$, $2\pi^2U'_{33}$. These U's represent the mean-square displacements of the atoms in the directions of the three principal axes of the ellipsoid. The direction of each of these axes, however, is generally oblique to each of the three crystal axes, so that a complete description of the thermal motion must include the three angles between each ellipsoid axis and the three crystal axes.

Least-squares refinement of thermal parameters

One of the interesting results observed in difference electron-density functions was that the maps of these functions showed anomalies due to an underestimation or overestimation of the Debye temperature coefficient B which had been used in the computed amplitude. As pointed out earlier in this chapter, the temperature coefficient of the atom in question could then be adjusted until the anomaly was alleviated, and in this way a better temperature coefficient could be found for each atom.

When least squares came into use in the refinement of crystal structures, it became evident that, in addition to the coordinates x, y, and z for each atom, an additional parameter, namely, the temperature coefficient B for each atom, could also be refined. As computing facilities were improved, it became common practice eventually to seek all six β's of (23) or the six B_{ij}'s of (22) or the six U_{ij}'s of (21) in order to find the best fit of variables to observed data. By adding six thermal parameters to the three coordinates of each atom, nine parameters per set of equivalent atoms had to be found by least-squares refinement. This is ordinarily successful provided the number of observed intensity data exceeds, by a reasonable factor, the number of variables to be refined. Usually it is sufficient if the intensity data are several times more numerous than the number of parameters to be fixed, that is, if there is a reasonable overdetermination of the variables.

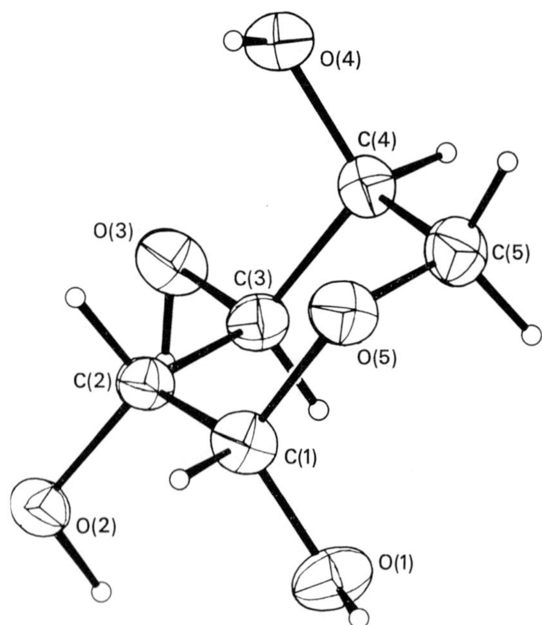

Fig. 5. Thermal ellipsoids in β-DL-arabinose. [*From S. H. Kim and G. A. Jeffrey, Acta Cryst.* **22** (1967) 537–545.]

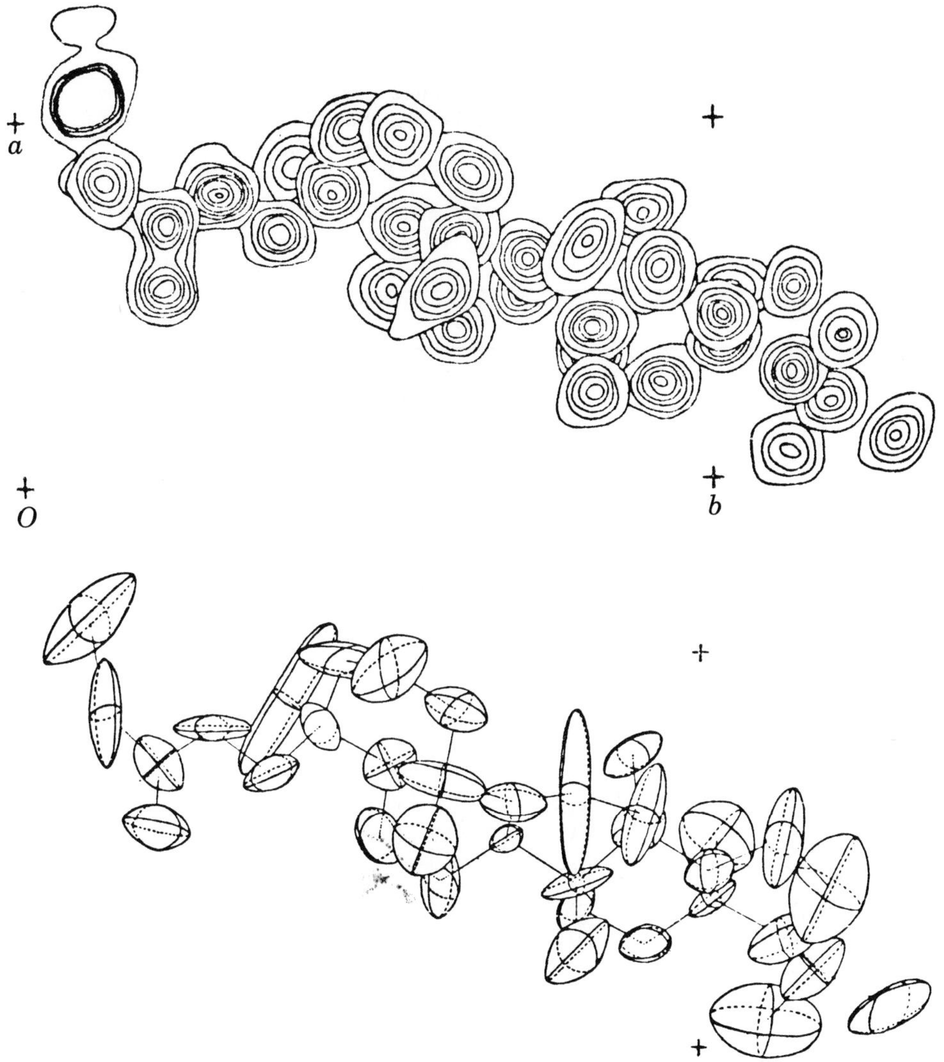

Fig. 6. Atoms in a molecule of davallol iodoacetate: Fourier synthesis (*above*) and thermal ellipsoids (*below*). [*From Yow-Lam Oh and E. N. Maslen, Acta Cryst.* **20** (1966) 852–864.]

An example of the thermal information which can be obtained about the atoms in a crystal is the thermal data for β-DL-arabinose, as presented by Kim and Jeffrey. Their thermal results for this crystal are listed in Table 2. The crystals of this compound have symmetry $P\ 2_1/c$, with a cell defined by $a = 5.925$, $b = 7.820$, $c = 13.357$ Å, $\beta = 99°45'$. This cell contains 4 molecules of composition $C_5H_{10}O_5$. The thermal data for each atom are neatly listed in Table 2 in two forms. The ellipsoids based upon the magnitudes of the Debye temperature coefficients in the

Table 2
Characteristics of the thermal ellipsoids for the various atoms in β-DL-arabinose, shown in Fig. 5
[After S. H. Kim and G. A. Jeffrey. Acta Cryst. 22 (1967) 537–545.]

The root-mean-square displacement U_i corresponds to the ith principal axis of the ellipsoid, and θ_{ia}, θ_{ib}, θ_{ic} are angles between the ith axis and the crystallographic axes a, b, c.

	i	B_i†	U_i	θ_{ia}	θ_{ib}	θ_{ic}
C(1)	1	1.720Å²	0.1476Å	146.2°	79.8°	48.7°
	2	2.215	0.1675	86.0	155.0	66.5
	3	2.694	0.1847	56.5	67.5	50.4
C(2)	1	1.570	0.1410	83.9	19.1	108.9
	2	1.919	0.1559	50.4	107.7	143.7
	3	2.062	0.1616	40.2	83.0	60.3
C(3)	1	1.478	0.1368	131.0	50.5	59.2
	2	2.088	0.1626	126.8	140.0	70.8
	3	2.279	0.1699	62.7	95.3	37.5
C(4)	1	1.827	0.1521	76.6	105.0	27.8
	2	2.186	0.1644	80.7	160.1	108.8
	3	2.328	0.1717	16.4	77.2	109.6
C(5)	1	1.815	0.1516	108.1	86.3	9.2
	2	2.519	0.1786	62.8	149.8	82.6
	3	3.089	0.1978	33.6	60.1	84.6
O(1)	1	1.755	0.1491	122.9	53.9	47.1
	2	2.247	0.1687	130.3	138.4	75.1
	3	4.051	0.2265	57.6	107.8	46.7
O(2)	1	1.621	0.1433	94.2	140.5	50.6
	2	2.155	0.1652	173.8	89.6	86.5
	3	2.349	0.1725	94.6	50.5	39.6
O(3)	1	1.372	0.1318	138.7	66.7	49.6
	2	2.155	0.1652	93.1	40.4	128.9
	3	3.718	0.2170	48.8	59.1	64.5
O(4)	1	1.863	0.1536	86.9	3.6	88.6
	2	2.279	0.1699	170.0	87.3	70.7
	3	2.817	0.1889	80.5	92.4	19.4
O(5)	1	1.615	0.1430	133.2	94.0	33.8
	2	2.274	0.1697	78.1	166.7	86.3
	3	2.993	0.1947	45.7	77.4	56.5

† $B_i = 8\pi^2 U_i^2$.

direction of each ellipsoid axis 1, 2, and 3 are labeled B_1, B_2, and B_3. The root-mean-square displacements in these directions are labeled U_1, U_2, and U_3. The angle between the first principal axis of the ellipsoid and the *a* axis is labeled θ_{1a}, etc., so that the orientations of the three axes of the thermal ellipsoid of each atom with respect to the crystallographic axes are available. From these data the thermal ellipsoids plotted in Fig. 5 can be constructed.

In Table 2 it can be seen that root-mean-square displacements of an atom from its equilibrium position vary from about 0.13 to about 0.22 Å. For organic crystals these magnitudes are normal, but for inorganic crystals they are somewhat smaller. The ellipsoids shown in Fig. 5 are not markedly eccentric. An example of greater anisotropies in the motions of atoms, as displayed by the eccentricities of their ellipsoids, is shown in Fig. 6.

Summary

The chief obstacle to be overcome in the analysis of a crystal structure is the determination of the rough structure. This part of the problem may require real ingenuity, but once a rough structure has been found, there are several standard routines for improving the precision and details of the structure. The degree to which a structure has been improved is commonly measured by a discrepancy index

$$R = \frac{\Sigma\big||F_{\text{obs}}| - |F_{\text{cal}}|\big|}{\Sigma|F_{\text{obs}}|}.$$

This is a crude measure of how much the model departs from the actual structure, as expressed by the differences in their diffraction amplitudes. The value of this index is limited by the quality of the data (as well as by ideas concerning the model), and may be as small as 2 percent. Well-refined structures commonly have R in the neighborhood of 5 to 6 percent. For a completely wrong structure the expected value of R is 82.8 percent if the crystal is centrosymmetrical, or 58.6 percent if noncentrosymmetrical. Models with R values of 45 percent or lower are worth trying to improve by refinement.

A structure may be improved by a trial-and-error attempt to adjust parameters. This procedure is difficult or impossible with structures having many atoms, and is seldom used in currect practice.

A series of successive Fourier syntheses provides a way of producing a moderate improvement of an essentially correct structure model. Even in cases where the locations of all atoms have not been established, cautious use of successive Fourier syntheses usually reveals the locations of the previously unlocated atoms. The chief precaution is that no atom whose location is unknown or doubtful should be used in computing the structure factors whose phases are to be attributed to the F_{obs}'s used as coefficients of the Fourier series. Refinement of a structure by the method of successive Fourier syntheses leads to only a moderate degree of refinement because of series-termination errors. These result from including only a limited number of measured F_{hkl}'s in the synthesis, instead of all F's with h, k, and l each ranging from $-\infty$ to $+\infty$.

Refinement can also be achieved by a difference Fourier synthesis; this amounts to a subtraction of a Fourier synthesis based upon the computed F_{hkl}'s from one based upon the observed F_{hkl}'s. Such a Fourier synthesis provides the difference of the electron densities of these two structures. This produces a map of the anomalies of the model. From the nature of the anomaly it can be ascertained whether an atom has been completely misplaced, whether it should be shifted a little, and by what amount and in what direction, and whether the temperature coefficient has been underestimated or overestimated. When the difference-density map is featureless, the model and the actual structure are the same. Difference syntheses are nearly free from series-termination effects and so lead to well-refined structures.

Difference Fourier syntheses are also useful in revealing the locations of light atoms which would otherwise escape detection in a Fourier synthesis. This feature is very useful in locating hydrogen atoms, especially in organic structures.

With the development and improvement of high-speed electronic computers, it is now common practice to refine structures by the method of least squares. This method seeks the best values of the variable parameters for each atom to fit the observed intensities. It is used not only to find the best coordinates of each atom but also to find the thermal parameters of each atom. If the atoms vibrate in a field with linear restoring forces in various directions, the mean-square displacement marks out a triaxial ellipsoid. Least-squares refinement is typically used to obtain parameters from which the dimensions and orientations of these ellipsoids for each atom in the structure can be computed.

Notes on history

In the early days of crystal-structure analysis, crystals with few variable parameters were investigated. The structures of these simple crystals were refined by adjusting the coordinates of the atoms to give the best fit of computed to observed intensities. When Fourier syntheses came into use in the 1930s, it became evident that the coordinates of the atoms could be improved by successive Fourier syntheses. But it was recognized that the final coordinates obtained by this scheme were in error due to the series-termination effect, and various devices were tried to correct for this.

The discrepancy index $R = (\Sigma||F_{obs}| - |F_{cal}||)/\Sigma|F_{obs}|$ has been used since the 1930s. It has been commonly used to estimate the departure of the model from the actual structure ever since.

During the Second World War Bunn devised an "error synthesis," which was a difference Fourier synthesis using as coefficients the terms $(F_{obs} - F_{cal})_{hkl}$, but only those for which $F_{obs} = 0$. For these terms the phase of the term $(F_{obs} - F_{cal})_{hkl}$ is that of F_{cal}. This Fourier synthesis has the property of having negative values at the positions of atoms which are improperly placed. Unfortunately, this Fourier series has so few terms that it is not very perfect, and is consequently not easy to interpret. This was the first version of a difference Fourier synthesis. The number of terms can be increased to all, or nearly all, *hkl*'s, however, if the structure is known. This form, which was first suggested for use in refinement by Booth in 1948, has been used to test the correctness of final or near-final versions of crystal-structure models ever since. It has been used not only for the refinement of structures but also for finding the positions of very light atoms such as hydrogen.

The least-squares method for determining the best parameters to fit the amplitude data is actually one of the older methods of refining crystal structures, for it was introduced by Hughes in 1941. It gained no great popularity at that time, because of the tedious computations involved. This drawback was removed with the availability of high-speed electronic computers in the 1950s. Since then the method, originally used for the refinement of the coordinates of atoms, has been extended to the refinement of thermal parameters and other variables (such as the amount of substitution of one atom by another in a particular site) which arise in particular cases.

Additional reading

Martin J. Buerger. *Crystal-structure analysis.* (Wiley, New York, 1960) 585–628.

H. Lipson and W. Cochran. *The determination of crystal structures.* (G. Bell, London, 1966) rev. ed. 300–310, 317–357.

Significant literature

E. W. Hughes. *The crystal structure of melamine.* J. Am. Chem. Soc. **63** (1941) 1737–1752.

A. D. Booth. *An expression for following the process of refinement in x-ray structure analysis using Fourier series.* Phil. Mag. **36** (1945) 609–615.

Andrew D. Booth. *A new Fourier refinement technique.* Nature **161** (1948) 765–766.

D. Crowfoot, C. W. Bunn, B. W. Rogers-Low, and A. Turner-Jones. *The x-ray crystallographic investigation of the structure of penicillin. The chemistry of penicillin.* (Princeton, Princeton, N.J., 1949) 310–367.

W. Cochran. *The effect of anisotropic thermal vibration on the atomic scattering factor.* Acta Cryst. **7** (1954) 503–504.

David R. Davies and J. J. Blum. *The crystal structure of parabanic acid.* Acta Cryst. **8** (1955) 129–136.

Jürg Waser. *The anisotropic temperature factor in triclinic coordinates.* Acta Cryst. **8** (1955) 731.

D. W. J. Cruickshank. *The determination of the anisotropic thermal motion of atoms in crystals.* Acta Cryst. **9** (1956) 747–753.

Index

Index